Axel Bronstert

Jesus Carrera

Pavel Kabat

Sabine Lütkemeier

Coupled Models for the Hydrological Cycle

Integrating Atmosphere, Biosphere, and Pedosphere

Axel Bronstert
Jesus Carrera
Pavel Kabat
Sabine Lütkemeier
(Editors)

Coupled Models for the Hydrological Cycle

Integrating Atmosphere, Biosphere, and Pedosphere

With 92 Figures, 19 in colour and 20 Tables

Springer

Axel Bronstert
University of Potsdam
Institute for Geoecology
Chair for Hydrology and Climatology
P.O. Box 60 15 53
14415 Potsdam
Germany

Pavel Kabat
Wageningen University and
Research Centre
ALTERRA Green World Research
Droevendaalsesteeg 3
6708 PB Wageningen
The Netherlands

Jesus Carrera
Technical University of Catalonia (UPC)
Department of Geotechnical Engineering
and Geosciences
School of Civil Engineering
Campus Nord, Edif. D-2
08034 Barcelona
Spain

Sabine Lütkemeier
Potsdam Institute for
Climate Impact Research
Telegrafenberg
14473 Potsdam
Germany

Library of Congress Control Number: 2004108318

ISBN 3-540-22371-1 **Springer Berlin Heidelberg New York**

Springer is a part of Springer Science+Business Media
springeronline.com

Printed in Germany

Cover design: Erich Kirchner
Production: Almas Schimmel
Typesetting: PR-Graphix, Karl C. Stahl, Heidelberg
Printing: Mercedes-Druck, Berlin
Binding: Stein + Lehmann, Berlin

Printed on acid-free paper 30/3141/as - 5 4 3 2 1 0

Preface

Freshwater resources are an indispensable part of all terrestrial ecosystems. The Dublin Statement on Water and Sustainable Development states: "Scarcity and misuse of fresh water pose a serious and growing threat to sustainable development and protection of the environment. Human health and welfare, food security, industrial development and the ecosystems on which they depend are at risk, unless water and land resources are managed more effectively in the present decade and beyond than they have been in the past". Similarly, the first principle defined at the Dublin Conference states "Fresh water is a finite and vulnerable resource, essential to sustain life, development and the environment" (UNESCO 1996). Strategies for sustainable development must be based on an accurate assessment of the Earth's carrying capacity and resilience to human activity. A deeper understanding of the interconnections between water, nutrient and biogenic cycles, and the energy flows of land, oceans and atmosphere, is crucial (UNCED 1992). AGENDA 21 calls for an "Integrated Approach to Land-Resource Use, and Protecting and Managing Freshwater Resources". Increasing demand for land – a finite resource – and for the natural resources land supports is creating competition and conflict that results in land degradation.

The Framework and General Outline of IHP-V (www.unesco.org/water/ihp/) states that if future water resource development and management schemes are to be sustainable, they will have to deal effectively with environmental and social consequences; land-water linkages; allocation of water among competing uses and users; and achieving effective implementation. To achieve these goals, four areas of concern need addressing:

1. the role of scales in hydrological processes;

2. vulnerability of the environment;

3. integrated water resource management; and

4. education, training and transfer of technologies.

IHP Project 1.2 addresses the first two, in particular, fostering the development of coupled hydroecological/atmospheric models, synthesising hydrological data relevant to the calibration of those models and identifying the effects of enhanced atmospheric carbon dioxide. However, it is important to keep the last two goals in mind in order to stimulate stronger interactions between research, education and application and to emphasise integrated resource management and science.

The International Dialogue on Water and Climate (DWC) aims to improve interactions between climate scientists and the water resource management. DWC (www.waterandclimate.org) bridges the information gaps between the water and climate sectors to improve our capacity to cope with the impacts on water management of increasing climate variability and change. The goal of the Dialogue is to develop a knowledge base, generate widespread awareness, identify policy and management options that build such capacities, learn from the experiences throughout the world, and make this knowledge available to the most affected communities.

In general, coupled models are developed in pursuit of a particular objective and will therefore always have bias towards particular processes and phenomena. Similarly, this document is deliberately biased to place greater emphasis on hydrological processes and less emphasis on processes elsewhere in the coupled system. Later, for instance, when the interrelationships between processes is explored, processes and phenomena in the hydrological system are diagnosed in greater detail than are atmospheric, land-cover and transport processes. Further, to focus more on land-surface hydrological/atmospheric coupling, the domain of interest considered is limited to coupling processes relevant at the spatial scale of regions or large catchments. It is assumed, therefore, that either the incoming and outgoing atmospheric fields around the continental domains are measured or, if they are calculated elsewhere in a global coupled model, that the coupling between oceans and the atmosphere has been represented adequately.

When modelling the hydrological system, it is not always necessary that all (or indeed any) coupling processes be represented. There are, however, at least four general situations where the use of coupled models is essential, as follows.

- It is not feasible or economic to measure the boundary between compartments in the modelled hydro-geo-biosphere system and thereby isolate them. This will be common when models are being used in truly predictive mode.

- Observations of two or more state variables that are coupled cannot be modelled simultaneously unless the processes that couple them are included in the model. Sometimes, for instance, this is the case for models that describe the flows not only of water quantity but also of the chemical composition of that moving water simultaneously. Representing both with a coupled model can reveal hitherto unrecognised weaknesses in the uncoupled description of water flow.

- The science that is being addressed is beyond the normal range of activity and interest of specialists in individual coupled processes. In this case, the discipline of developing a coupled model can often produce totally new scientific understanding that would not otherwise have been possible.

- Vulnerability assessments of the whole system are undertaken.

This book is intended primarily for applied scientists and graduate students. It should also become a valuable source of information for hydrology and water resource management professionals who wish to advance their knowledge of the key principles of the coupled climate-hydrology system, and to use water system management models. As such, this book will be a valuable help in education and training.

Mike Bonell, Axel Bronstert, Pavel Kabat, Jim Shuttleworth

Agenda 21 (1993) Programme of action for sustainable development. Rio declaration on environment and development. The final text of agreements negotiated by governments at the United Nations Conference on Environment and Development (UNCED), 3 - 14 June 1992, Rio de Janeiro, Brazil. New York, NY: United Nations, p 294

UNCED (1992) Report of the United Nations Conference on Environment and Development: Rio de Janeiro, 3 - 14 June 1992, New York, United Nations publication

Acknowledgements

The Editors would like to express their sincere thanks to the International Hydrology Programme (IHP) of UNESCO, to the former Core Project of the International Geosphere-Biosphere Programme "Biospheric Aspects of the Hydrological Cycle (BAHC)" and to the Dialogue on Water and Climate (DWC). All these three international programmes helped in convening workshops for the authors or in providing funding to finalise the book for publication.

We are grateful to the German Federal Ministry of Education and Research (BMBF) for funding the BAHC Project Office at the Potsdam Institute for Climate Impact Research (PIK). The Dutch Ministry of Development Cooperation provided support through the DWC. Additional funding and support have been provided by the German IHP-HWRP (Hydrology and Water Resources Programme of the WMO) Committee.

We also wish to thank Ursula Werner (PIK) for converting most of the figures and Andreas Bauer (University of Potsdam) for additional technical support.

Cover design:

River Spree in Brandenburg, Germany © M. Zebisch, Technical University Berlin/ Potsdam Institute for Climate Impact Research, Germany.
Bushes in the Sahel, West Africa: courtesy of H.-J. Pachur ©, Free University Berlin, Germany.
Hillslopes after burning harvest remains, semi-arid Northeast Brazil, State of Ceará © Axel Bronstert, University of Potsdam, Germany.
Elbe flood near Havelberg, Germany, on 21 August 2002 © M. Zebisch, Technical University Berlin/Potsdam Institute for Climate Impact Research, Germany.

Content

List of Contributors

Elena Abarca
School of Civil Engineering, Technical University of Catalonia (UPC)
Campus Nord, Edif. D2, 08034 Barcelona, Spain
E-mail: Elena.Abarca@upc.es

Ramón Arandes
Municipality of Barcelona
Torrent de l'Olla, 218-220, 6º, 08012 Barcelona, Spain

Lawrence E. Band
Dept. of Geography, University of North Carolina
200S Elliott Rd., Chapel Hill, NC 27514-5830, USA
E-mail: lband@email.unc.edu

Luis A. Bastidas
Dept. of Civil and Environmental Engineering and Utah Water Research Laboratory, Utah State University
Old Main Hill 4110, Logan UT 84322-4110, USA
E-mail: luis.bastidas@usu.edu

Mike Bonell
Division of Water Sciences, International Hydrological Programme
United Nations Educational Scientific and Cultural Organization (UNESCO)
1, rue Miollis, F-75732 Paris, France
E-mail: m.bonell@unesco.org

Axel Bronstert
Institute for Geoecology, University of Potsdam
P.O. Box 60 15 53, 14415 Potsdam, Germany
E-mail: axelbron@rz.uni-potsdam.de

Jesus Carrera
School of Civil Engineering, Technical University of Catalonia (UPC)
Campus Nord, Edif. D2, 08034 Barcelona, Spain
E-mail: Jesus.Carrera@upc.es

Thomas N. Chase
CIRES/CSES, University of Colorado
Campus Box 216, Boulder CO 80309, USA
E-mail: tchase@terra.colorado.edu

Martin Claussen
Potsdam Institute for Climate Impact Research
Telegrafenberg, 14473 Potsdam, Germany
E-mail: claussen@pik-potsdam.de

Victor N. Demidov
Water Problems Institute of Russian Academy of Sciences
117735, Gubkin 3, Moscow, Russia
E-mail: demidov@aqua.laser.ru

Bas Eickhout
National Institute of Public Health and the Environment (RIVM)
P.O. Box 1, 3720 BA Bilthoven, The Netherlands
E-mail: bas.eickhout@rivm.nl

Andrey Ganopolski
Potsdam Institute for Climate Impact Research
Telegrafenberg, 14473 Potsdam, Germany
E-mail: Andrey.Ganopolski@pik-potsdam.de

Alexander N. Gelfan
Water Problems Institute of Russian Academy of Sciences
117735, Gubkin 3, Moscow, Russia
E-mail: hydrowpi@aqua.laser.ru

Alexander Georgiadi
Institute of Geography, Russian Academy of Sciences
Staromonetny per., 29, 109017 Moscow, Russia
E-mail: galex@online.ru

Lars Gottschalk
Dept. of Geosciences, University of Oslo
P.O. Box 1047, Blindern, NO-0316 Oslo, Norway
E-mail: lars.gottschalk@geo.uio.no

Sven-Erik Gryning
Wind Energy Department, Risø National Laboratory
P.O. Box 49, DK-4000 Roskilde, Denmark
E-mail: sven-erik.gryning@risoe.dk

Forrest G. Hall
Goddard Space Flight Center
National Aeronautic and Space Administration (NASA)
Code 923, Greenbelt MD 20771, USA
E-mail: fghall@ltpmail.gsfc.nasa.gov

Sven Halldin
Department of Earth Sciences, Uppsala University
Villavägen 16, SE-75236 Uppsala, Sweden
E-mail: sven.halldin@hyd.uu.se

Ekkehard Holzbecher
Institute of Freshwater Ecology and Inland Fisheries
Müggelseedamm 310, 12587 Berlin, Germany
E-Mail: holzbecher@igb-berlin.de

Ronald W.A. Hutjes
Wageningen University and Research Centre
ALTERRA Green World Research
Droevendaalsesteeg 3, 6708 PB Wageningen, The Netherlands
E-mail: ronald.hutjes@wur.nl

Annekathrin Jaeger
Hydrology Research Group, Department of Geography
University of Bonn, Meckenheimer Allee 166, 53115 Bonn, Germany
E-mail: jaeger@giub.uni-bonn.de

Pavel Kabat
Wageningen University and Research Centre
ALTERRA Green World Research
Droevendaalsesteeg 3, 6708 PB Wageningen, The Netherlands
E-mail: Pavel.Kabat@wur.nl

Maarten S. Krol
University of Twente, Faculty of Engineering Technology
Discipline Group Water Engineering and Management
P.O.Box 217, 7500 AE-Enschede, The Netherlands
E-mail: m.s.krol@ctw.utwente.nl

Lev S. Kuchment
Water Problems Institute of Russian Academy of Sciences
117735, Gubkin 3, Moscow, Russia
E-Mail: kuchment@hotmail.com

George Leavesley
USGS, Denver Federal Center, Lakewood, CO 802250046, USA
E-mail: george@usgs.gov

Rik Leemans
Environmental Systems Analysis Group, Wageningen University
6700 AA Wageningen, The Netherlands
E-mail: Rik.Leemans@wur.nl

Glen E. Liston
Department of Atmospheric Science, Colorado State University
Fort Collins, CO 80523, USA
E-mail: liston@atmos.colostate.edu

Sabine Lütkemeier
Potsdam Institute for Climate Impact Research
Telegrafenberg, 14473 Potsdam, Germany
E-mail: Sabine.Luetkemeier@pik-potsdam.de

Walter A. Lyons
FMA Research, Inc.,Yucca Ridge Field Station
46050 Weld County Road 13, Fort Collins CO 80524, USA
E-mail: walyons@frii.com

Curtis H. Marshall
NCEP Environmental Modeling Center
5200 Auth Road, Room 207, Camp Springs MDÿ 20746
E-mail: Curtis.Marshall@noaa.gov

Nicole Mölders
Geophysical Institute, University of Alaska Fairbanks
903 Koyukuk Drive, P.O. Box 99775-7320Fairbanks, AK 99775, USA
E-mail: nicole.molders@gi.alaska.edu

Yuri G. Motovilov
Russian State Institute for Applied Ecology
Pyatnitskaya St., 44/3, 109017 Moscow, Russia
E-mail: motol@siae.ru

Roger A. Pielke Sr.
Department of Atmospheric Sciences, Colorado State University
Fort Collins CO 80523, USA
E-mail: pielke@atmos.colostate.edu

Xavier Sánchez-Vila
School of Civil Engineering, Technical University of Catalonia (UPC)
Campus Nord, Edif. D-2, 08034 Barcelona, Spain
E-mail: xavier.sanchez-vila@upc.es

Jim Shuttleworth
Dept. of Hydrology and Water Resources, University of Arizona
Tucson, AZ 85721, USA
E-mail: shuttle@hwr.arizona.edu

Everett P. Springer
Los Alamos National Laboratory, MS J495, Los Alamos,
NM 87545, USA
E-mail: everetts@lanl.gov

Louis T. Steyaert
U.S. Geological Survey, EROS Data Center, and
NASA Goddard Space Flight Center, Code 923
Greenbelt, MD 20771 USA
E-mail: steyaert@ltpmail.gsfc.nasa.gov

Christina L. Tague
Dept. of Geography, San Diego State University
San Diego, CA 92182-4493
E-mail: ctague@mail.sdsu.edu

Oleg F. Vasiliev
Institute for Water and Environmental Problems (IWEP)
Morskoy Prosp. 2, 630090 Novosibirsk 90, Russia
E-mail: vasiliev@ad-sbras.nsc.ru

Enric Vázquez-Suñé
School of Civil Engineering, Technical University of Catalonia (UPC)
Campus Nord, Edif. D-2, 08034 Barcelona, Spain
E-mail: enric.vazquez-sune@upc.es

Pier-Luigi Vidale
Climate Research, Swiss Federal Institute of Technology (ETH)
Winterthurerstrasse 190, 8057 Zürich, Switzerland
E-mail: pier-luigi.vidale@ethz.ch

Robert L. Walko
Department of Civil and Environmental Engineering
Duke University, Durham, NC 27708-0287, USA
E-mail: Robert.Walko@duke.edu

CHAPTER 1

Cycles, Feedbacks, and Main Issues

1.0 Global Cycles: Overview and Introduction

Jim Shuttleworth, Jesus Carrera, Pavel Kabat

Global climate change has become a prominent issue recognised by scientists, policy makers, and the public in general. Most environmental scientists are aware of the factors controlling climate (Fig. 1.0-1), at least at a qualitative level. Changes in atmospheric composition, such as increases in CO_2 and other trace gases, are changing the energy balance of the Earth. Specifically, part of the long wave radiation emitted by the Earth's surface is absorbed more strongly lower in the atmosphere, leading to warming of the air near the ground and, it is believed, to changes in atmosphere circulation and precipitation patterns. In turn, such changes in precipitation will affect vegetation covers. Some plants accustomed to receiving a great deal of water might not to survive a decrease in water availability and be replaced by drought-tolerant species, and vice versa where rainfall increases. Plants may also react directly to increases in CO_2 by either growing faster (increasing photosynthesis) and, in this way, become more widespread, or by reducing transpiration (increased CO_2 may allow them to benefit from closing their stomata to some extent), or both. In turn, changes in rainfall and soil cover are known to affect surface runoff and infiltration. Increased surface runoff may lead to increases in the frequency and severity of floods, thus potentially changing geomorphology and land use. Decreases in infiltration will cause a decrease in the base flow of rivers and many wetlands to disappear, and vice versa. Consequently, there may be significant changes in the landscape, causing further changes in evaporation that, in turn, affect atmospheric circulation and the water cycle.

Changes in temperature and the spatial distribution of inflows to the ocean also change ocean circulation patterns. The Atlantic Ocean, which receives a disproportionate amount of surface water, will likely continue to feed the Pacific Ocean, which has a larger surface area to support evaporation. However, ocean currents are unstable and even slight changes in temperature or salinity may change their trajectories, leading to further changes in local climate. To compound the complexity of the problem, humans, who have forced many of these changes, will react to them in a manner that is at least as difficult to predict as the physical and biological changes in the environment. The central point is that the water, energy, and biogeochemical cycles through which the dynamic Earth system operates are coupled together amongst themselves and to the way we live. Moreover, they act over a wide range of spatial and temporal scales and their interactions are complex.

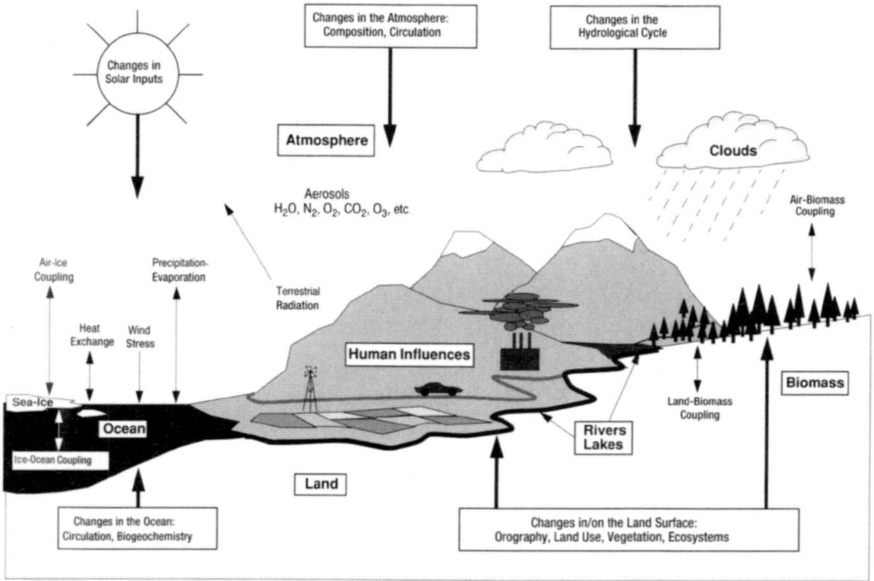

Fig. 1.0-1 The global climate system and the interactions of its subsystems (Intergovernmental Panel on Climate Change 1996)

Scientists seek to quantify the coupled global cycles through which the Earth system operates. Fig. 1.0-2 shows the carbon cycle, which controls the level of CO_2 in the atmosphere and thus affects photosynthesis, the main process by which inorganic carbon becomes organic carbon. The largest fluxes are those linking atmospheric carbon dioxide to vegetation. About half of the CO_2 dissolved in oceans is taken up by marine biota, while the CO_2 concentration in surface ocean water is controlled by Henry's law (its concentration is proportional to that of the atmosphere). Dissolution of carbonate acts as a buffer. Consequently, marine biota activity is expected to be less sensitive to changes in the carbon cycle than might be anticipated. The carbon cycle shown in Fig. 1.0-2 is not in balance: the inputs to the atmosphere are larger than the outputs. Although human input to the atmosphere is a relatively small portion of the total, it is the cause of this imbalance. This is why scientists are recommending restricting emissions and it is the motivation for the Kyoto agreements. It is also the motivation for much research on carbon sinks. Analysing Fig. 1.0-2 allows identification of some of these sinks. For example, one might expect that increased CO_2 in ocean water will lead to an increase in carbonate dissolution. However, uncertainties in reaction kinetics as well as in the dynamics of oceanic circulation and the availability of nutrients (carbonate precipitation is often catalysed by microorganisms) make it difficult to establish this. Soil uptake is a potential sink, but increased temperatures could increase the rate of decomposition of soil organic matter and increase what is now a relatively small portion of the total input to the atmosphere. Also, rock weathering is expected to increase, thus fostering Ca and Mg circulation, and changes in landscape, which imply further changes in the hydrological cycle. It is clear that the carbon cycle is more complex than implied by Fig. 1.0-2 and that it is coupled to several other biogeochemical cycles.

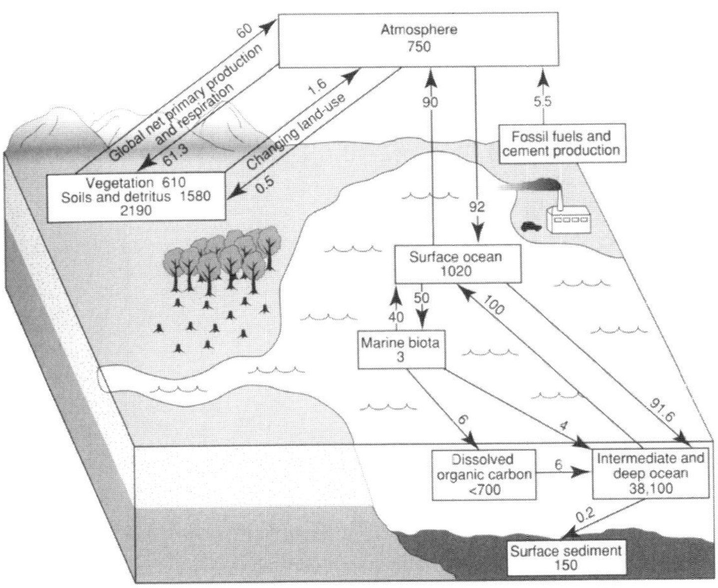

Fig. 1.0-2 The global carbon cycle (Intergovernmental Panel on Climate Change 1996). Pools (in boxes) are expressed in 10^{15} g, fluxes (in arrows) are expressed in 10^{15} g y^{-1}. Notice that the system is in imbalance and the atmosphere is receiving 3.3 10^{15} g y^{-1} more than is returning, hence the build-up in atmospheric CO_2

The main subject of this book is the hydrological cycle, which is represented at a global scale in Fig. 1.0-3. The atmospheric stock (approximately 13 000 km^3) is a small portion of the yearly flux (of approximately 505 000 km^3)

and is negligible when compared to the total stock of approximately 1400 million km³. This is why potential changes in the hydrological cycle are so important. It is likely that the hydrological cycle – or at least parts of it – is sensitive to changes in climate. Warming of the lower atmosphere will cause evaporation to increase, causing an increase in rainfall. However, spatial variability in the hydrological cycle is so large that, at any given location, it is very hard even to say whether it is accelerating or slowing down.

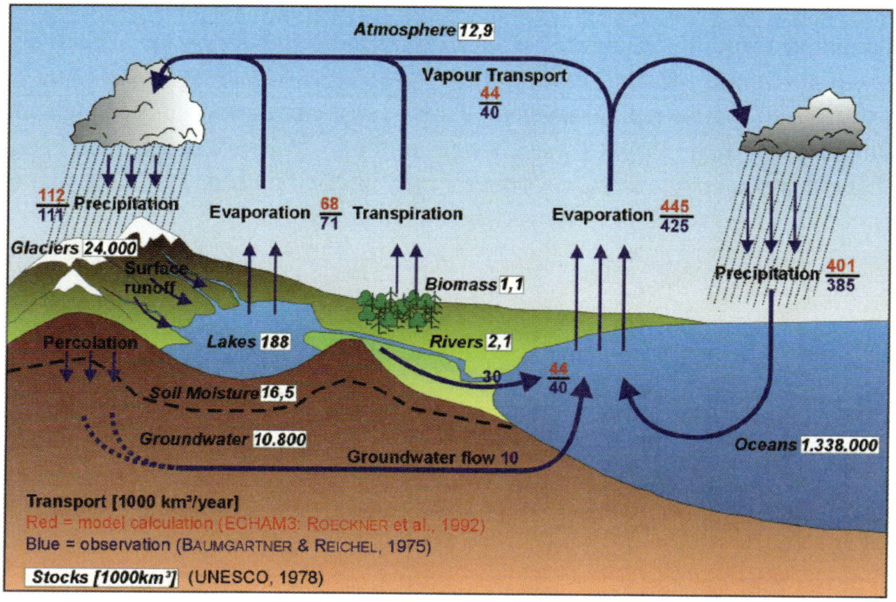

Fig. 1.0-3 The global water cycle: stocks and fluxes. The groundwater stock includes fresh groundwater and water stored in saturated, permanently frozen soil. The lake water stock includes fresh water lakes, saline lakes and swamps. The number for runoff from land to ocean provided by the GCM ECHAM3 does not distinguish between runoff from rivers and direct groundwater exfiltration into the ocean. All numbers are associated with considerable uncertainty. Different sources present rather different numbers, due to varying assumptions of categorising different stocks. An overview of different sources on the global water cycle is included in Gleick (1993)

Some of the uncertainty in the response of the water cycle to climate change is because of its complex interaction with other global cycles while some, as previously stated, is because of its temporal and spatial variability. The different energy balances of temperate and boreal forests (Fig. 1.0-4) and of temperate agricultural and forested land cover (Fig. 1.0-5) illustrate this last point. Forests absorb more of the incoming sunlight (their albedo is smaller). Because of energy, water, and nutrient availability, temperate forests typically return two thirds of the incoming radiation as latent heat through evapotranspiration. Boreal forests, on the other hand, are thriftier; their stomatal resistance is higher, so they return less energy through evapotranspiration. This implies more runoff and also implies that more energy is returned to the atmosphere as sensible heat, leading to a thicker atmospheric boundary layer. In general, the water cycle is necessarily intimately linked to the energy and biochemical cycles and the nature of these links is complex. This argues for viewing and modelling the Earth as a coupled system.

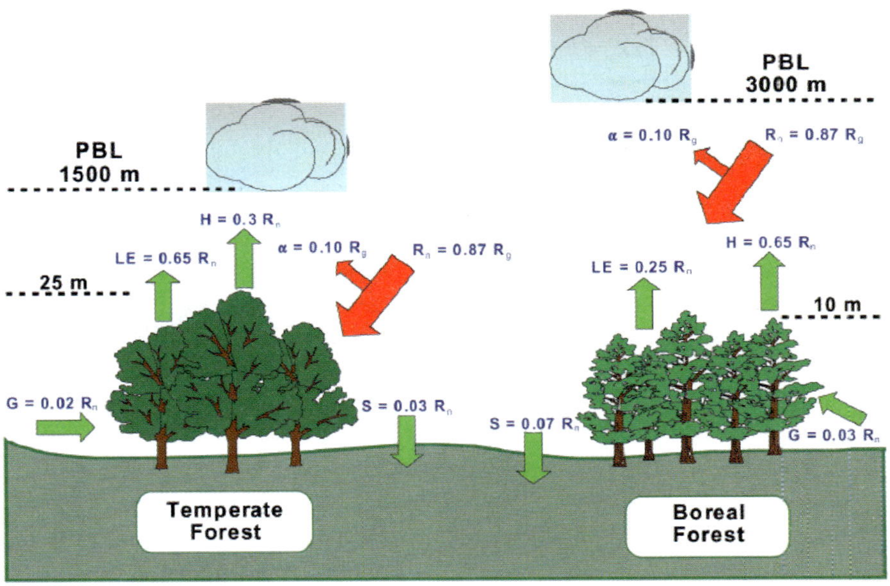

Fig. 1.0-4 Illustration of the typical difference in energy balance between forests in temperate and boreal climates. Most striking is the different partitioning of available energy over latent and sensible heat. Tight stomatal control by boreal forests is induced by water scarcity resulting from still frozen soils in spring, and from free drainage on coarse textured soils (boreal forests are therefore sometimes referred to as "green deserts"). This leads to a diversion of energy to sensible heat resulting in (very) deep boundary layers. Symbols used: H – sensible heat, LE – latent heat (evapotranspiration), S – soil heat flux, G – heat stored in canopy (air and biomass). α – albedo, R_n – net radiation, R_g – incoming shortwave radiation, PBL – planetary boundary layer

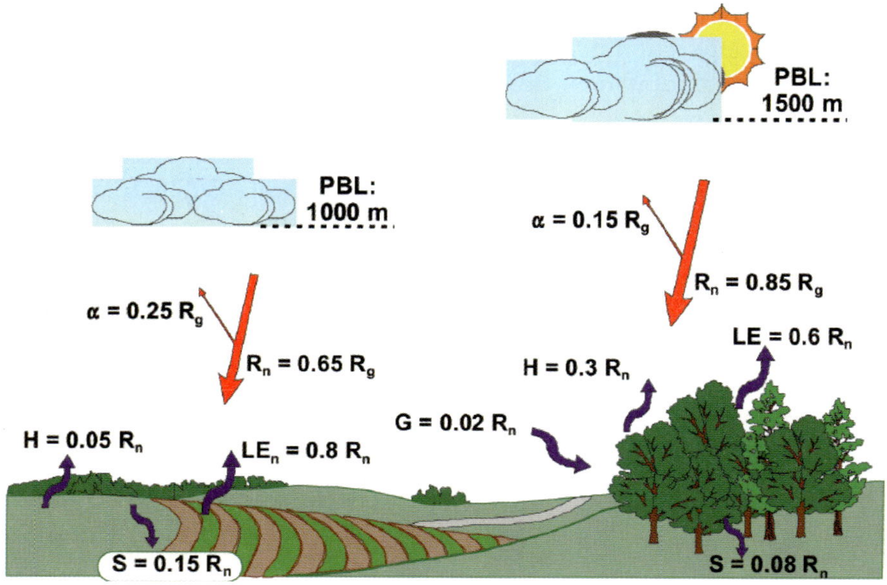

Fig. 1.0-5 Illustration of the typical difference in energy balance between temperate forests and agricultural lands. Most striking is again the different partitioning of available energy over latent and sensible heat, though for different reasons. Whereas evapotranspiration from both forest and crops can be comparable in absolute terms, available energy over crops is generally much smaller due to its larger albedo and soil heat flux. As a result, less energy is left for sensible heat, leading to shallower boundary layers. (Symbols as in Fig. 1.0-4)

Acting together within confines determined by the Earth's planetary motion in the solar system, the coupled water, energy, and biogeochemical cycles generate spatial and temporal patterns in the atmosphere, biosphere, hydrosphere and geosphere that have the general appearance of a stable equilibrium. Planners and resource managers assume that, at any point in time in the Earth's history, such patterns have quantifiable average values that can be specified at the time scales of decades to centuries, though they may exhibit noticeable variations around these average values (also coupled) which are assumed temporary. Further, it is often assumed that, unless there are changes in external forcing or substantial accidental or human-imposed changes in internal feedback processes, these multi-scale, coupled variations in the Earth system will be self-correcting and that they can themselves be quantified in statistical terms. However, this hypothesis may not be true, but planning, design, and management in general, and of hydrological systems in particular, remain in large measure based on the assumption that mean values and variations in hydrological variables within the coupled Earth system can be quantified locally and then applied.

However, the processes that generate equilibria in the atmosphere, biosphere, hydrosphere and geosphere are coupled, so what happens if there is a substantial change in one or more of the coupled components at local, regional, or global scale? What will happen to the other components? Most important in the context of this book, what will happen to the hydrological variables that are coupled to that changed component? If there are changes in the carbon balance of the atmosphere as a whole, for example, or regional or local human induced changes in vegetation cover, or enhanced nitrogen deposition to the ground from the atmosphere, how will the hydrology of a particular catchment respond? The only way to address such questions is to identify, understand, and model the coupled processes that occur over continental surfaces that involve or impact water in its vapour, liquid and solid forms. This is why coupled models have become so important for hydrology, and this book provides a state-of-the-art review of progress towards this goal.

In this first chapter, we consider some of the more important aspects of this issue in general terms. The first section is an overview of the most important water-related coupled systems. It is structured around phenomena that induce some aspect of "memory" into coupling processes. Topography is a feature of the landscape that induces persistent patterns in variables important in hydrology, such as precipitation and runoff, which are, in this sense, remembered. Water is stored over land surfaces as ice and snow, and also in the ground as soil moisture coupled to the atmosphere via vegetation. Within the coupled Earth system, water stored in both these two ways provides a memory of the combined effect of the past influence of coupling processes such as precipitation, evaporation and sublimation. The strong coupling between the carbon and water cycles through vegetation processes also involves memory through the nature, amount, and vigour of the vegetation. This is also described in this section.

The second section of the chapter is concerned with the problematic issue of scale. Coupled processes occur over a broad range of temporal and spatial scales and in this section the relevant scales at which coupling between the water, energy, and biogeochemical cycles occurs are defined. Over the last decade, there has been much progress towards bridging between process representation at small and large spatial scales using alternative upscaling and downscaling methods. Finally, in the third section, attention is given to the important discovery that, in some situations and locations, models capable of representing coupled interactions predict the existence of distinct alternative equilibria if significant change occurs or is imposed in one of the factors involved in determining the equilibrium state. This surprising and important feature of coupled interactions is illustrated by two very different examples, one relating to the hydroclimate of the Sahelian region of Africa, the other to the salinisation of water bodies.

1.1 The Nature of Land-Biosphere-Atmosphere Feedbacks in the Hydrological Cycle

Luis A. Bastidas, Mike Bonell, Pavel Kabat, Jim Shuttleworth

Land use and land cover are crucial factors in climate and global change. Human activity has transformed most of the Earth's land cover during the last 200 years, with prominent influence on several subsystems of the total Earth system (Turner et al. 1992). In some parts of Europe, for example, most of the forest has been replaced by agricultural land. The *terrestrial biosphere* itself has been modified fundamentally by land-cover conversions for agriculture and other purposes (industrialisation, urbanisation), as well as by forest and rangeland management. The *atmosphere* is affected by these changes, both through an altered energy balance over a significant part of the land surface (i.e. the more intensively managed areas) where physical and physiological properties such as albedo, roughness and canopy conductance are modified, as well as through changed fluxes of CO_2, CH_4 and other trace gases between soil/vegetation and the troposphere (Solomon et al. 1993; Woodwell et al. 1983).

The fundamental research question, from a "coupled system" perspective, is the degree of sensitivity of feedbacks between the terrestrial biosphere and the atmosphere, to land-use change (Tans and Wallace 1999, Pielke 2001). Answering this question must involve processes at the appropriate spatial and temporal scales, and the non-linearities involved. Ideally, investigation of the past dynamics of the interaction, for example for the last 200 years, should be explored in tandem with the development of predictive models that can be extrapolated into the future.

The land surface and the atmosphere form a tightly coupled system that evolves together on both diurnal and seasonal time scales (Betts 2000). The existing understanding says that storage of water near the surface, as soil moisture, and its storage on the surface, as snow and ice cover, are land-memory processes that can affect both the overlying atmosphere and runoff to streams or to ground water. Moreover, research suggests that the nature and seasonal progression of the vegetation growing at the ground also represents a land memory, because the current nature and growth status of plants reflects not only past climate but also controls current surface energy, moisture, and momentum exchanges.

Topography, snow cover, soil moisture and vegetation all form an interactive system. For example, a large winter snowfall may produce more spring soil moisture as it melts, delaying the spring warming of the surface. This in turn will affect the timing and rate of initiation of growth of plants. Denser vegetation will provide more surface to intercept snow but at the same time shade snow on the ground. Higher peaks receive more snowfall. Thus it is not possible to study each of these elements of land-surface hydrology in isolation. Details of the impacts and interactions of these elements can be found in e.g. McBean and Hantel 1993, or Schultz et al. 1994.

1.1.1 Topographic Influences

Subscale variability of topography has a marked influence on atmospheric convective processes and a strong control on surface hydrological flows. The interaction between atmospheric and hydrological processes and topography necessarily involves questions of both temporal and spatial scale, as well as identification of significant, controlling topographic parameters, such as elevation, slope and aspect. Such topographic parameters can be readily derived from digital elevation models at various spatial scales. On the other hand, precipitation data in mountainous areas are often poorly sampled by ground stations and poorly measured by radar.

Indirect and remote effects, that are more complex than direct orographic forcing, arise through the effects on convection initiation of heat-generated mesoscale circulations in mountainous terrain. Remote (far-field) effects involve travelling mesoscale systems not represented by existing (single-column) parameterisations. These issues can be addressed using a hierarchical approach based on (parameterised) regional-scale and (explicit) cloud-resolving models (Liu et al. 1999, 2001a,b) that stem from oceanic convection studies (Grabowski et al. 1998; Wu et al. 1999). They provide statistically meaningful results that contribute to the development of both GCMs and statistical models.

Statistical analysis of the interaction of precipitation with topographic parameters involves calibrating appropriate, single-point statistical models to the observed temporal variability (Hutchinson 1991a, 1995a). Such models can separate long-term average precipitation patterns conveniently from anomaly patterns, and these separate components may well have different spatial scales and different topographic dependencies. This can be advantageous when using measured data to calibrate topographic dependencies to support spatial interpolation and statistical simulation of precipitation patterns. It can also help to identify key statistical parameters associated with longer-term change and predictability.

So far, most statistical analyses of the interaction between precipitation and topography have focused on spatial interpolation of long-term monthly mean precipitation (Daly et al. 1994; Hutchinson 1995b). This can be used to describe long-term average seasonal variability, this being closely associated with patterns of natural vegetation. Moreover, interpolated monthly mean precipitation can also be considered one of the important parameters in a temporal-statistical model of precipitation. Monthly mean precipitation patterns are modulated strongly by topography. Elevation is usually the primary factor but its influence varies spatially, thus invalidating the use of simple regression between precipitation and elevation (Chua and Bras 1982; Hutchinson 1995b). However, spatially coherent elevation dependency can be calibrated by using robust multivariate spatial analysis techniques such as thin plate smoothing splines (Wahba and Wendelberger 1980; Hutchinson 1991b, 1995b).

These techniques and others indicate that the relative impact of elevation on precipitation patterns is two orders of magnitude greater than the impact of horizontal position (Hutchinson 1995a, 1998; Running and Thornton 1996). Thus, precipitation patterns usually reflect a topographic landscape that is exaggerated in the vertical, leading to a significant influence on precipitation patterns by relatively modest topographic features (Bindlish and Barros 1996; Barros and Kuligowski 1998). The spatial resolution of this dependence has been estimated as 4–10 km (Daly et al. 1994, Hutchinson 1998). Slope and aspect also affect precipitation patterns, at a similar spatial resolution (Hutchinson 1998). Further study is needed to clarify the scales of the topographic dependency of precipitation. These may differ for different precipitation averaging periods and for windward and leeward precipitation-generating processes (Barros and Kuligowski 1998; Buzzi et al. 1998). More studies are needed to develop and calibrate robust spatial statistical models capable of representing the interrelationship between precipitation and complex topography.

Statistical analysis shows that precipitation anomaly patterns that are due to large-scale synoptic weather patterns display relatively broad spatial coherence and are relatively insensitive to topography (Hutchinson 1995a). Such patterns are more readily represented in regional, mesoscale meteorological models and are amenable to analysis in relation to broad-scale circulation patterns (Walsh et al. 1982; Klein and Bloom 1987; Lyons 1990). It is fortunate that the statistical properties of such patterns can be more readily determined from less dense networks operated during a restricted time period. Once determined, these anomaly patterns can be added to the background (long-term mean) topography-dependent patterns to generate complex statistical models of precipitation patterns. Study is needed to identify the relative sensitivity of long-term mean patterns and anomaly patterns to the factors that affect long-term change and predictability. This will require close attention to the statistics of precipitation and due recognition of appropriate physical constraints where these can be identified.

1.1.2 Snow, Ice and Frozen Soil

Snow processes are important in climate and weather prediction models because of the unique characteristics of snow, specifically its high albedo, low thermal conductivity, and the fact that snow and ice cover often exhibit considerable spatial and temporal variability (Liston et al. 1999). The control exerted by snow and ice on energy and water exchanges between the atmosphere and the underlying soil are therefore markedly different to other surfaces. In addition, the timing of snowmelt and the subsequent fate of melted water play an extremely important role in the hydrological response of catchments.

Several studies have investigated the development and validation of snow submodels in climate models (e.g. Loth et al. 1993; Lynch-Stieglitz 1994; Yang et al. 1997; Schlosser et al. 2000; Slater et al. 2001; Liston 1999). These stand-alone model evaluations reveal that there are significant observational problems when validating snow/ice models. There are, for instance, challenges in specifying the model-forcing data accurately. Measuring snowfall is difficult, especially in windy conditions. It is also difficult to specify a threshold temperature to characterise when precipitation falls as snow rather than rain; it is also difficult to measure downward long-wave radiation but it is important to do so because this is the dominant wintertime radiation flux (Yang et al. 1997).

When coupled to atmospheric models, snow/ice models do seem to be able to capture the broad features of the seasonal snow regime such as seasonal variations in the snow line. However, a convincing explanation of the still significant discrepancies in simulations of snow cover remains illusive because of the complex feedbacks between precipitation, air temperature, radiation, topography, vegetation and snow. In the case of global models, for instance, the reasons for the often modelled delays in snowmelt at high latitude in Eurasia and North America are unknown (Yang et al. 1999), and similar unexplained weaknesses are also observed in regional model simulations of, for instance, the climate of the Pacific Northwest U.S. (Leung and Ghan 1999).

Thus, there are many difficult challenges remaining in providing adequate models of snow accumulation and snowmelt processes before predictive coupled hydrologic/atmospheric models can be expected to reproduce successfully the reported observed relationship between spring snow pack and subsequent summer rainfall in the southwest U.S. (Gutzler and Preston 1997). It is necessary to investigate the development of both uncoupled and coupled snow/ice models in regions with and without strong topography and to study the potential climate predictability associated with the snow/ice memory at catchment to regional length scales and at time scales up to seasonal.

Although there have been many studies of the effect of liquid soil moisture on climate, the effect of frozen soils on land-surface processes and the climate system have received little attention. However, about 70% of the Earth's land surface experiences seasonal freezing (Kinoshita 1982) which often lasts several months and may reach a depth of 2–3 m, while 24% of northern hemisphere continents are underlain by permafrost (Zhang et al. 1999). Because of the latent heat of fusion, freezing and thawing wet soil involves a very substantial uptake or release of energy so soils that freeze and thaw have, in effect, a large heat capacity. Freeze-thaw cycles influence the thermal and hydrological properties of the soil and this has a significant impact on surface energy and moisture balance (Kinoshita 1982; Williams and Smith 1989; Yershov 1998) and, hence, the climate system. Freezing soil increases the thermal conductivity and hence the soil heat flux, but it reduces the hydraulic conductivity and infiltration, thus increasing runoff; near-surface soil moisture may still increase due to restricted deep drainage. The existence of a thin frozen layer at the surface essentially decouples the moisture exchange between the land surface and the atmosphere.

There are substantial inter-seasonal fluctuations in the area of snow in the Northern Hemisphere and recent studies suggest there has been a decrease over the last 20 years (Armstrong and Brodzik 1999; Pielke et al. 2000) as air temperature has increased (by almost 1° C in the case of western North America). If the area with snow cover is reduced, the areal extent and thickness of frozen soils might well increase. On the other hand, increased temperature may itself result in less frozen soil. Additional data on snow cover extent, depth, and snow water extent are needed to aid study of this phenomenon.

1.1.3 Soil Moisture

The land, biosphere, atmosphere and oceans are coupled together in an Earth system in which there is variability at a wide range of time and space scales (Pielke et al. 2000). Variability and memory in this system are due to the cycling of water between reservoirs and may be strengthened by the development of feedbacks in the linkages between the reservoirs. Although the land fraction of the Earth is fairly small (30%), its distribution into large contiguous areas, the prevalence of thunderstorms over land (Lyons 1999) and its distinctive hydrothermal inertia cause significant variations in regional climatic systems.

Hydrological states that have a long memory, such as soil moisture, may serve to integrate past atmospheric forcing and enhance prediction skills for regional climates. Fennessey and Shukla (1999), Atlas et al. (1993), Bounoua and Krishnamurti (1993a,b), Xue and Shukla (1993), and Oglesby (1991) presented examples of numerical experiments, based on General Circulation Models (GCMs), which indicate the sensitivity of climate simulation to initialization of surface soil moisture. Early studies by Delworth and Manabe (1993) showed that the presence of an interactive soil-moisture reservoir acts to increase the variance and add memory to near-surface atmospheric variables such as humidity, while Milly and Dunne (1994) identified and analysed shifts in atmospheric general circulation and hydrological cycle in response to soilwater storage capacity. Koster and Suarez (1996, 1999) introduced statistical measures to distinguish between inherent climate variability and variability due to the presence of land memory in the form of soil moisture.

The presence of feedback mechanisms can enhance land-memory phenomena. Thus, if positive feedback mechanisms are present in the coupled land-atmosphere system, an initial anomaly can persist through reinforcement at both climate and weather time scales. Cook and Ganadeskian (1991) and Cook (1994), for instance, showed that precipitation and tropical general circulation following an initial soil-moisture anomaly are altered in ways that tend to reinforce the perturbed surface conditions. Brubaker et al. (1993) and Entekhabi et al. (1992) identified one such feedback mechanism that can reinforce surface anomalies, that is, when local precipitation is partially derived from local evaporation, reduced evaporation leads to reduced precipitation that, in turn, leads to further drying. Scott et al. (1997) and Vinnikov et al. (1996) demonstrated the importance of both soil-moisture reservoir size and precipitation recycling which “refers to how much evaporation in an area contributes to the same precipitation in the same area" (Trenberth 1999). However, there are conditions under which such simple feedback loops are not established. Thus, the relative roles of local versus external forcing for hydroclimatic anomalies over North America depends on factors that may either favour or counter drought or flood conditions (Giorgi et al. 1996). In addition, Trenberth (1999) noted for the tropics that the fractions of moisture flowing through a region which is precipitated out, exceed 40% over the areas of the African monsoon; the "maritime continent" of SE Asia; the southern part of the Amazon basin; and the western Ghats of India. Principally, these high fractions coincide with the strong low-level maritime transports (i.e. synoptic-scale moisture advection) associated with the seasonal movement of the precipitation fields.

On weather and storm event time scales, there is also evidence that initial soil conditions can reinforce the development of precipitating weather systems (Chen et al. 2001). At regional scale, soil-moisture availability has substantial influence on elevated mixed layers and on associated "lids" on atmospheric instability that act to focus the release of convective instability and hence determine the distribution of the regional precipitation in time and space (Benjamin and Carlson 1986; Clark and Arritt 1995). Such coupling was clearly demonstrated by numerical modelling in the GCIP–GEWEX Continental Scale International Project, that covers the entire Mississippi River Basin (Paegle et al. 1996; Beljaars et al. 1996; Betts et al. 1996; Pielke et al. 2000; Liu and Avissar 1999a,b). Such mechanisms are, for instance, believed to have played a significant role in the Mississippi River floods in 1993. Castelli and Rodriguez-Iturbe (1993) showed that growth of baroclinic instability could be enhanced by anomalies in surface fluxes. Using numerical mesoscale atmospheric models, Chang and Wetzel (1991), Ziegler et al. (1997) and Fast and McCorcle (1991) showed that the evolution of summertime weather systems in the Midwestern U.S. is critically dependent on so-called "dryline" conditions where sharp gradients in soil moisture are present.

Elsewhere, in the Sahel, Taylor and co-workers (Taylor et al. 1997; Taylor and Lebel 1998) observed a positive feedback between the land surface and rainfall in semi-arid conditions. These authors noted that in the absence of significant topographic features, there was an association between local intensification of convective instabilities, which delivered more intense rain from deeper convection, and a priori mixed layer anomalies in specific humidity within the planetary boundary layer (PBL). These latter anomalies were created by antecedent rain patterns which enhanced the specific humidity over the previously wetter (rained-on) areas, by way of a positive feedback of higher evaporation and available energy compared with areas of lower rainfall. Consequently there exists a persistence of preferred areas of higher precipitation, thus reinforcing soil moisture patterns at a convective scale of about 10 km.

The nature of the land-surface cover and antecedent moisture also has implications for the maintenance of easterly perturbations during their progression westwards over the more sparse, moisture limited vegetation typical of the Sahelian region, compared with the better watered, more dense vegetation further south. Modelling by Taylor et al. (2000) showed that the surface land cover and antecedent moisture has a major influence on the PBL depth and potential temperature, with the depth of the moist layer being deeper and warmer over sparse compared to dense vegetation within the high pressure ridge (24 hours ahead of the passage of the easterly trough). The larger sensible heat and long-wave heating for sparse vegetation causes the depth of the PBL to increase; this enables a near-surface parcel of air to more likely reach the lifting condensation level. A consequence of this is the higher convective available potential energy (CAPE) over sparse vegetation, thus triggering deep convection sooner over sparse vegetation compared with more dense vegetation. As a result, the subsequent passage of an easterly perturbation does not necessarily require the additional forcing of daytime heating to be convectively very active from the previous build-up of large conditional instabilities to support travelling easterly perturbations.

Further south towards the humid tropics of west Africa, well-watered, dense vegetation provides a higher evaporative fraction, and triggers more frequent, short-lived, daytime convection rain. Thus the relative contribution of squall lines to the total rainfall over a wet season tends to reduce over the more dense vegetation further south (Taylor et al. 2000). Moreover, heavy rainfall ahead of an easterly trough only develops during daylight hours due to the requirement of additional forcing from diurnal (afternoon) heating (Taylor et al. 2000). As indicated, this is not a requirement for the Sahelian thermodynamic environment. Thus the implications of the modelling work of Taylor et al. (2000) are that the nature of the surface land cover and antecedent moisture have a significant influence on the maintenance of easterly trough squall lines.

Therefore, it is apparent that, in certain conditions, land memory in the form of the soil-moisture store, perhaps reinforced by positive feedback mechanisms such as recycling of precipitation, has significant effects on atmospheric variability and predictability and can lead to greater persistence of weather and climate anomalies. Delineation of the conditions under which soil-moisture state is important to the evolution of weather and climate, coupled with ways of estimating the initial soil-moisture state based on in situ and satellite observations and the realistic simulation of the subsequent evolution of that soil-moisture state in predictive models, should extend atmospheric forecast skills.

1.1.4 Vegetation and Land-cover Dynamics

Vegetation plays a major role in determining the surface energy partition and the removal of moisture from the soil by transpiration. Representation of the vegetation's response (i.e. the change in live biomass) to atmospheric and hydrological influences is currently weak in models used to give monthly to seasonal predictions of precipitation and hydrological variables. Aspects of interest relate both to representing heterogeneous vegetation cover in models and to representing the seasonal evolution of vegetation.

There has been substantial progress in representing heterogeneous vegetation by specifying area average parameters on two fronts, one being essentially empirical and the other theoretical. The empirical approach (e.g. Mason 1988; Blyth et al. 1993; Noilhan and Lacarrere 1995; Avissar and Schmidt 1998; Arain et al. 1996, 1997; Walko et al. 2000) is to create a coupled surface-atmosphere model, to postulate and test hypothetical rules (often called "aggregation rules", Shuttleworth 1991) to give parameters applicable at larger scales by combining the parameters that control surface exchanges for small plots of uniform land cover. The theoretical approach (e.g. Lhomme 1992; McNaughton 1994; Raupach 1995; Raupach and Finnigan 1995, 1997) is to adopt the equations that are accepted as reasonable descriptions of surface-atmosphere exchanges for small plots of uniform land cover, to assume that such equations can also be used to describe the area-average behaviour of heterogeneous cover, and to derive theoretical equations that link the parameters required at large scales with those that apply for individual small plots. Ongoing research under GCIP (GEWEX Continental Scale International Project) is investigating the sensitivity of model predictions to improved representation of heterogeneous vegetation cover.

Most meteorological models either prescribe a seasonal evolution in vegetation parameters or assume that they are constant. Exceptions include Eastman et al. 2001a,b; Lu et al. 2001; Kiang and Eltahir 1999 and Wang and Eltahir 1999, 2000a, b. Assimilating satellite observations is one way to provide a more realistic representation of current vegetation status in model simulations, and there is now great opportunity to develop this approach with data from the recently launched Earth Observing System (Stewart et al. 1996). However, when using this approach, care is needed to avoid creating inconsistencies between the space-time distribution of soil moisture and the assimilated vegetation biomass growth pattern.

As part of the coupling of the land surface with the atmosphere, there is now wide acceptance that forests make a significant contribution towards the mechanism of precipitation recycling through wet canopy (interception) evaporation and dry canopy transpiration. Over the Amazon forest the basin-wide average of precipitation recycling is within the range of 25–34% from various modelling estimates (e.g. Eltahir and Bras,1994; Trenberth,1999). For the West African region, Gong and Eltahir (1996) predicted that 27% of the annual precipitation would be derived from local recycling.

Sensitivity studies with changes in the land surface, such as complete deforestation of tropical or boreal forests, and realistic existing changes (Chase et al. 2000) may lead to altered climatic conditions (e.g. Lean and Rowntree 1997). GCM studies by Dirmeyer and Shukla (1996) have shown how, for example, increasing land degradation in (semi-) arid areas affected precipitation locally in the degraded areas, as well as modifying rainfall regimes outside of these degraded areas.

Such new climatic and related hydrological conditions may be inadequate to support the vegetation that existed prior to the perturbation (daRocha et al. 1996; Claussen 1996; Kutzbach et al. 1996). On removal of the forest, for example, the surface fabric of selected tropical soils "collapse" on disturbance; this results in a major change in the infiltration and the corresponding preferred pathways of hillslope storm runoff and vertical percolation (Bonell 1998). Infiltration-excess overland flow now occurs with a corresponding reduction in surface infiltration and recharge to deeper groundwater bodies by vertical percolation. The associated reduction in subsurface soil moisture and, in particular, in the deeper soil moisture, arrests the ability of forest to regenerate. These new conditions also place limits on the ability of evergreen forests to maintain their transpirational requirements through reduced access to deeper groundwater during seasonal and inter-annual drought (Nepstad et al. 1994; Bonell 1998).

At smaller spatial and time scales, land-surface variability can influence the amount of precipitation and its spatial organisation (see review by Pielke et al. 1998, 1999). Factors involved include changes in energy budget, frictional effects, changes in horizontal convergence and associated vertical velocities (Copeland et al. 1996; Bonan 1997). Thus changes in land cover can affect both single weather events, and their time integral accumulating to seasonal pattern changes, mostly for (sub-areas of) the continental USA. For example, Pielke et al. (1999) noted an 11% decrease in July–August rainfall in south Florida which could be attributed to progressive conversion of everglades to urban and agricultural areas between 1900—1993; see also Sect. 4.8.

Elsewhere, Blyth et al. (1994) were able to demonstrate for a specific case of boundary conditions in humid temperate Southwest France that there was a sensitivity of mesoscale rainfall to forest cover as against bare soil. They concluded that complete forest cover in a 400 km × 400 km domain could increase frontal rainfall by 30% compared with a 3-D mesoscale model simulation for bare soil. About half of this increase in rainfall was caused by moisture convergence, the remainder by enhanced, rapid re-evaporation of intercepted rainfall (wet canopy evaporation) over the forest. In the latter case, two mechanisms were involved. First, a negative sensible heat flux provided the energy to sustain evaporation of intercepted water and second, the greater surface roughness to airflow over the forest canopy which enhanced the physical and dynamical effect of turbulent transport. Blyth et al. (1994) concluded that this effect was also strengthened by frictional effects associated with the passage of weather fronts from the sea to land along coastal areas. The extension and generality of this phenomenon to other environments with varied surface cover needs to be investigated

Because land-use change is currently most rapid in the tropics (O'Brien 2000), the land-surface sensitivity studies have been performed mostly for the tropics. However, Dirmeyer (1994) showed that a change in vegetation dynamics affected the prediction of weather in the mid-latitudes of the continental U.S. This result is of considerable importance as it suggests that even for mid-latitude conditions with advection-dominated weather patterns, the land surface can still play a key role in the weather. Similarly for western Europe, early results with the RACMO model suggest that the land surface forcing of weather is stronger in the summer than in the winter (van den Hurk, *pers. comm.*). Growth of vegetation depends to a large extent of the total environmental stress undergone during the summer period, making studies of the interaction of vegetation dynamics and weather particularly important for this season. It is also worth noting that some of the largest, ancient land-use changes have occurred at the mid-latitudes (Reale and Dirmeyer 2000).

The *direct (energy water exchange) and indirect (carbon dioxide exchange) land-cover/climate interactions are intrinsically coupled.* It is, for example, hypothesised that vegetation regulates stomatal opening to maximise CO_2 uptake while minimising water loss. To simulate the interaction of plant physiology with carbon dioxide realistically, models have been developed which relate photosynthesis and canopy conductance (Ball et al. 1987). The importance of the global carbon dioxide feedback on plant stomatal conductance was elucidated by Sellers et al. (1996) and Randall et al. (1996) in a GCM sensitivity study. Inclusion of this feedback increased surface temperatures through a reduction in evaporation and reduced rainfall over land areas. The magnitude of this reduction in rainfall and evaporation depends critically on the parameterisation employed in their land-surface scheme. However, those studies did not include the biogeochemical effect of vegetation on the subsequent atmospheric response; incorporating dynamic vegetation into a land-surface model is a fairly new endeavour, but research in this area has already provided important insight. Claussen (1995), for instance, used an interactively coupled global atmosphere-biome model to assess the dynamics of deserts and drought in the Sahel. He found that the comparison of atmospheric states associated with these equilibria corroborates Charney's (1975) hypothesis that deserts may, in part, be self-inducing through albedo enhancement. Ji (1995) developed a climate-vegetation interaction model to simulate the seasonal variations of biomass, carbon dioxide, energy and water fluxes for temperate forest ecosystems in northeastern China. Foley et al. (1998) directly coupled the GENESIS GCM and IBIS Dynamic Global Vegetation Model through a common treatment of land-surface and ecophysiological processes. They found that the atmospheric portion of the model simulates the basic zonal distribution of temperature and precipitation correctly (albeit with several important regional biases) and that the biogeographic vegetation model was able to capture the general placement of forests and grasslands reasonably well. On a regional scale, Eastman et al. (2001b) showed that the biogeochemical effect of double atmospheric CO_2 concentrations may have a more immediate, and quite substantial, effect on seasonal growing season weather than the radiative effect of doubled CO_2.

An interactive canopy model (Dickinson et al. 1998) has been added to the Biosphere-Atmosphere Transfer Scheme (BATS: Dickinson et al. 1986, 1993) to describe the seasonal evolution in leaf area needed in atmospheric models and to estimate carbon fluxes and net primary productivity. This scheme differs from that used in other studies by focusing on short time-scale leaf dynamics. Tsvetsinskaya (1999) introduced daily crop growth and development functions into BATS and coupled it to the National Center for Atmospheric Research (NCAR) Regional Climate Model to simulate the effect of seasonal crop development and growth on the atmosphere/land-surface heat, moisture, and momentum exchange. She found that the coupled model was in better agreement with observations than the earlier, non-interactive model. Lu et al. (2001) developed and implemented a coupled RAMS/CENTURY modelling system and successfully applied it in the central U.S. to study the two-way interactions between the atmosphere and land surface at seasonal-to-inter-annual time scales (Lu 1999). All these early attempts promote the value of including two-way feedbacks between the atmosphere and biosphere in meteorological models to create the soil-vegetation-atmosphere transfer schemes (SVATS) with "dynamic vegetation" (IV), hereafter called IV-SVATS.

Need for Understanding of Coupling between Pedosphere-Biosphere and Hydrology: Example of Carbon

It has become increasingly clear that questions relating to the uptake or release of greenhouse gasses by terrestrial vegetation cannot be seen separately from the direct interaction of vegetation with the atmosphere, through heat and water exchange, with possibly global and regional effects (Steffen et al. 1998). This calls for scientific studies addressing not only the possible options for long-term carbon sequestration but also possible climate change as a result of changing land use, and with explicit accounting for the feedbacks within hydrological cycle.

The terrestrial carbon balance is the sum of the amount of carbon absorbed by vegetation and released through heterotrophic respiration. In contrast with industrial emissions, fluxes to/from terrestrial vegetation are part of an active *biological and hydrological cycle*. Carbon stored in plants and soil may return to the atmosphere on the time scale of decades (plants) to centuries (soils). Two-thirds of the terrestrial carbon is found in the soil. Land and water management, for example, may be aimed at strengthening the sequestration of carbon into below-ground, rather than above-ground storage.

For example, Cox et al. (2000a,b) present the results of a numerical experimentation with a carbon-climate model indicating that carbon-cycle feedbacks could significantly accelerate climate change. In several studies, the role of European forests in the global carbon cycle have been estimated. One of these estimates, based on a review of *static conversion* (without accounting for the feedbacks) of national forest inventory data of European countries, gives a sink strength for the European forest of 101 Tg C y^{-1} for the mid-eighties (Nabuurs et al. 1997). This large sink would compensate 9.5% of total European fossil fuel emissions. However, Martin et al. (1998) provide an estimate based on the EUROFLUX data of 120 to 280 Tg C y^{-1}. The difference is clearly due to the fact that the latter estimate represents the total ecosystem flux (indicating a large sink in the soils) and therefore including the feedback mechanism.

The global picture is even more contradictory. Studies using the NOAA flask network suggest that a significant portion of the uptake of the terrestrial biosphere is located at mid-latitudes (Denning et al. 1995). A recent study suggests that most of this uptake is in North America, Eurasia being almost in equilibrium and the tropics being a small source (Fan et al. 1998). The tower flux data from EUROFLUX cited above (Valentini et al. 2000) suggest, however, that European forests also take up significant amounts of carbon (1-5 t ha^{-1} y^{-1}), as do the Amazonian forests (from 0.6 t ha^{-1} y^{-1}, according to Phillips et al. (1998) up to 5.9 t ha^{-1} y^{-1} according to Mahli et al. (2002). Changes in the global signal of seasonal CO_2 concentrations (Keeling et al. 1996) show the sensitivity of climate variability and the consequences for source/sink relations of the biosphere.

Possible causes of these differences in carbon uptakes, and whether they are transient in nature, are heavily discussed at present and are clearly linked to the level of our understanding of the feedbacks involved. Possibilities include climatic variations (ENSO), reforestation, and carbon dioxide fertilisation. Schimel et al. (2000) recently questioned the latter possibility for a larger uptake of CO_2 by the biosphere under higher CO_2 concentrations and a warmer climate, because that would also mean at least a doubling of nitrogen fluxes into ecosystems. Current nitrogen deposition trends do not indicate such increases, suggesting that the future C-cycle will probably be limited by nitrogen. Therefore it is of utmost importance to integrate biogeochemical and water cycles.

1.2 Scale Issues

Axel Bronstert, Jesus Carrera, George Leavesley, Nicole Mölders

1.2.0 Introduction

In the development and application of coupled process models, the effects of space and time scales on process interactions and interpretations are major. It is important to have in mind, therefore, that the concept of different scales can be applied on various levels, in particular see Blöschl (1996) for example:

- *Process scale*: spatial and/or temporal extent for which a certain process occurring in nature can be called representative;
- *Measurement scale*: spatial and/or temporal extent which is determined by the measurement device or sampling principles;

- *Modelling scale*: spatial and/or temporal resolution of a model. Usually, variability which is smaller than the modelling scale is not treated by the model approach, at least not directly.

It is important to understand that process, measurement and modelling scales differ from each other in most hydrological and atmospheric studies. In some cases, the differences between theses scales can be of different orders of magnitude (example: rainfall formation, measurement and modelling). These differences are a major source of uncertainty in the description and modelling of hydrological, ecological and atmospheric processes.

Furthermore, one should be aware that the question which is the appropriate scale to be applied in a modelling study also depends on the overall question under study. Therefore, it is not advisable to give a general applicable guideline on what is "the best" scale for a model. This statement is even more relevant for coupled hydrological, ecological and atmospheric models.

Scale issues are important for almost every scientific discipline, including all types of natural and sociological sciences. Traditionally, most of the scientific disciplines have developed their own specific terminology for different scales. This applies also for hydrologists and climatologists, which makes understanding rather difficult when developing or applying coupled hydrological-atmospheric models. To simplify understanding further, we give below a summary of the terminology of space and time scales used in hydrology and atmospheric sciences. A comprehensive overview of scale issues in hydrology has been given by Feddes (1995). Stewart et al. (1996) summarised the possible use of remote sensing data for upscaling in hydrology.

Besides clear identification of the relevant space and time scales, it is important to realise that in many cases information has to be transferred between different scale levels. This information transfer is termed *scaling*. If information is transferred to a more detailed level (in space or time), we speak of disaggregation or *downscaling*. In contrast, if detailed information is lumped together and thus made available at a coarser level, we speak of aggregation or *upscaling*. Going from one spatial scale to another scale does not only require aggregation or disaggregation, it may also require the consideration of processes which are of minor importance at one scale but of high importance at another (see also the discussion in the next paragraph).

Scale issues have to be considered in particular each time a certain modelled process requires information input (measured or calculated) from an interacting process. So, in the case of coupled models with many interacting processes, scaling methodologies are of particular importance.

1.2.1 Why is Scale an Issue for Modelling Hydrological Processes?

Watching a movie of boats in the ocean, alert spectators immediately and intuitively identify whether the boats were filmed in a pool or in the actual ocean. Pool waves may look as large as ocean ones, but they never look as foamy. The reason is that surface tension, which is an important force for small water bodies, becomes negligible when compared with the inertial forces of true ocean waves. While it is evident that dominant processes may change with scale, scientific education has not emphasised that. Instead, the conventional continuum mechanics approach is built into the way of quantifying things. This approach essentially consists of several steps. First, observe processes at the laboratory scale. These observations are then generalised into constitutive laws. The third step is to recall some conservation law to derive a partial differential equation, which can then be used to predict the behaviour of the system at different scales and under different conditions to those of the laboratory observations.

The continuum mechanics approach is fine as long as the main processes remain unaltered. However, one of the points of this book is that such is never the case in hydrology or indeed for most of the related scientific disciplines. Viscosity is the main force for slowly moving water through narrow water paths, but again becomes negligible when compared to inertial forces in large channels. The former leads to laminar flow, the latter to turbulent flow. Heat conduction may be the main energy transfer mechanism at small distances, but becomes negligible at large distances when compared to convection.

One might wonder why the space and time scales are so closely linked (the examples we pointed out above illustrate what we have termed process scale and they only deal with "large" or "small", but not with time). To illustrate time dependence, let us consider the one-dimensional diffusion-advection equation (variants of this equation control the transport of vapour in the atmosphere, heat by ocean currents or pollutants in rivers and aquifers):

$$\frac{\partial c}{\partial t} = D\frac{\partial^2 c}{\partial x^2} - v\frac{\partial c}{\partial x} \tag{1.2-1}$$

where c represents the quantity being transported per unit volume of fluid (e.g. concentration), v is the mean velocity of the fluid and D is the diffusion (dispersion) coefficient, which represents mixing caused by Brownian motion (or by fluctuations of the fluid velocity around its mean). The solution of this equation for constant D and v and a unit input is shown in Fig. 1.2-1.

The main point in this figure is that the input is displaced and diffused. Therefore, one can define an advection scale as the distance travelled by the centre of mass and a diffusion scale as the distance over which this mass has diffused. As shown in Fig. 1.2-1, the expressions for these two scales are:

$$L_A = vt \tag{1.2-2}$$

$$L_D = \sqrt{2Dt} \tag{1.2-3}$$

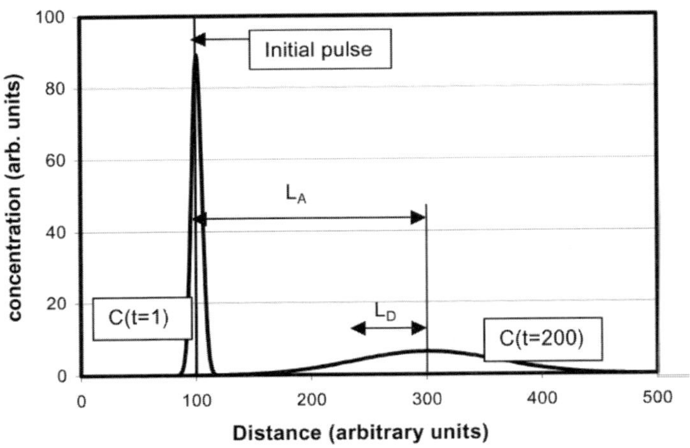

Fig. 1.2-1 Concentrations after 1 and 200d in response to a pulse injection of 1000g mass at x = 100 m in a fluid moving to the right at 1 m d^{-1} (actually, units are arbitrary). The diffusion coefficient is 10 m^2 d^{-1}. Notice that advection is negligible after 1 day, while diffusion displays a sizable effect (that is, the centre of gravity does not appear to change, but concentrations define a Gaussian bell, quite different from the initial point pulse). After 200 d, the pulse has moved 200 m to the right. Diffusion, still relevant, is overcome by advection, so that concentration is negligible at the injection point

These two scales are represented versus time in Fig. 1.2-2. Simple inspection of the figure makes it apparent that, for very short time intervals one might as well ignore advection because L_D is much larger than L_A. On the other hand, for long time intervals diffusion becomes negligible. Actually, things can be a bit more complex, because D itself may grow with scale, leading to dispersion in response to the increase of velocity fluctuations around its mean (e.g. the size of eddies in the atmosphere or heterogeneities in an aquifer that a contaminant may encounter grow as it disperses). Still, the fact remains that (1) the spatial scale of each process grows with time, and (2) the dominant process is not always easy to anticipate. In summary, processes that are dominant at one scale may become negligible at large scales, while new processes (e.g. dispersion) may emerge in response to changes in scale.

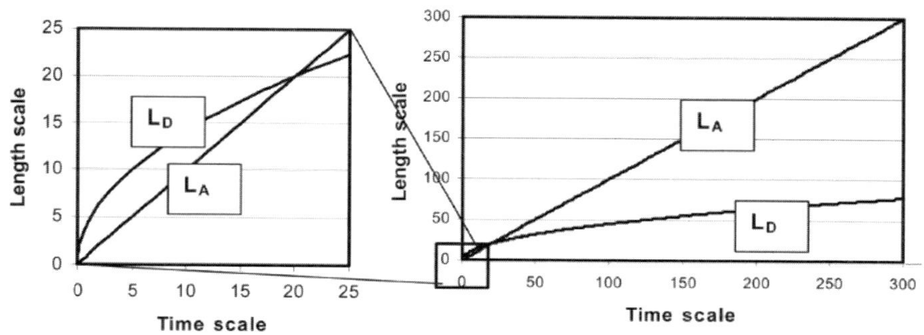

Fig. 1.2-2 Advection and diffusion length scales (L_A and L_D, respectively) versus time in the example of Fig. 1.2-1 (v=1 m d^{-1}, D=10 m^2 d^{-1}). Notice that, before 20 days, diffusion dominates advection, which is negligible before 1d

This example is restricted to a single subsystem (transport of solutes in a river or vapour in the atmosphere). The problem becomes eerie when coupling several subsystems. Each of them will have its own time and spatial scales and their corresponding dominant processes, which may not be easy to anticipate. Worse, the relative importance of each subsystem is problem-dependent. For example, regional groundwater flow may not be important when dealing with floods but may control river base flow. This is why it is important to understand the process, measurement and modelling scales that are relevant for each subsystem. Thus, it is important to define the basic hydrological scales, which is the topic of the following section.

1.2.2 Terminology of Scales in Hydrology and Atmospheric Sciences

The terminology concerning temporal and spatial scales used in different science disciplines is far from uniform. Thus, many disciplines derived their own notion for typical scales for that particular scientific area. Even within one scientific area, there may be several terms for a specific spatial or temporal extent. In the case of coupling processes, concepts or models of two or more scientific disciplines, it is of particular importance to understand the meaning of the terminology at the scales of the corresponding discipline.

In Fig. 1.2-3 the time scale terminology applied in hydrology and atmospheric sciences is presented and related to actual times or durations. One can see that the terminology of the hydrological time scales are derived from the distinction of rainfall/non-rainfall periods ("event scale"), or the diurnal and the seasonal periodicities, respectively. In atmospheric sciences, terminology is mainly derived from or related to the synoptic time scale, a term which has been used for a long time in weather forecasting.

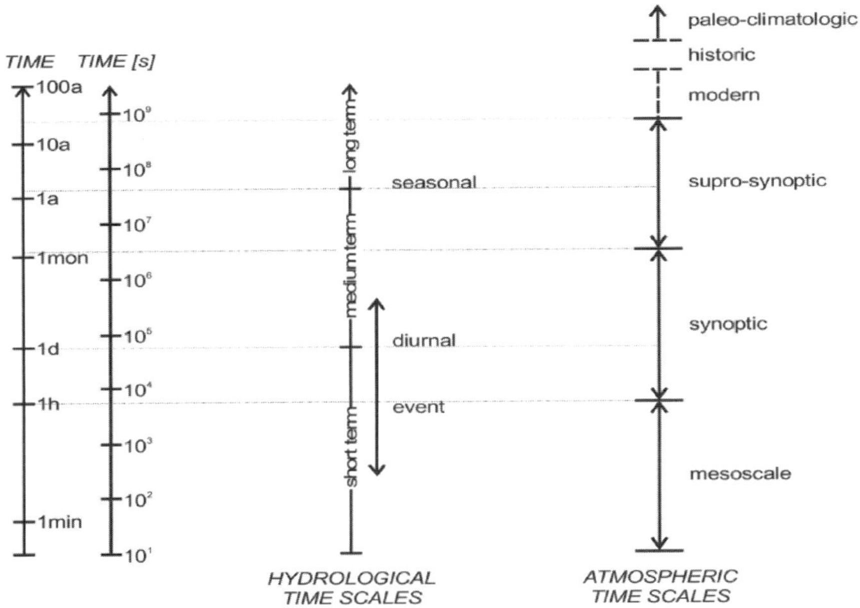

Fig. 1.2-3 Time scales in hydrology and climatology

In Fig. 1.2-4 the terminology on spatial scales applied in hydrology and atmospheric sciences is presented and related to actual spatial lengths or areas. In hydrology, two concepts are mainly applied: first, the distinction of the typical size of different landscape fractions (such as local/point scale, hillslope, catchment, region, continent). Second, the terms micro-, meso- and macroscales are applied to distinguish between the dimensionality of the processes considered: the hydrological microscale is often (but not always!) treated as one-dimensional in the vertical direction and relates to small spatial extensions of usually not more than a few metres (e.g. a soil column in the laboratory, a lysimeter, an experimental field plot). The hydrological mesoscale is mostly attributed to more or less uniform landscapes or catchments where, on the one hand, many different hydrological processes are occurring, but – on the other hand – might be described by so-called "lumped" concepts, aggregating the hydrological cycle of the whole area in a straightforward systems approach. Finally, the hydrological macroscale covers large areas with very different landscape features (e.g. mountains and lowlands) and hydrological systems (e.g., river systems and lakes), which cannot be aggregated by a lumped approach. Having this in mind, it is obvious that the interface between the hydrological meso- and macroscale is a fuzzy one. The definition of the spatial scale in atmospheric sciences is more straightforward, distinguishing the climatological micro-, meso-, and macroscales with clear defined interfaces (2 km length and 2000 km length respectively), with subdivisions noted alpha, beta and gamma, where the alpha, beta, gamma stands for the large, medium, small subscale, respectively.

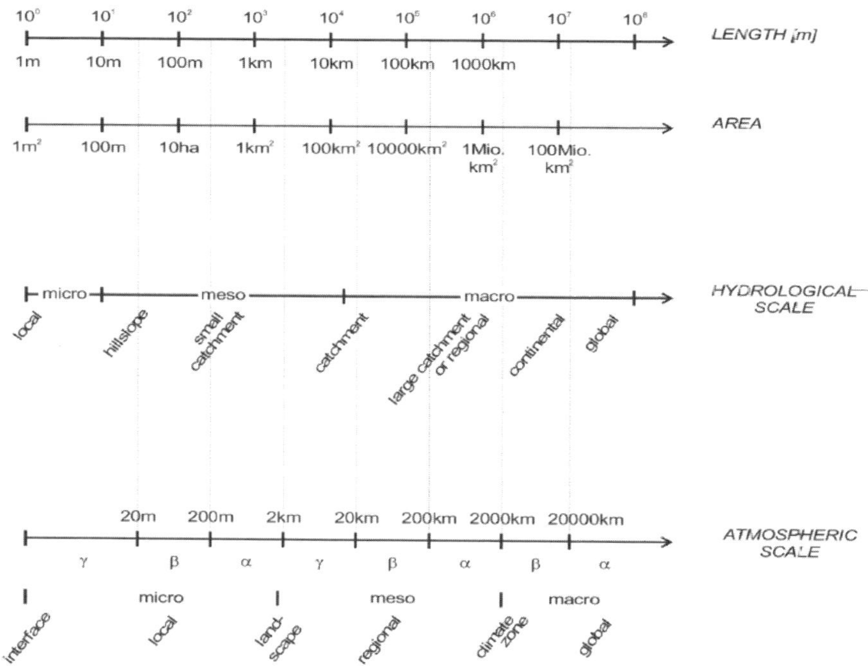

Fig. 1.2-4 Spatial scales in hydrology and climatology

There is a typical relationship between space and time scales. For example, the smaller the spatial extent investigated, the shorter the appropriate time (Hugget 1991).

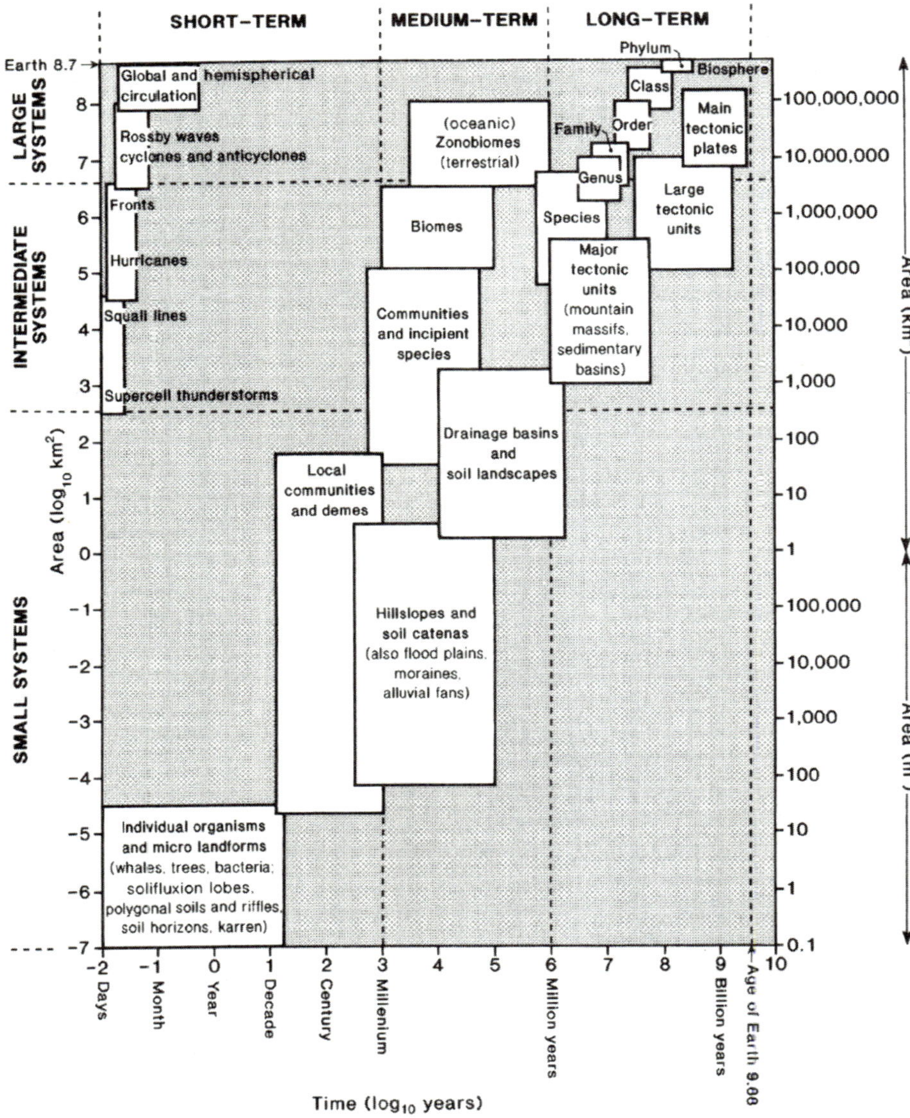

Fig. 1.2-5 Schematic relationship between spatial and temporal process scales for a variety of Earth system processes (Hugget 1991)

Fig. 1.2-5 gives a very comprehensive example of this relationship putting all kinds of Earth system processes in a framework of space and time scales, ranging between durations of about an hour to a billion years and areas of a few dm^2 to the complete surface of the globe.

This concept has been adopted by Blöschl (1996) to relate the space and time scales of rather different hydrological and atmospheric processes (Fig. 1.2-6). Precipitation processes are shown to range from spatial scales of tens of metres to more than 1000 km and over time scales from minutes to days. Hydrological processes are shown to have a wide range of scale effects depending on the process of interest. Infiltration and surface runoff processes occur at scales from metres to 100s of metres at time scales of minutes. Subsurface and groundwater processes, however, range from 10s of metres to a 1000 km at time scales from hours to much more than a 100 years. Coupling this range of process variability with spatial variations in topography, soils and vegetation presents a complex problem in resolving the linkages and interactions among these processes across the wide range of space/time scales of interest to hydrologist and atmospheric scientists.

One can see from Fig. 1.2-6 that the relationship between characteristic space and time scales (i.e. the gradient of the regions of particular processes, which is the ratio of characteristic space and time scales for a given process over a range of scales) is pretty constant for the specific processes but varies significantly between those processes. This relationship has been noted as characteristic velocity by Haltiner and Williams (1980) and Blöschl (1996). Additionally, one can see, if a time/space scale is fixed and the space/time scale is varying, the processes of relevance are changing. This varying relevance of processes depending on scales has been illustrated before by the convection-dispersion equation (Eq. 1.2-1). The different time scales of importance for a number of parameters in biosphere-atmosphere interactions are summarised in Box 1.2-1.

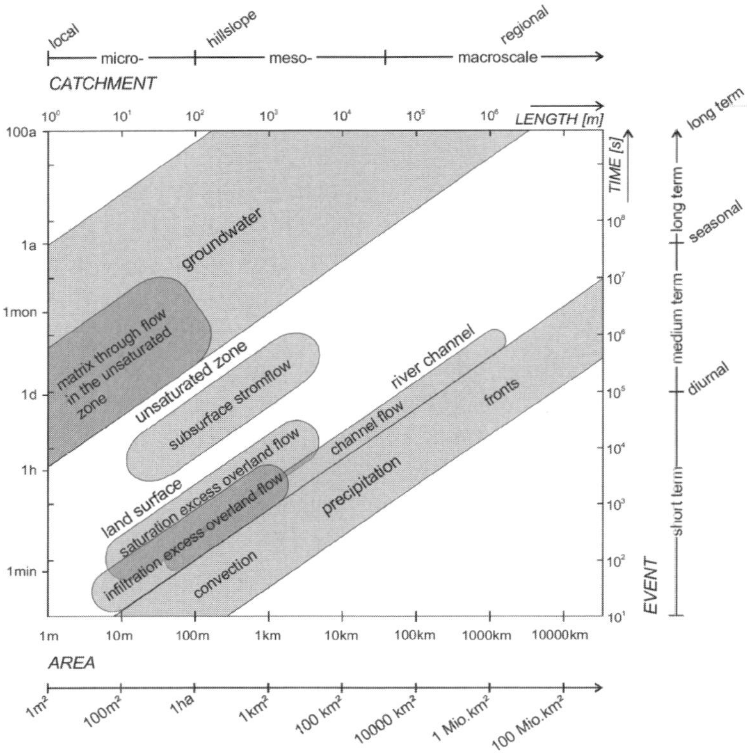

Fig. 1.2-6 Schematic relationship between spatial and temporal process scales for a variety of hydrological and atmospheric processes, from Niehoff (2002), extended from Blöschl (1996)

1.2.3 Scaling of Hydrological Processes to be used in Atmospheric Models and vice versa

Coupling the range of process variability presented, e.g. in Fig. 1.2-6, with spatial variations in topography, soils and vegetation, presents a complex problem in resolving the linkages and interactions among these processes across the wide range of space/time scales of interest to hydrologists and atmospheric scientists.

This is of particular importance if hydrological and atmospheric models are to be coupled in order to provide information about the hydrological or atmospheric processes, including their feedbacks, to the corresponding model. Atmospheric model scales are spatially too coarse to provide reliable assessments for smaller scale basins where most of the concerns of hydrologists and resource managers occur. At the same time, the hydrological process models typically used in basin studies operate at scales that are too small for their direct incorporation into atmospheric models. As discussed by Hostetler (1994), there is a problem of discordant scales between atmospheric and hydrological models which is being addressed by scaling up or aggregating hydrological models and downscaling or disaggregating atmospheric models. Below, we summarise the current methods applied to aggregate (upscaling) or disaggregate (downscaling) information between hydrological and atmospheric models and vice versa.

Aggregation (Upscaling)

Aggregation of subscale hydrological or land-surface processes for the purpose of describing land-atmosphere interactions is accomplished using a variety of methods. Several different strategies have been developed to parameterise subgrid-scale heterogeneity of land-surface processes. For instance, an "effective" parameter approach attempts to describe the heterogeneities or distribution of a given process using a single aggregate or weighted value by averaging surface properties (e.g. Lhomme 1992; Dolman 1992) or by a statistical-dynamic weighting approach (e.g. Wetzel and Chang 1988; Entekhabi and Eagleson 1989). Computationally more expensive procedures to consider patchy surface properties are the mosaic approach (Avissar and Pielke 1989), the explicit subgrid strategy (Seth et al. 1994), or the mixture strategy, wherein, for the different surface types, tightly coupled energy balances are determined (e.g. Sellers et al. 1986; Dickinson et al. 1986). Several authors comparing the results provided by simulations with and without consideration of subgrid-scale surface heterogeneity found that for very patchy surfaces large differences in the predicted atmospheric fluxes can occur (e.g. Avissar and Pielke 1989; Seth et al. 1994; Mölders and Raabe 1996). A review of methods to treat heterogeneity is given by Giorgi and Avissar (1997).

Another approach uses a probability distribution function to describe heterogeneities of land-surface processes. The *variable infiltration capacity* model (Wood et al. 1992) couples a distributed function of soil moisture storage values in a catchment to account for subgrid variability of soil moisture. This also allows for the simulation of the effects of changing probability density functions of soil moisture on runoff generation and other hydrological processes. The topographic index of Beven and Kirkby (1979) has also been used to describe soil moisture distribution and to aggregate subgrid variability of a soil-vegetation-atmosphere transfer processes (Famiglietti and Wood 1994).

Delineation by elevation zones has also been used to account for subgrid variability in mountainous regions. Leung et al. (1996) developed a subgrid orographic precipitation model that was then included in the Pacific Northwest Laboratory (PNL) regional atmospheric model. The scheme partitions a grid cell into elevation classes and radiative transfer, turbulent mixing, convection and land-surface physics are computed for each elevation class. A full set of hydrological variables, including precipitation, snow water equivalent, soil moisture and surface runoff, are simulated for each elevation class as well.

The process of aggregating small-scale hydrological process model formulations to macroscale formulations suitable for incorporation in a General Circulation Model (GCM) has been a focus of a number of climate research programs such as FIFE, BOREAS and GCIP. Research is also being conducted independently of these large projects and the extent of this research is demonstrated in the large number of land-surface models participating within PILPS. References to these models are available in Henderson-Sellers et al. (1995).

Subscale variability is not only of importance for coupling hydrological and atmospheric processes, i.e. land-atmosphere interactions. Many authors (e.g. Dooge 1995) have stressed the importance of small-scale variability on soil moisture dynamics and on infiltration conditions. Looking on runoff generation processes, Bronstert and Bárdossy (1999) presented an example on how strongly the spatial subscale variability of soil moisture can influence the runoff generation process. In Bronstert and Bárdossy (2003) the problem has been elaborated for the temporal subscale variations of precipitation, i.e. short-term fluctuations of rainfall intensity, which proved to have an enormous impact on the generation of surface runoff, in particular if the runoff is only a small fraction of the precipitation, i.e. the soil has a large storage potential.

Disaggregation (Downscaling)

While downscaling attempts to address the heterogeneities within a model grid, the simulation of regional climate variations at the scales required for environmental impact assessments is unreliable at individual grid and subgrid box scales (Intergovernmental Panel on Climate Change 1996). This mismatch, between what the climate impacts community requires and what the GCMs are able to supply, has been a confounding issue affecting the confidence placed in impact scenarios at the hydrological mesoscale (Hostetler 1994; Xu 1999). Two techniques have been developed that attempt to counter this deficiency: semi-empirical (statistical) downscaling

(SDS) of GCM outputs, and regional climate models (RCMs) nested within a GCM (Giorgi and Mearns 1991; Wilby and Wigley 1997).

Statistical downscaling (SDS) bridges the two different scales by establishing empirical (statistical) relationships between large-scale features (such as geopotential height fields) and regional or local climate variables (such as temperature and precipitation at a certain location). It is analogous to the "model output statistics" and "perfect prog" approaches used for short-range numerical weather prediction (Klein and Glahn 1974) which both use correlations with atmospheric variables at the synoptic scale (such as geopotential height fields) to simulate weather at the local scale (such as single site precipitation). Common SDS procedures involve weather-type classification, linear and non-linear regression, or modifications to stochastic weather generators (see Wilby and Wigley 1997). A key strength of SDS is the low computational demand which facilitates the generation of ensembles of climate realisations. However, realistic SDS scenarios are contingent on strong/stationary empirical relationships, and on the choice of predictor variable(s) and transfer function(s) used for the downscaling (see Winkler et al. 1997).

Although many studies have discussed the theory and practice of statistical downscaling (e.g. Bárdossy and Plate 1992; Hay et al. 1992; Karl et al. 1990; Kim et al. 1984; von Storch et al. 1993; Wigley et al. 1990; Bürger 1996), relatively few have considered explicitly the limitations of such techniques (Giorgi and Mearns 1991; Wilby and Wigley 1997). However, sensitivity analyses have demonstrated the susceptibility of downscaled scenarios to season definitions, the choice of data standardisation technique, length of calibration period, function form and predictor variable(s) (e.g. Winkler et al. 1997). It has also been shown that different circulation schemes (Buishand and Brandsma 1997) and downscaling methodologies (Wilby et al. 1998b) yield markedly different regional climate change scenarios, even when common sets of GCM predictors are used. Finally, there is scepticism regarding the assumed stationarity of predictor-predictand relations (Wilby 1997), and the reproduction of low-frequency surface climate variability in downscaling schemes continues to be problematic (Katz and Parlange 1996).

Dynamic downscaling is performed by applying regional atmospheric models. They simulate subscale climate features (relative to global models) dynamically at resolutions of 20-50 km, given time-varying atmospheric conditions supplied by the GCM bounding a specified domain (see reviews by McGregor 1997; Giorgi and Mearns 1999). The main advantage of regional atmospheric models (RCMs) is their ability to respond through their lateral boundary conditions to different external forcings as well as to such regional effects as land-surface or atmospheric chemistry changes. RCMs can also resolve important atmospheric processes such as orographic precipitation better than the driving GCM (Jones et al. 1995). However, RCMs are computationally demanding and require orders of magnitude more computer time than SDS to compute equivalent scenarios.

There are other issues with RCMs. The use of a mesoscale model forced by a GCM which still remembers its initial conditions can add value through its lateral boundary conditions (i.e. dynamical downscaling) in all cases. This is the numerical weather prediction mode of dynamical downscaling. A GCM forced by observed SSTs also should insert skill into a RCM through the lateral boundary conditions. In the RCM area, detailed knowledge of surface forcing will add value when using the data from a larger scale model or reanalysis grid (with the resolution of those grids).

However, atmospheric structure in the RCM cannot be improved from a GCM which is not constrained by observed data and that is dependent on the lateral boundary conditions from the GCM. This conclusion has been reached in Castro and Pielke (2004), and is supported by the work of von Storch et al (2000; see their Table 1). Surface forcing can add value, however, its skill is dependent on the extent its effect is dependent on the lateral boundary condition information. If it is weakly dependent, than driving the RCM with the coarse resolution GCM results and the detailed surface forcing, for example, will add significant value.

Spectral nudging (von Storch et al. 2000; Miguez-Macho et al. 2004) appears the optimal way to include some "initialisation" yet still leave some freedom for the RCM to create smaller scale structure in response to that nudging, and from the surface forcing. Any spectral skill in the smaller scale features is a result of surface forcing and the spectral nudging, not the lateral boundary conditions. Of course, with spectral nudging, the RCM is then constrained in its larger scale features to what the larger model provides, and the ability for the RCM to alter this large-scale structure is lost.

A simplified method to downscale hydrologically-relevant information from atmospheric models is the explicit subgrid-scheme: Here, a higher resolution grid is defined at the land-atmosphere interface consisting of several subgrid cells per cell of the atmospheric model (see also Sect. 3.3). According to the mixture approach, these subgrid cells may be covered by at least one vegetation- and/or soil-type. Thus, coupled energy (for soil and vegetation) and hydrological budgets are maintained for each subgrid cell using the subgrid cell surface characteristics and the micro-climate at the representative location. Soil water content, soil temperature, near-surface air-temperature and humidity are simulated dynamically for each subgrid cell. The coupling of the subgrid cells to the atmospheric grid cell – which is actually an aggregation procedure – is realised by one of the upscaling strategies mentioned before, e.g. in a simple manner by the arithmetic average of individual subgrid cell fluxes (e.g. Avissar and Pielke 1989, or Mölders et al. 1996).

The explicit subgrid scheme belongs to the class of mosaic approaches. The main advantages of the subgrid scheme, as compared to a simple (non-explicit) mosaic approach, are that by explicitly breaking down the grid cells of the atmospheric model (1) the spatial location of each subgrid flux is known, (2) precipitation can be disaggregated explicitly, and (3) the coupling can be realised at the resolution and the surface parameters of the subgrid cells. The disadvantage of the explicit subgrid scheme is that it is much more computationally expensive, especially when the surface conditions are relatively homogeneous (see Mölders et al. 1996).

A particular challenge is the downscaling of precipitation as when applying the above-mentioned explicit subgrid scheme, the precipitation provided by a cloud module within the atmospheric model has to be downscaled (e.g. von Storch et al. 1993; Leung and Ghan 1995) to the subgrid resolution. In most regions, long-duration precipitation increases with elevation, due to the orographic uplift and cooling of moist air leading to precipitation. Assuming that precipitation increases with elevation and taking into account the direction of wind, stratiform precipitation can be disaggregated (e.g. Leung and Ghan 1995; Leung et al. 1996). For convective cases, the disaggregation of rainfall cannot easily be related to surface characteristics because of the more random character of the appearance of convective cells.

Box 1.2-1 ***Important Time Scales for Biosphere-Atmosphere Interactions***

Pavel Kabat, Roland W.A. Hutjes

The following tables give an indication of the most relevant time scales of importance / variance for a number of parameters in biosphere-atmosphere interactions.

Atmospheric Processes / Energy Exchange

Seconds	Hours	Days	Week/ synoptic	Season	Year
u	H/LE	P	Θ_v	P	Θ_v
R_g	CO_2	R_g	P	T_{air}	P
	T_{air}	T_{air}		T_{soil}	
	R_g			R_g	

(u: wind velocity; R_g: global radiation; H/LE: heat fluxes; CO_2: flux of carbon dioxide; T_{air}: air temperature; P: precipitation; Θ_v: potential virtual air temperature; T_{soil}: soil temperature)

CO_2 Exchange

Seconds	Hours	Days	Week/ synoptic	Season	Year
P_s	g_s	R_{soil}	LAI	P_s capacity	disturbance
	P_s		Vegetation height	g_s max	decomposition
				Leaf nitrogen	stand structure
				LAI	species
				Phenology	P_s capacity
					Soil carbon and nitrogen

(P_s: photosynthesis; g_s: stomatal conductance; R_{soil}: soil respiration; LAI: leaf area index)

Water Vapour Exchange

Seconds	Hours	Days	Week/ synoptic	Season	Year
Transpiration from canopy	z_{PBL}	ET	surface wetness	z_{PBL}	ET
	G	R_n	G_c	R_n	
	G_c			ET	
	E_{soil}			G	
	H			G_c	
	T			T	

(z_{PBL}: height of the planetary boundary layer; G: soil heat flux; G_c: heat flux in the canopy; E_{soil}: soil evaporation; H: sensible heat flux; T: surface temperature; ET: evapotranspiration; R_n: net radiation)

Box 1.2-2 ***Comparison of Dynamically and Statistically Downscaled Global Atmospheric Model Output***

George Leavesley

The ability to simulate adequately the spatial and temporal distribution of meteorological variables such as precipitation, temperature and radiation within an atmospheric model grid cell is of major importance to the adequate simulation of the subgrid heterogeneity of hydrological and ecosystem processes. Differences in daily precipitation and temperature for the Animas River basin, southwest Colorado, were examined using raw NCEP (National Centers for Environmental Prediction) data, statistical downscaling (SDS) and regional atmospheric model (RCM) simulations for current climate conditions (Wilby et al. 2000). The simulated surface climate variables were used to drive a distributed hydrological model. Since the hydrological response of the basin is an integration of the regional climate (in time and space), the results provide insights into the overall "value" added (or lost) to hydrological model skill due to the choice of downscaling technique.

Area-average daily precipitation (P) and maximum and minimum temperatures (T_{max}, T_{min}) were computed for the water years (WYs) 1980 to 1995 using two Snow Telemetry and one National Weather Service station. Area-average data for WYs 1987-95 were used to calibrate the SDS method and WYs 1980-86 to evaluate all models. Both the SDS and RCM were driven by gridded (approx. 200 km grid spacing) variables obtained from the NCEP/NCAR re-analysis (Kalnay et al. 1996). The SDS method (see Wilby et al. 1999) uses step-wise multiple linear regression to identify parsimonious sets of NCEP atmospheric variables – at the grid-point nearest the Animas basin – to predict local P, T_{max} and T_{min}. Separate regression equations were produced for each climatological season (i.e. DJF, MAM, JJA, SON) and surface variable (i.e. P, T_{max} and T_{min}).

RCM output was produced by RegCM2 (Giorgi et al. 1996), employing the continental U.S. domain of the Project to Intercompare Regional Climate Simulations (Takle et al. 1999). Initial and boundary conditions from the NCEP/NCAR re-analysis were supplemented by observations of water-surface temperature in the Gulf of California and the Great Lakes which are under-resolved in the re-analysis. Model grid spacing equates to 52 km on a Lambert conformal projection of the mid-latitudes.

The mean elevation of the RegCM2 and NCEP grid cells were adjusted on a monthly basis by solving for elevation using observed monthly *z*-climate relations. The fictitious elevations were then used to distribute P, T_{max} and T_{min} as in the case of observed and SDS data (for more details see Hay et al. 2000). This correction is necessary because of the coarseness of the RegCM2 and NCEP grids, which cause the model terrain heights to depart from the actual elevation of the basin.

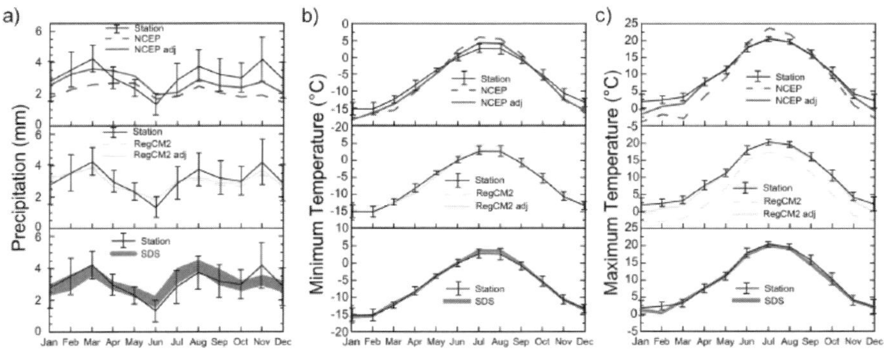

Fig. 1.2-7 Downscaled monthly mean daily (a) P, (b) T_{min} and (c) T_{max} for WYs 1980-86, compared with area-averaged station data for the Animas River basin. The error bars for station data correspond to 2 SE. The dashed lines for NCEP and RegCM2 represent the uncorrected model output. The grey shading for the SDS results shows the range of values produced by an ensemble of twenty members

Fig. 1.2-7a shows that the NCEP output captures the timing of the June minimum and March maximum wet-day amounts but underestimates total rainfall by 47% in the uncorrected case and by 36% in $NCEP_{adj}$. In comparison, RegCM2 output for both the corrected and uncorrected cases is closer to the observed seasonal cycle, with mean-bias values of –6 and –5% respectively. The SDS ensemble spans the observed rainfall regime in all months except for June (too high) and November (too low), but underestimates the rainfall total by 6%. It is noteworthy that relative to temperature and runoff (see below), the explained variance for daily P is low for all models, ranging between 14% (NCEP) and 26% ($RegCM2_{adj}$).

In contrast, T_{min} (Fig. 1.2-7b) is well represented by all methods. NCEP output shows a warm bias in summer and cold bias in winter that is reduced by elevation correction in $NCEP_{adj}$. The comparable monthly biases are smaller in RegCM2 and the corrected output of $RegCM2_{adj}$ has a mean bias of –0.1° C. However, NCEP and RegCM2 show large biases in T_{max} (Fig. 1.2-7c) with values of –2.0 and –4.6° C respectively. NCEP has a significant cold bias in winter and spring that is partially offset by correction, whereas RegCM2 has a cold bias throughout the year that is removed by $RegCM2_{adj}$ except in November to March. In comparison, SDS has a mean bias of –0.5° C for T_{max}.

The Precipitation-Runoff Modelling System (PRMS) (Leavesley and Stannard 1995) was used to simulate daily runoff (Q) given time series of surface climate variables (P, T_{max} and T_{min}). Fig. 1.2-8 compares the relative skill of the downscaling methods at simulating daily runoff using annual values of the coefficient of efficiency (CE) (Nash and Sutcliffe 1970). The station data provided the best simulation results with the majority of the years having CE scores of 0.8 or higher. The magnitude of these values indicates that, even though parameters were not optimised, the performance of the hydrological model is still quite good.

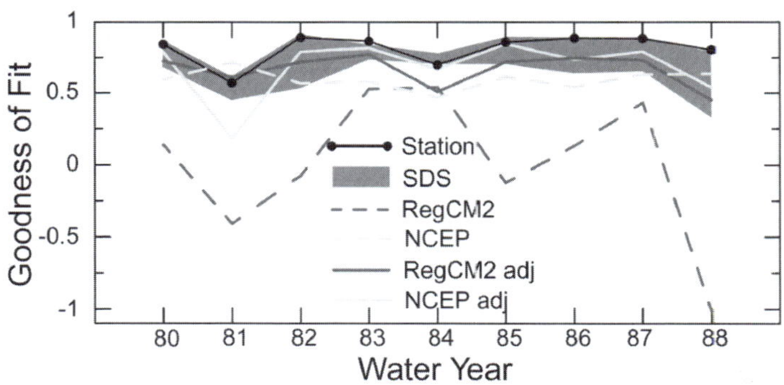

Fig. 1.2-8 Nash-Sutcliffe scores of simulated versus observed daily Q computed on a water year basis for different downscaling methods. The goodness of fit scores for the station output provide a measure of errors due to the hydrological model and/or choice of stations since observed P, T_{max} and T_{min} were used in this case

Overall, the CE scores for station data fall within the ensemble range of the SDS method. The CE scores for $RegCM2_{adj}$ and $NCEP_{adj}$ also lie within the bounds of the SDS ensemble for all years except WY1984 and WY1981 respectively, when the dynamic models had lower skill. In comparison, the skill of NCEP was lower than that of SDS in all years apart from WY1981 when the modelled runoff from NCEP was closest to observed. The least skilful model output was obtained from RegCM2: the negative CE scores in WYs 1981, 1985 and 1988 indicate that the observed mean of Q is a better predictor of daily Q than the model (Wilcox et al. 1990).

From the single-basin study it was concluded that the SDS and RCM methods have greater skill (in terms of modelling hydrology) than the coarse resolution data used to drive the downscaling. The SDS has the advantage of requiring very few parameters – an attribute that makes this procedure attractive for many hydrological applications. The RCM output, once elevation-corrected, provides better estimates of the water balance than the raw and corrected NCEP output. However, since the methods provide varying results, care must be taken in interpreting scenarios of basin-scale hydrology under both present and future climate forcing.

The reasonable simulations of daily runoff obtained using SDS and RCM output in the Animas River basin is in part related to the fact that this is a snowmelt-dominated basin. Daily variations in winter precipitation are less important than the volume of precipitation over the accumulation season. As noted by Wilby and Dettinger (2000) in a similar study in the Sierra Nevada mountains, hydrological "skill" arises from the fact that the snowpack acts as an integrator of the hydrological processes. Alternatively, daily variations in a rainfall-dominated basin are much more important in the simulation of basin hydrological response. Application of RCM output to the Alapaha River basin in Georgia showed a very low accuracy in daily runoff simulation for the period 1979-1988 (Hay et al. 2002). This is related to errors in the seasonality of simulated precipitation and the large errors in the magnitude and frequency of simulated storms when compared to observed storms.

Box 1.2-3 ***Upscaling Example: Aggregation of Evapotranspiration***

Nicole Mölders

In meteorological modelling, computational capacity, parameterisation limitations, or reasonable simulation times[1] require coarser grid resolutions

[1] In the case of introducing a finer resolution, all meteorological processes have to be calculated on that fine resolution for which not only the amount of grid points increases, but also the time step has to be reduced to fulfill the Courant-criteria.

than are desirable to describe realistically the exchange of momentum, heat and matter at the earth-atmosphere interface. The task to be solved is to aggregate information of the exchange from smaller to larger scale so as to consider various vegetation and soil types.

Let us assume that the land-use and soil-type data are given in a resolution of 1 x 1 km^2 and that they represent the land-use or soil type dominating in this one square kilometre area. The meteorological model is intended to be run at a coarser resolution but the fine resolution of the land-use and soil data is to be taken into account by a mosaic approach (see Avissar and Pielke 1989; Mölders et al. 1996). A fundamental assumption of this approach is that the local-scale near-surface meteorological forcing at the patch, which is experienced by the surface of the patch, is important in determining the net exchange of heat, moisture and momentum at the earth-atmosphere interface. For each patch of equal land-use/soil type, unique energy and hydrological budgets are maintained using the grid cell forcing of wind, temperature, moisture, pressure and radiation predicted by the meteorological model. For each patch, its own soil temperatures, soil wetness, and near-surface meteorological forcing in the immediate vicinity of the Earth's surface is used to determine the fluxes of the respective patch. These individual quantities result as a consequence of the consideration of the individual land-use and soil type at the patch level. The fluxes of net radiation, Q, sensible, H, and latent heat, L_vE, as well as the soil heat flux, G, for the ith patch of the jth grid cell are written as (e.g. Mölders et al. 1996)

$$Q_j^i = R_{S\downarrow,j}^i(1-\alpha_j^i) - \varepsilon_j^i R_{L\downarrow,j}^i + \varepsilon_j^i \sigma T_{g,j}^{i\ 4} \tag{1.2-4}$$

$$H_j^i = \rho_j^i c_p C_{h,j}^i u_{R,j}(\Theta_{g,j}^i - \Theta_{R,j}) \tag{1.2-5}$$

$$L_v E_j^i = \rho_j^i L_v C_{q,j}^i u_{R,j}[q_{g,j}^i(T_{g,j}^i) - q_{R,j}]w_{e,j}^i \tag{1.2-6}$$

$$G_j^i = -\lambda_j^i \frac{\partial T_{s,j}^i}{\partial z} \tag{1.2-7}$$

where Θ and q represent the potential temperature and specific humidity at the surface (index g) and the reference height (index R) located at the first half level of the model. Furthermore, α, ε, λ, σ, $R_{S\downarrow}$ and $R_{L\downarrow}$ stand for the albedo, the emissivity of the surface, the soil thermal conductivity, the Stephan-Boltzmann constant, the short-wave and long-wave radiation, respectively. T_g and T_s denote the surface and soil temperature, and u_R is the wind speed at the reference height. The density of air is denoted as ρ,

c_p and L_v are the specific heat at constant pressure and the latent heat of condensation, C_h and C_q are the transfer coefficients for heat and water vapour. For bare soil, the wetness factor, w_e, equals the soil surface wetness while for vegetated surfaces it considers the canopy conductivity, g_s, which depends on the maximal evaporative conductivity, the insolation, the water vapour deficit, the air temperature and the soil wetness (e.g. Deardorff 1978).

In the simplest approach, the n individual fluxes $F_j^{i,k}$, ocurring at the (local) patch scale within the jth grid cell are upscaled to the flux, F_j^k, of this j^{th} grid cell by the area weighted mean (e.g., Avissar and Pielke 1989; Mölders et al. 1996)

$$F_j^k = \sum_{i=1}^{n} a_j^i F_j^{i,k} \quad . \tag{1.2-8}$$

Here, the index k refers to net radiation and soil heat flux as well as the fluxes of sensible and latent heat. The letter n is the number of patches occurring within the jth grid cell, and a_j^i is the relative area covered by these subgrid-scale patches of equal land-use/soil type i where

$$\sum_{i=1}^{n} a_j^i = 1 \quad . \tag{1.2-9}$$

Note that the variables of state (e.g. soil temperature, soil moisture, etc.) are upscaled in the same manner. To investigate the effect of upscaling, the following simulations have been performed: (1) a simulation with a resolution of 4 x 4 km^2 using distributions of land-use and soil types that were derived from the fine resolution data by assuming the dominant soil/land-use type of the grid cell to be representative (HOM4; Fig.1.2-9), (2) same as HOM4, but for a grid resolution of 8 x 8 km^2 (HOM8), (3) a simulation with a resolution of 4 x 4 km^2 applying the upscaling procedure (HET4), and (4) same as HET4, but for a grid resolution of 8 x 8 km^2 (HET8). Note that in this example land use is linked to a typical soil type. A change in dominant landuse might lead to a drier surface if urban area increases, and also by an increase in dunes, or coniferous forest growing on sandy soil. Moreover, the different land use yields to other aerodynamic roughness for which wind speed changes. The altered surface fluxes lead to different moisture and temperatures states of the near-surface atmosphere that again affects the fluxes. Thus, differences may increase with time. Other authors reported similar results (e.g., Shao et al. 2001).

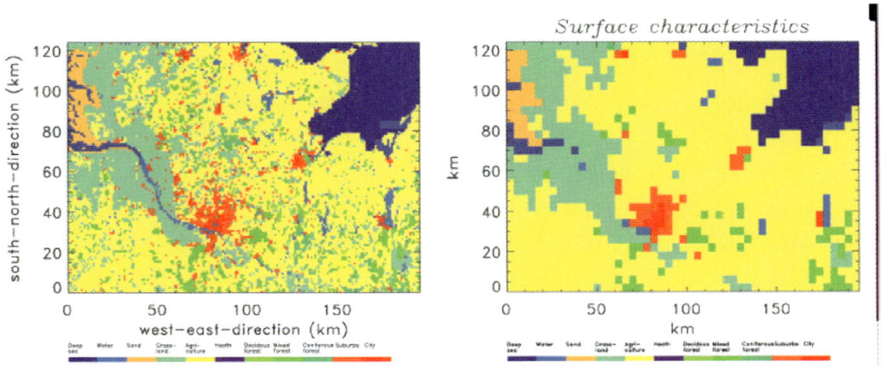

Fig. 1.2-9 Distribution of land use on the fine resolution of 1 x 1 km^2 used in the upscaling (HET4, HET8) and the coarse resolution of 4 x 4 km^2 used under the assumption of dominant land-use type in HOM4. Note that the information loss is even worse for a resolution of 8 x 8 km^2 (HOM8) (after Mölders and Raabe 1996)

Here, upscaling of evapotranspiration is discussed exemplarily. The model results showed that upscaling influences predicted evapotranspiration, which could be derived by comparing HOM4 and HOM8. The detailed consideration of land use (by direct simulation of the 1 x 1 km^2 resolution, or by the mosaic approach with an enlarged resolution, e.g. HET4 and HET8) allows consideration of the often subgrid-scale cities, dunes and coniferous forests that have a low evapotranspiration as compared to the surrounding agricultural land, grassland or deciduous forests. Thus, upscaling without accounting for such subscale features by considering only the dominant land use (HOM4 and HOM8), yields to an over-prediction of evapotranspiration. Note that in semi-arid areas, upscaling without accounting for subscale features may lead to an under-prediction of evapotranspiration because subscale wet spots, such as irrigated fields and plots, river riparian zones, or oases, which provide more water to the atmosphere than their environments, are disregarded with respect to the resolution of the meteorological model. However, the mosaic approach (coarse resolution but accounting for subgrid variations) can account for such effects. It also allows for consideration of lake surface temperatures different from the terrestrial environment (e.g. colder in spring and warmer in autumn), where evaporation can be rather different from that of the surrounding fields. Mölders and Raabe (1996) have shown that upscaling by applying the mosaic approach (HET4, HET8) provides nearly the same domain-averaged evapotranspiration, independent of the grid resolution.

1.3 Multiple Equilibria

Ekkehard Holzbecher, Martin Claussen, George Leavesley

System equilibrium refers to a stable state of a complex system and the ability of interactive system components to maintain their current states over extended periods of time. Variations in the magnitudes of feedbacks among the components, in terms of energy and mass, are sufficiently small to maintain this balance. For example, the natural distribution of specific vegetation types, over large regions of a continent, reflects, in part, the effects of regional precipitation and temperature regimes. Conversely, the types and distribution of the regional vegetation control, in part, regional precipitation and temperature regimes.

Major changes in the state of system components and the magnitude of their mass and energy fluxes can cause a shift in the state of the system towards an alternative level of stability. These changes can be the result of natural causes such as forest fires, or human causes such as deforestation. The ability to simulate these alternative levels of stability, or multiple equilibria, necessitates the use of models that adequately describe the atmospheric, hydrological and biological feedbacks among system components. This section provides an overview of the concepts of system equilibria and specific examples of multiple equilibria in ecosystem and geochemical systems.

Parts of complex non-linear systems, as they are discussed here, may interact in different ways. Generally one can distinguish between positive and negative feedback (the term feedback is introduced more formally in Sect. 2.1 and ways to deal with them in Sect. 2.2). Positive feedback drives the system away from equilibrium, while it is taken back towards the equilibrium in the case of negative feedback.

Equilibria are most commonly defined in terms of energy to change the status quo. If the energy available to change the status quo is less than the energy required to make a change, then the status quo will be maintained. If little energy is required to change the current state, then that state is likely to move to a new equilibrium, often more stable, but not necessarily the most stable state. This perspective assumes that the energy input is a trigger to change the status quo, and that no further energy input is required to maintain the new state. The most common examples are illustrated by a marble on a hill where a small input of energy will move the marble from the top of the hill to a lower level (unstable), the case where the marble is on top of the hill, but there is an intermediate valley where the marble will stop (conditionally stable), or where the marble is in a valley and will always return to the valley (stable). This simplified mechanistic view has no feedbacks.

Watson and Lovelock's (1983) adaptive control system with negative feedbacks (often referred to as the Daisyworld model) has oscillations about a stable solution and negative feedbacks govern the outcome at any given instant. Black and white daisies compete in an ideally simplified artificial world model. The environmental feedback may lead to different steady states. Watson and Lovelock show that a hysteresis representing multiple equilibria is obtained, depending on whether the perturbing parameter, solar luminosity, is increased or decreased.

Let us consider a dynamical system which, for example, describes the dynamics of subtropical, water limited vegetation (e.g., Brovkin et al. 1998) in a very simple form:

$$dV/dt = f(V, P, t) \qquad (1.3\text{-}1)$$

where, V and P refers to vegetation cover and precipitation, respectively. V and P are functions of time t. If a solution

$$dV/dt = 0 \qquad (1.3\text{-}2)$$

exists, then it is called an equilibrium solution. Generally, dynamical systems can exhibit several equilibrium solutions[1], or multiple equilibria. In the case of multiple equilibria, there are different types of equilibria which can conveniently be categorised by discussing the so-called potential function F(V) being defined by

$$dF/dV = dV/dt \qquad (1.3\text{-}3)$$

By definition, any extreme value $dF/dV = 0$ indicates an equilibrium solution. The maximum $d^2F/dV^2 > 0$ of the potential refers to an unstable equilibrium, and the minimum $d^2F/dV^2 < 0$ to a stable one. Any infinitesimally small perturbation will kick the system from an unstable solution, say V_3, to a stable one. If the potential F at two stable equilibria V_1 and V_2 differ, say $F_1 < F_2$, than the equilibrium with the smaller potential is said to be globally stable, and the other locally stable. I.e., it needs a stronger perturbation to move the system from F_1 to F_2 than from F_2 to F_1. The graph in Fig. 1.3-1 nicely demonstrates this behaviour (the effect of the nature of f and F on the time evolution of the system will be discussed further in Sect. 2.2.2).

[1] These may also be called steady-states.

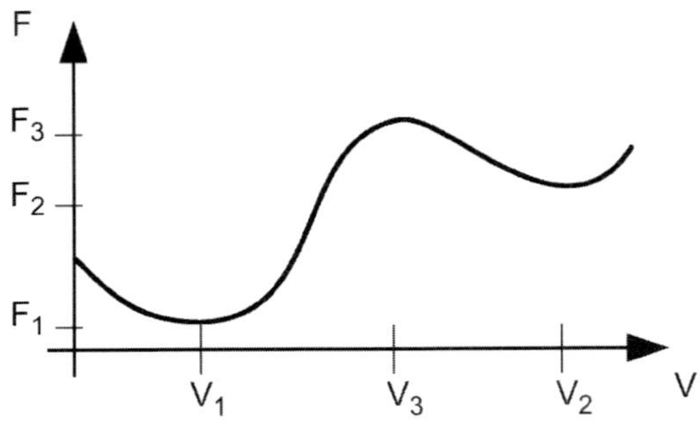

Fig. 1.3-1: Potential as a function of a state variable (e.g., vegetation cover), illustrating local (V1) and global (V2) equilibrium states

In complex systems, involving the hydrological, biogeochemical and energy cycles of the atmosphere, biosphere and pedosphere, and one in which disturbances from both natural and human activity are common, multiple stable equilibrium states may exist. Such systems contain characteristics of resilience, adaptability and diversity. Each of these characteristics can bring about stable environments. As example, in the Sahel, equal energy, biogeochemical and water exchanges are obtained through two diverse ecotypes – Savannah and bushlands (Goutorbe et al. 1997). Similarly, a constant disturbance can bring about a new equilibrium in which feedbacks to the atmosphere from the biosphere have resulted in a stable environment – deforestation of the Mediterranean basin as example, where changes in energy and water fluxes between the atmosphere and the biosphere have provided an environment for competitive, disturbance-tolerant plant species to dominate.

A coupled model of the atmosphere, biosphere and pedosphere, addressing the processes of the hydrological, biogeochemical and energy cycles, must be sufficiently robust to capture theses characteristics of resilience, adaptability and diversity, and must contain the trigger mechanisms of soil saltation, transport of nutrients across space and most importantly, contain the impacts of human activities.

Two quite different examples, concerning scale and field of application, are presented in the following side-boxes:

- Box 1.3-1: Large-scale multiple equilibria: two Sahelian vegetation covers under two different precipitation schemes;
- Box 1.3-2: Multiple equilibria for the chemisty of salinised water.

Box 1.3-1 ***Multiple Equilibria in the Vegetation-Climate System***

Martin Claussen

The interaction between components of the climate system is non-linear. Hence one might expect multiple equilibrium solutions, not only in the atmosphere-ocean system (which has been amply documented, e.g. Rahmstorf 1995), but also in the vegetation-atmosphere system. Zonally-averaged models using rather simple coupling between vegetation and atmosphere (see Gutman et al. 1984; Gutman 1984, 1985) revealed only unique, steady-state solutions. The possibility of multiple equilibria in the 3-dimensional atmosphere-vegetation system was discovered later by Claussen (1994) and subsequently analysed in detail by Claussen (1997, 1998) and Brovkin et al. (1998).

Two solutions of the atmosphere-vegetation system seem to be possible: the arid, present-day climate and a humid solution resembling more the mid-Holocene climate, some 6000 years ago, i.e. with a Sahara greener than today, albeit less green than in the mid-Holocene (see Fig. 1.3-2). The two solutions differ mainly in the subtropical areas of North Africa and, but only slightly, in Central East Asia. Interestingly, experiments with mid-Holocene vegetation yield only one solution, the green Sahara (Claussen and Gayler 1997), while for the Last Glacial Maximum (LGM), some 21,000 years ago, two solutions exist (Kubatzki and Claussen 1998). Levis et al. (1999) seek multiple solutions of the atmosphere-vegetation-sea-ice system at high northern latitudes. Their model converges to one solution in this region corroborating the earlier assertion (Claussen 1998) that multiple solutions emerge in the subtropics, mainly in North Africa.

Why do multiple solutions appear in the subtropics, but not at high latitudes – and why for present-day and LGM climate, but not for mid-Holocene climate? Claussen et al. (1998) analyse large-scale atmospheric patterns in present-day, mid-Holocene, and LGM climate. They find that velocity potential patterns, which indicate divergence and convergence of large-scale atmospheric flow, differ between arid and humid solutions mainly in the tropical and subtropical regions. Apparently the Hadley-Walker circulation slightly shifts to the west. This is consistent with Charney's (1975) theory of albedo-induced desertification in the subtropics. Moreover, changes in surface conditions directly influence vertical motion, and thereby large-scale horizontal flow, in the tropics (Eltahir 1996) but hardly at all at middle and high latitudes (e.g. Lofgren 1995a, b). For the mid-Holocene climate, the large-scale atmospheric flow is already close to the humid mode, even if one prescribes present-day land-surface conditions. This is caused by differences in insolation: in the mid-Holocene boreal summer, the Northern Hemisphere received up to 40 W m^{-2} more energy than today,

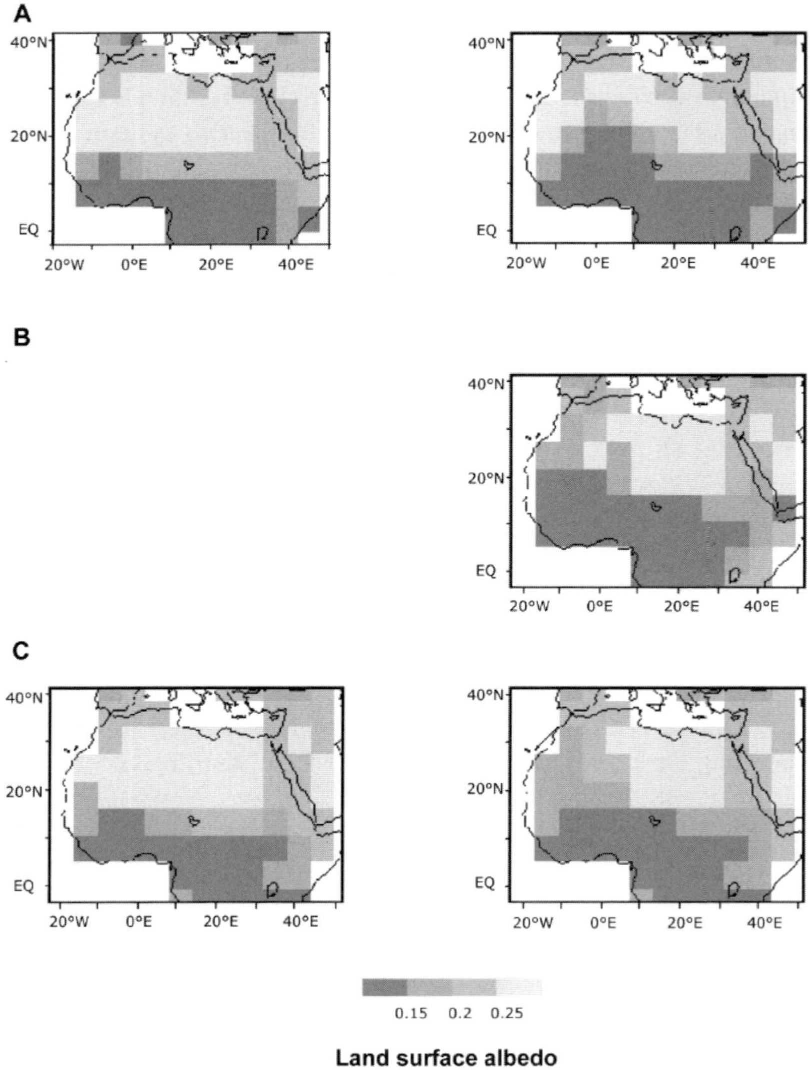

Fig. 1.3-2: Multiple equilibria computed for present-day climate (A), and for the climate of the last glacial maximum (C). For mid-Holocene conditions, only one solution is obtained (B). Fig. 1.3-2 summarises the results of Claussen (1997) (A), Claussen and Gayler (1997) (B) and Kubatzki and Claussen (1998) (C) using an asynchronously coupled atmosphere-biome model. The figure is taken, with modifications, from Brovkin et al. (1998)

thereby strengthening African and Asian summer monsoon (Kutzbach and Guetter 1986). During the LGM, insolation was quite close to present-day conditions. A more ecological interpretation of multiple equilibria is given by Brovkin et al. (1998). They develop a conceptual model of vegetation-precipitation interaction in the western Sahara which is applied to interpret the results of comprehensive models. The conceptual model finds three solutions for present-day and LGM climate; one of these, however, is unstable to infinitesimally small perturbations. The humid solution is shown to be less probable than the arid solution, and this explains the existence of the Sahara desert as it is today. For mid-Holocene climate, only one solution is obtained. Application of the conceptual model to biospheric feedbacks at high latitudes (Levis et al. 1999) yields only one solution for present-day conditions.

Are multiple equilibria just a matter of the atmosphere-vegetation system, or do they occur also in the atmosphere-ocean-vegetation system? So far, multiple solutions have not yet been found in the model of Ganopolski et al. (1998), a fully coupled atmosphere-ocean-vegetation model. (This model attains, however, multiple solutions associated with multiple states of the thermohaline convection.) Presumably the lack of multiple solutions is related to the coarse resolution of this model, because North Africa is represented by just three grid boxes, Sahara, Sudan and tropical North Africa. Subsequently, Saharan precipitation in the coarse model of Ganopolski et al. (1998) appears to be less sensitive to changes in land-surface conditions than the West-Saharan precipitation in the model used by Claussen (1997, 1998). On the other hand, the study of Ganopolski et al. (1998) shows that the biogeophysical feedback in North Africa is mainly a vegetation-atmosphere feedback. Therefore one can assume that the conclusion drawn from coupled vegetation-atmosphere models should generally be valid, i.e., vegetation-atmosphere-ocean models (with finer horizontal resolution) should also exhibit multiple equilibria in the North African region.

The discussion of multiple equilibria seems to be somewhat academic. However, the existence of these could explain abrupt transitions in vegetation structure (Claussen et al. 1998; Brovkin et al. 1998). If global stability changes in the sense that one equilibrium solution becomes less stable to finite amplitude perturbations than the others, than an abrupt change of the system from the less stable to a more stable equilibrium is to be expected. Brovkin et al. (1998) find in their conceptual model that the green solution becomes less stable around 3.6 ka BP. However, these arguments are based on studies of the system at or in the vicinity of an equilibrium state. Only with fully coupled, dynamic vegetation models one can explore the time evolution of biogeophysical feedbacks. This has been undertaken by Claussen et al. (1999) who analyse the transient structures in global vegetation pattern and climate using the coupled atmosphere-ocean-vegetation model of Ganopolski et al. (1998), but with a dynamic vegetation module. Their simulations clearly show (not just suggest) that subtle changes in orbital forcing triggered changes in North African climate which were then strongly amplified by biogeophysical feedbacks in this region, leading to a rather fast desertification within a few centuries starting around 5500 years ago. This seems to be in at least qualitative agreement with palaeogeological reconstructions (e.g. Pachur and Wünnemann 1996).

Box 1.3-2 ***Multiple Equilibria and Salinisation***

Ekkehard Holzbecher

Salinisation in general is a phenomenon that is initiated either by contact with a salt source, which may be a saline water reservoir or solid salt precipitates, or by shrinking water volumes. Here the last case – salinisation as effect of shrinking water volume – is examined in some detail. Salinisation becomes a problem when salt concentration rises above a certain limit, which makes the water unsuitable for a special purpose, as for irrigation, for drinking water or for industrial use.

The major cause for shrinking water volumes is evaporation. When actual evaporation exceeds the source terms of whatever type, the volume shrinks and concentrations of all species would accumulate if reactions of any type would be neglected. Various chemical processes accompany the accumulation, causing a complex change in water chemistry leading to specific types of brines.

Here, in a simplifying approach, a precipitation/dissolution type of reaction is taken into account in addition to accumulation. Precipitation and dissolution, as shown below, interact in a complex way with the process of salt accumulation. Hanor (2001) describes that with increasing salinity divalent cations increase at a much greater rate than monovalent cations and can attribute this phenomenon to reactive transport involving rock buffering.

The phenomenon is relevant in salt lakes, where salt concentrations tend to increase until there is a balance between precipitation, surface and subsurface sources on the one hand and sinks on the other. Highest salt concentrations can be expected for terminal lakes, for which evaporation is the only "sink" for water. The situation is similar in soil columns, where seepage water is extracted by evaporation or by root uptake. While in the latter case dissolved components are transferred to the biosphere compartment, there is an accumulation of concentrations in the former case. Measurements of dissolved tracers in the soil clearly show rising concentrations with depth from the surface.

In their classical contribution on salt-lakes Eugster and Hardie (1978) describe the chemistry accompanying the accumulation of salt. Considering different species, they introduce the concept of a *chemical divide*, which will be outlined here briefly following basically the description given by Drever (1997).

Two components with fluid phase concentrations c_1 and c_2 forming a binary salt are considered. If initial concentrations are c_{10} and c_{20} the increase of concentrations can be described by:

$$\begin{aligned} c_1 &= nc_{10} - e \\ c_2 &= nc_{20} - e \end{aligned} \qquad (1.3\text{-}4)$$

The factor n quantifies the accumulation of salt due to the loss of water as a result of evaporation. The sink term e measures the losses of the fluid phase due to precipitation of the binary salt. The precipitation process is determined by the solubility product which, using Eq. 1.3-4 in a simplified form, is stated as:

$$(nc_{10} - e)(nc_{20} - e) = K \qquad (1.3\text{-}5)$$

K is the equilibrium constant which is typical for the reaction. From both Eqs. 1.3-4 and 1.3-5 follow two solutions for e:

$$e_{1,2} = n\frac{c_{10} + c_{20}}{2} \pm \sqrt{K + n^2\left(\frac{c_{10} - c_{20}}{2}\right)^2} \qquad (1.3\text{-}6)$$

For the concentrations holds:

$$\begin{aligned} c_1 &= n\left(\frac{c_{10} - c_{20}}{2} + \sqrt{\frac{K}{n^2} + \left(\frac{c_{10} - c_{20}}{2}\right)^2}\right) \\ c_2 &= n\left(\frac{c_{20} - c_{10}}{2} + \sqrt{\frac{K}{n^2} + \left(\frac{c_{10} - c_{20}}{2}\right)^2}\right) \end{aligned} \qquad (1.3\text{-}7)$$

The formulae show that the sign of the first term determines the development of c_1 and c_2. When c_{10} is greater than c_{20}, then at all times c_1 will exceed c_2 as both terms in the brackets have to be summed for component 1, while they have to be subtracted for component 2. The role of the components is changed when c_{10} is smaller than c_{20}. The difference between both concentrations obviously fulfils the simple equation:

$$|c_1 - c_2| = n|c_{10} - c_{20}| \tag{1.3-8}$$

The concentration difference increases as long as n increases. The behaviour can be described as a chemical divide: the component which dominates in the beginning, will dominate during the further development. Finally the dominated component will vanish out of the system. Without restriction of generality one may assume $c_{10} > c_{20}$ and find for $n \to \infty$:

$$\begin{aligned} c_1 &\to n\frac{c_{10} - c_{20}}{2} + n\frac{|c_{10} - c_{20}|}{2} = n(c_{10} - c_{20}) \\ c_2 &\to n\frac{c_{20} - c_{10}}{2} + n\frac{|c_{10} - c_{20}|}{2} = 0 \end{aligned} \tag{1.3-9}$$

Analytically, both asymptotic solutions represent two steady equilibrium solutions of the system. In one the concentration of component 2 vanishes, while in the other component 1 disappears from the fluid phase. Which of the solutions will prevail is already determined by the initial condition, in which the smallest difference determines the future fate of the components.

In a real system, several other influences and disturbances will have to be taken into account, so that there may be a fluctuation between the paths leading to the two equilibrium states. This will be shown by the following example, in which additional sources for one component are considered. For a water body in the field one may imagine only one of the components added by runoff or groundwater. The starting point is the extended set of differential equations:

$$\begin{aligned} \frac{\partial c_1}{\partial t} &= n_t c_1 - e + q_1 \\ \frac{\partial c_2}{\partial t} &= n_t c_2 - e + q_2 \end{aligned} \tag{1.3-10}$$

in which n_t is a temporary accumulation rate and q_1 and q_2 quantify additional sources. In lakes, these sources could represent material transported by surface or subsurface water into the lake. Subtraction of both equations delivers:

$$\frac{\partial}{\partial t}(c_1 - c_2) = n_t (c_1 - c_2) + q_1 - q_2 \qquad (1.3\text{-}11)$$

which is a differential equation for $d=c_1-c_2$. Once the difference d is determined, using Eq. 1.3-7 the concentrations can be computed according to:

$$c_1 = \tfrac{1}{2}\left(d + \sqrt{4K^2 + d^2}\right) \qquad c_2 = c_1 + d \qquad (1.3\text{-}12)$$

The figure below shows three solutions of this system. The accumulation regime in all three cases is the same and is shown with the function N[1], wich decreases linearly down to zero at a time instant, after which there is no further accumulation. In all simulations the initial concentration of the first component is slightly higher than that of the second component.

According to the classical chemical divide concept, c_1 increases monotonically, while c_2 decreases; both reaching a finite asymptotic value when concentration accumulation ceases. This behaviour is shown by the curves in Fig. 1.3-3a. The transient development and the final steady state are significantly determined by the initial condition. Note that the different roles of component 1 and 2, can be taken differently by the components of a binary salt: in some natural environments calcium may play the dominant role, while in others CO_3 takes the lead.

The difference between the concentration of both components rises to a constant positive value. In a second run of the model it is assumed that component 2 enters the system by some additional process, but component 1 has no additional source (q_1=0). The solution for that situation is depicted by the curves in Fig. 1.3-3b. It turns out that the concentration of component 2 remains significantly high compared to the first run, but does not reach the concentration c_1. In the third run, represented by curves in Fig. 1.3-3c, the source term for component 2 is increased, so that after some time c_2 exceeds c_1. Now the system converges to a solution with zero concentration for c_1.

Note that in the simulation shown in Fig. 1.3-3b and 1.3-3c no constant asymptotic value for c_2 is approached for $t \to \infty$, because the source term is not switched off. If, at a certain time instant the source goes to zero, the

[1] In the figure legend capital letters are used throughout: D for d, N for n_t

principle of the chemical divide can be applied: the component, which is dominant at that time instant will dominate further in the fluid. To keep the modelling work easy and comprehensible, the source term has been kept constant, which is realistic only for a limited period of time. In real systems the source term will vary spatially and temporarily. Thus the fluctuation between paths to the two equilibria will probably be even more pronounced than in the simplified examples presented here.

The proposed method could be applied to steady-state salinisation in soil columns. Without considering any chemical reactions, Ginn and Murphy (1997) propose the differential equation

$$\theta\frac{\partial c}{\partial t}+q\frac{\partial c}{\partial z}=q_{ex}c \qquad (1.3\text{-}13)$$

(with volumetric moisture θ, interior flux q and boundary flux q_{ex}. After the introduction of a reaction and source term, for steady-state this leads to the system of differential Eq. 1.3-10, in which the space variable z replaces the time variable t (and with suitable new parameter n).

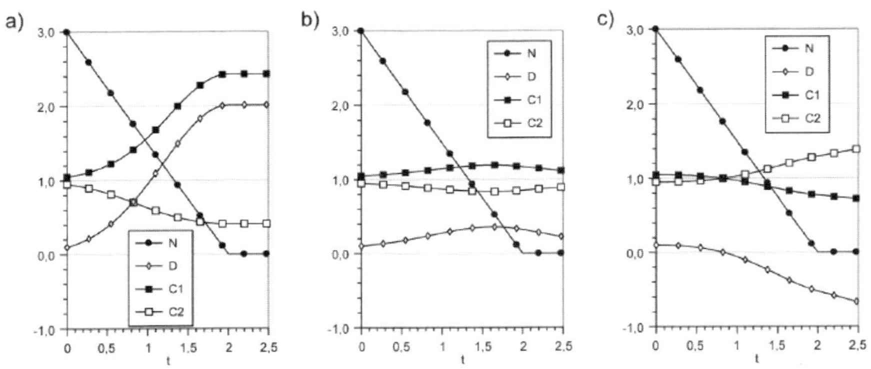

Fig. 1.3-3 Fluctuation of salinisation as function of time between paths leading to two different equilibrium states (for details see text)

In the preceding example the temporal change of two chemical components of a binary salt, for example $CaCO_3$, was examined. As the three runs show, the transient development and the steady state are crucially determined by the source term for the components. The concept of the chemical divide analogously describes the crucial dependence on the initial conditions, more precisely on the relative size of c_{10} and c_{20}. As long as

$c_{10}>c_{20}$ or $c_{20}>c_{10}$, the equilibrium does not change drastically. But the components exchange their roles when the relation changes. For $c_{10}\approx c_{20}$, tiny differences can determine which of the components finally plays the dominant role. As the last two scenarios show, changes in source rates may become similarly important additionally.

Eugster and Hardie (1978) summarise the major point concerning the carbon characteristic of natural waters: "clearly, then, calcite precipitation is a critical evolutionary step: it will immediately determine whether an evaporating water will become carbonate-rich or carbonate-poor".

References Chapter 1

Arain AM, Michaud JD, Shuttleworth WJ, Dolman AJ (1996) Testing of vegetation parameter aggregation rules applicable to the Biosphere-Atmosphere Transfer Scheme (BATS) at the FIFE site. J Hydrol 177:1-22

Arain AM, Shuttleworth WJ, Yang Z-L, Michaud JD, Dolman AJ (1997) Mapping surface cover parameters using aggregation rules and remotely sensed cover classes. Q J Roy Meteor Soc 123(B):2325–2348

Armstrong RL, Brodzik MJ (1999) A twenty year record of global snow cover fluctuations derived from passive microwave remote sensing data. Proceedings of American Metorological Society 5th Conference on Polar Meteorology and Oceanography, Dallas, Texas

Atlas R, Wolfson N, Terry J (1993) The effects of SST and soil moisture anomalies on GLA model simulations of the 1988 U.S. summer drought. J Climate 6(11):2034-2048.

Avissar R, Pielke RA (1989) A parameterization of heterogeneous land surface for atmospheric numerical models and its impact on regional meteorology. Mon Weather Rev 117:2113-2136

Avissar R, Schmidt T (1998) An evaluation of the scale at which ground-surface heat flux patchiness affects the convective boundary layer using large-eddy simulations. J Atmos Sci 55:2666-2689

Ball JT, Woodrow IE, Berry JA (1987) A model predicting stomatal conductance and its contribution to the control of photosynthesis under different environmental conditions. In: Biggins J (ed) Progress in photosynthesis research. Martinus Nijhoff, Dordrecht, pp221-224

Bárdossy A, Plate EJ (1992) Space-time model for daily rainfall using atmospheric circulation patterns. Water Resourc Res 28:1247-1259

Barros AP, Kuligowski RJ (1998) Orographic effects during a severe wintertime rainstorm in the Appalachian Mountains. Mon Weather Rev 126:2648-2672

Baumgartner A, Reichel E (1975) The world water balance. Mean annual global, continental and maritime precipitation, evaporation and runoff. Elsevier, Amsterdam

Beljaars ACM, Viterbo P, Miller MJ, Betts AK (1996) Anomalous rainfall over the U.S. during July 1993: Sensitivity to land surface parameterization. Mon Weather Rev 124:364-383

Benjamin SG, Carlson TN (1986) Some effects of surface heating and topography in the regional severe storm environment. Part I: Three-dimensional simulations. Mon Weather Rev 114:307-329

Betts AK (2000) Idealized model for equilibrium boundary layer over land. J Hydrometeorol 1(6):507-523

Betts AK, Ball JH, Beljaars ACM, Miller MJ, Viterbo P (1996) The land-surface-atmosphere interaction: A review based on observational and global modeling perspectives. J Geophys Res 101D:7209-7225

Beven KJ, Kirkby MJ (1979) A physically based, variable contributing area model of basin hydrology. Hydrol Sci Bull 24:43-69

Bindlish R, Barros AP (1996) Aggregation of digital terrain data using a modified fractal interpolation scheme. Comput Geosci 22(8):907-917

Blöschl G (1996) Scale and scaling in hydrology. Habilitationsschrift, Technische Universität, Wien, p346

Blyth E, Dolman AJ, Wood N (1993) Effective resistance to sensible and latent heat in heterogeneous terrain. Q J Roy Meteor Soc 119:423-442

Blyth EM, Dolman AJ, Noilhan J (1994) The effect of forest on mesoscale rainfall; An example from HAPEX-MOBILHY. J Appl Meteorol 33(4):445-454

Bonan GB (1997) Effects of land use on the climate of the United States. Climatic Change 37:449-486

Bonell M (1998) Possible impacts of climate variability and change on tropical forest hydrology. Climatic Change 39:215-272

Bounoua L, Krishnamurti TN (1993a) Influence of soil moisture on Sahelian climate prediction I. Meteorol Atmos Phys 52:183-203

Bounoua L, Krishnamurti TN (1993b) Influence of soil moisture on Sahelian climate prediction II. Meteorol Atmos Phys 52:205-224.

Bronstert A, Bárdossy A (1999) The role of spatial variability of soil moisture for modelling surface runoff generation at the small catchment scale. Hydrol Earth Syst Sci 3(4):505-516

Bronstert A, Bárdossy A (2003) Uncertainty of runoff modeling at the hillslope scale due to temporal variations of rainfall intensity. Phys Chemistry Earth 28: 283-288

Brovkin V, Claussen M, Petoukhov V, Ganopolski A (1998) On the stability of the atmosphere-vegetation system in the Sahara/Sahel region. J Geophys Res 103(D24):31613-31624

Brubaker KL, Entekhabi D, Eagleson PS (1993) Estimation of continental precipitation recycling. J Climate 6(6):1077-1089

Buishand TA, Brandsma T (1997) Comparison of circulation classification schemes for predicting temperature and precipitation in the Netherlands. Int J Climatol 17:875-889

Bürger G (1996) Expanded downscaling for generating local weather scenarios. Climate Res 7:111-28

Buzzi A, Tartaglione N, Malguzzi P (1998) Numerical simulations of the 1994 Piedmont flood: Role of orography and moist processes. Mon Weather Rev 126:2369-2383

Castelli F, Rodriguez-Iturbe I (1993) Large-scale spatial dynamics of soil moisture-atmosphere interaction. American Meteorological Society, Fourth International Conference on Precipitation, April 26-28, Iowa City, Iowa

Castro CL, Pielke RA Sr. (2004) Dynamical downscaling: Assessment of value restored and added using the Regional Atmospheric Modeling System (RAMS). J Geophys Res - Atmospheres, submitted

Chang J-T, Wetzel PJ (1991) Effects of spatial variations of soil moisture and vegetation on the evolution of a prestorm environment: A numerical case study. Mon Weather Rev 119(6):1368-1390

Charney JG (1975) Dynamics of deserts and drought in the Sahel. Q J Roy Meteor Soc 101:193202

Chase TN, Pielke RA, Kittel TGF, Nemani RR, Running SW (2000) Simulated impacts of historical land cover changes on global climate. Clim Dyn 16:93-105

Chen F, Pielke RA Sr., Mitchell K (2001) Development and application of land-surface models for mesoscale atmospheric models: Problems and promises. In: Lakshmi V, Alberston J, Schaake J (eds) Observations and modeling of the land-surface hydrological processes. Am Geophysic Union, pp107-136

Chua S-H, Bras RL (1982) Optimal estimators of mean areal precipitation in regions of orographic influence. J Hydrol 57:23-48

Clark CA, Arritt RW (1995) Numerical simulations of the effect of soil moisture and vegetation cover on the development of deep convection. J Appl Meteorol 34:2029-2045

Claussen M (1994) On coupling global biome models with climate models. Climate Res 4:203-221

Claussen M (1995) Modeling bio-geophysical feedback in the Sahel. Max-Planck-Institut für Meteorologie, No.163, p26

Claussen M (1996) Variability of global biome patterns as a function of initial and boundary conditions in a climate model. Clim Dyn 12(6):371-379

Claussen M (1997) Modeling biogeophysical feedback in the African and Indian Monsoon region. Clim Dyn 13:247-257

Claussen M (1998) On multiple solutions of the atmosphere-vegetation system in present-day climate. Global Change Biol 6:369-377

Claussen M, Gayler V (1997) The greening of Sahara during the mid-Holocene: Results of an interactive atmosphere - biome model. Glob Ecol Biogeogr Lett 6:369-377

Claussen M, Brovkin V, Ganopolski A, Kubatzki C, Petoukhov V (1998) Modeling global terrestrial vegetation - climate interaction. Phil Trans R Soc Lond B 353:53-63

Claussen M, Kubatzki C, Brovkin V, Ganopolski A, Hoelzmann P, Pachur HJ (1999) Simulation of an abrupt change in Saharan vegetation at the end of the mid-Holocene. Geophys Res Lett 26(14):2037-2040

Cook KH (1994) Mechanisms by which surface drying perturbs tropical precipitation. J Climate 7:400-413

Cook KH, Ganadeskian A (1991) Effects of saturated and dry land surfaces on tropical circulation and precipitation in a General Circulation Model. J Climate 4(9):873-889

Copeland JH, Pielke RA, Kittel TGF (1996) Potential climatic impacts of vegetation change: A regional modeling study. J Geophys Res 101:7409-7418

Cox PM, Betts RA, Jones CD, Spall SA, Totterdell IJ (2000a) Acceleration of global warming due to carbon-cycle feedbacks in a coupled climate model. Nature 408(6809):184-187

Cox PM, Betts RA, Jones CD, Spall SA, Totterdell IJ (2000b) Acceleration of global warming due to carbon-cycle feedbacks in a coupled climate model. Nature 408(6813):750-750

Daly C, Neilson RP, Phillips DL (1994) A statistical-topographic model for mapping climatological precipitation over mountainous terrain. J Appl Meteorol 33:140-158

daRocha HR, Nobre CA, Bonatti JP, Wright IR, Sellers PJ (1996) A vegetation-atmosphere interaction study for Amazonia deforestation using field data and a „single column" model. Q J Roy Meteor Soc Part A 122(531):567-594

Deardorff JW (1978) Efficient prediction of ground surface temperature and moisture, with inclusion of a layer of vegetation. J Geophys Res 84C:1889-1903

Delworth T, Manabe S (1993) Climate variability and land-surface processes. Adv Water Res 16:3-20

Denning AS, Fung IY, Randall D (1995) Latitudinal gradient of atmospheric CO_2 due to seasonal exchange with land biota. Nature 376(6537):240-243

Dickinson R, Henderson-Sellers A, Kennedy PJ, Wilson MF (1986) Biosphere-Atmosphere Transfer Schemes (BATS) for the NCAR community climate model. NCAR Techn. Note NCAR/TN-275+STR, Boulder, Colorado, p69

Dickinson RE, HendersonSellers A, Kennedy OJ (1993) Biosphere-Atmosphere Transfer Scheme (BATS) version 1e as coupled to the NCAR community climate model. NCAR Techn. Note NCAR/TN387+STR, p72

Dickinson RE, Shaikh M, Bryant R, Graumlich L (1998) Interactive canopies for a climate model. J Climate 11:2823-2836

Dirmeyer PA (1994) Vegetation stress as a feedback mechanism in mid-latitude drought. J Climate 7:1463-1483

Dirmeyer PA, Shukla J (1996) The effect on regional and global climate of expansion of the world's deserts. Q J Roy Meteor Soc 122:451-482

Dolman A (1992) A note on areally-averaged evaporation and the value of the effective surface conductance. J Hydrol 138:583-589

Dooge JCI (1995) Scale problems in surface fluxes. In: Feddes RA (ed): Space and time scale variabilities and interdependencies in hydrologic processes. Cambridge University Press, pp21-32

Drever J (1997) The geochemistry of natural waters. Prentice-Hall, Upper Saddle River, p436

Eastman JL, Coughenour MB, Pielke RA (2001a) Does grazing affect regional climate? J Hydrometeorol 2:243-253

Eastman JL, Coughenour MB, Pielke RA (2001b) The regional effects of CO2 and landscape change using a coupled plant and meteorological model. Global Change Biol 7(7):797-815

Eltahir EAB (1996) Role of vegetation in sustaining large-scale atmospheric circulation in the tropics J Geophys Res 101(D2):4255-4268

Eltahir EAB, Bras RL (1994) Precipitation recycling in the Amazon Basin. Q J Roy Meteor Soc 120:861-880

Entekhabi D, Eagleson P (1989) Land surface hydrology parameterization for atmospheric general circulation models including subgrid-scale spatial variability. J Climate 2:816-831

Entekhabi D, Rodriguez-Iturbe I, Bras R L (1992) Variability in large-scale water balance with landsurface-atmosphere interaction. J Climate 5(8):798-813

Eugster H, Hardie L (1978) Saline lakes. In: Lerman A (ed) Lakes - chemistry, geology, physics. Springer-Verlag, New York, pp237-293

Famiglietti JS, Wood EF (1994) Multiscale modeling of spatially variable water and energy balance processes. Water Resourc Res 30:3061-3078

Fan S, Gloor M, Mahlman J, Pacala S, Sarmiento J, Takahashi T, Tans P (1998) A large terrestrial carbon sink in North America implied by atmospheric and oceanic carbon dioxide data and models. Science 282(5388):442-446

Fast JD, McCorcle MD (1991) The effects of heterogeneous soil moisture on a summer baroclinic circulation in the central United States. Mon Weather Rev 119(6):2140-2167

Feddes RA (ed) (1995) Space and time scale variabilities and interdependencies in hydrologic processes. Cambridge University Press, p181

Fennessey MJ, Shukla J (1999) Impact of initial soil wetness on seasonal atmospheric prediction. J Climate 12(11):3167-3180

Foley JA, Lewis S, Prentice IC, Pollard D, Thompson SL (1998) Coupling dynamic models of climate and vegetation. Global Change Biol 4:561579

Ganopolski A, Kubatzki C, Claussen M, Brovkin V, Petoukhov V (1998) The influence of vegetation-atmosphere-ocean interaction on climate during the mid-Holocene. Science 280:1916-1919

Gleick PH (ed) (1993) Water in crisis. A guide to the world's fresh water resources. Oxford University Press

Ginn T, Murphy E (1997) A transient flux model for convective infiltration: Forward and inverse solutions for chloride mass balance studies. Water Resour Res 33(9):2065-2079

Giorgi F, Avissar R (1997) Representation of heterogeneity effects in earth system modeling: Experience from land surface modeling. Rev Geophys 35(4):413-437

Giorgi F, Mearns LO (1991) Approaches to the simulation of regional climate change. A review. Rev Geophys 29:191-216

Giorgi F, Mearns LO (1999) Introduction to special section: Regional climate modeling revisited. J Geophys Res 104:6335-6352

Giorgi F, Means LO, Shields C, Mayer L (1996) A regional model study of the importance of local versus remote controls of the 1988 drought and the 1993 flood over the central United States. J Climate 9:1150-1162

Gong C, Eltahir E (1996) Sources of moisture for rainfall in West Africa. Water Resourc Res 32:3115-3121

Goutorbe JP, Lebel T, Dolman AJ, Gash JHC, Kabat P, Kerr YH, Monteny B, Prince SD, Stricker JNM, Tinga A (1997) An overview of HAPEX-Sahel: A study in climate and desertification. J Hydrol 188/189:4-17

Grabowski WW, Wu X, Moncrieff MW, Hall WD (1998) Cloud-resolving modeling of tropical cloud systems during Phase III of GATE. Part II: Effects of resolution and the third dimension. J Atmos Sci 55:3264-3282

Gutman G (1984) Numerical experiments on land surface alterations with a zonal model allowing for interaction between the geobotanic state and climate. J Atmos Sci 41:2679-2685

Gutman G (1985) On modeling dynamics of geobotanic state-climate interaction. J Atmos Sci 43:305-306

Gutman G, Ohring G, Joseph JH (1984) Interaction between the geobotanic state and climate: A suggested approach and a test with a zonal model. J Atmos Sci 41:2663-2678

Gutzler DS, Preston JW (1997) Evidence for a relationship between spring snow cover in North America and summer rainfall in New Mexico. Geophys Res Lett 24:2207-2210

Haltiner GJ, Williams RJ (1980) Numerical prediction and dynamic meteorology. John Wiley & Sons, New York, p477

Hanor JS (2001) Reactive transport involving rock-buffered fluids of varying salinity. Geochim Cosmochim Acta 65(21):3721-3732

Hay LE, McCabe GJ, Wolock DM, Ayers MA (1992) Use of weather types to disaggregate General Circulation Model predictions. J Geophys Res 97:2781-2790

Hay LE, Wilby RL, Leavesley GH (2000) A comparison of delta change and down-scaled GCM scenarios for three mountainous basins in the United States. J Am Water Resourc 36(2):387-397

Hay LE, Clark MP, Wilby RL, Gutowski WJ Jr., Leavesley GH, Pan Z, Arritt RW, Takle ES (2002) Use of regional climate model output for hydrologic simulations. J Hydrometeorol 3(5):571-590

Henderson-Sellers A, Pitman AJ, Love PK, Irannejad P, Chen TH (1995) The project for intercomparison of land surface parameterization schemes (PILPS): Phases 2 and 3. B Am Meteorol Soc 74(7):1335-1349

Hostetler SW (1994) Hydrologic and atmospheric models: The (continuing) problem of discordant scales. Climatic Change 27:345-350

Huggett RJ (1991) Climate, earth processes and earth history. Springer-Verlag, Berlin, XIV, p281

Hutchinson MF (1991a) Climatic analyses in data-sparse regions. In: Muchow RC, Bellamy JA (eds) Climatic risk in crop production. CAB International, Wallingford, UK, pp55-71

Hutchinson MF (1991b) The application of thin plate smoothing splines to continent-wide data assimilation. In: Jasper JD (ed) Data assimilation systems. BMRC Research Report No. 27, Bureau of Meteorology, Melbourne, Australia, pp104-113

Hutchinson MF (1995a) Stochastic space-time weather models from ground-based data. Agr Forest Meteorol 73:237-264

Hutchinson MF (1995b) Interpolating mean rainfall using thin plate smoothing splines. Int J GIS 9:385-403

Hutchinson MF (1998) Interpolation of rainfall data with thin plate smoothing splines: II. Analysis of topographic dependence. J Geog Inf Decis Anal 2:168-185

Intergovernmental Panel on Climate Change (1996) Climate change 1995: Impacts, adaptations and mitigation of climate change: Scientific - technical analyses. Contribution of Working Group I to the Second Assessment Report of the Intergovernmental Panel on Climate Change. Cambridge University Press, Cambridge

Ji JJ (1995) A climate-vegetation interaction model: Simulating physical and biological processes at the surface. J Biogeochem 22:2063-2068

Jones RG, Murphy JM, Noguer M (1995) Simulation of climate change over Europe using a nested regional-climate model. I: Assessment of control climate, including sensitivity to location of lateral boundaries. Q J Roy Meteor Soc 121:1413-1449

Kalnay E, Kanamitsu M, Kistler R, Collins W, Deaven D, Gandin L, Iredell M, Saha S, White G, Woollen J, Zhu Y, Chelliah M, Ebisuzaki W, Higgins W, Janowiak J, Mo KC, Ropelewski C, Wang J, Leetmaa A, Reynolds R, Jenne R Joseph D (1996) The NCEP/NCAR 40-year reanalysis project. B Am Meteorol Soc 77:437-471

Karl TR, Wang WC, Schlesinger ME, Knight RW, Portman D (1990) A method of relating General Circulation Model simulated climate to the observed local climate. Part I: Seasonal statistics. J Climate 3:1053-1079

Katz RW, Parlange MB (1996) Mixtures of stochastic processes: Application to statistical downscaling. Climate Res 7:185-193

Keeling CD, Chin JFS, Whorf TP (1996) Increased activity of northern vegetation inferred from atmospheric CO2 measurements. Nature 382(6587):146-149

Kiang JE, Eltahir EAB (1999) Role of ecosystem dynamics in biosphere-atmosphere interaction over the coastal region of West Africa. J Geophys Res 104(D24):31173-31189

Kim JW, Chang JT, Baker NL, Wilks DS, Gates WL (1984) The statistical problem of climate inversion: Determination of the relationship between local and large-scale climate. Mon Weather Rev 112:2069-2077

Kinoshita S (1982) Physics of frozen soil. Morikita Publishing Company, Japan

Klein WH, Bloom HJ (1987) Specification of monthly precipitation over the United States from the surrounding 700 mb height field. Mon Weather Rev 115:2118-2132

Klein WH, Glahn HR (1974) Forecasting local weather by means of model output statistics. B Am Meteorol Soc 55:1217-1227

Koster RD, Suarez MJ (1996) The influence of land surface moisture retention on precipitation statistics. J Climate 9:2551-2567

Koster RD, Suarez MJ (1999) A simple framework for examining the interannual variability of land surface moisture fluxes. J Climate 12:1911-1917

Kubatzki C, Claussen M (1998) Simulation of the global biogeophysical interactions during the last glacial maximum. Clim Dyn 14:461-471

Kutzbach JE, Guetter PJ (1986) The influence of changing orbital parameters and surface boundary conditions on climate simulations for the past 18,000 years. J Atmos Sci 43:1726-1759

Kutzbach JE, Bonan G, Foley J, Harrison SP (1996) Vegetation and soil feedbacks on the response of the African monsoon to orbital forcing in the early to middle Holocene. Nature 384(6610):623-626

Lean J, Rowntree PR (1997) Understanding the sensitivity of a GCM simulation of Amazonian deforestation to the specification of vegetation and soil characteristics. J Climate 10:1216-1235.

Leavesley GH, Stannard LG (1995) The precipitation-runoff modeling system – PRMS. In: Singh VP (ed) Computer models of watershed hydrology. Water Resources Publications, Highlands Ranch, Colorado, pp281-310

Leung LR, Ghan SJ (1995) A subgrid parameterization of orographic precipitation. Theor Appl Climatol 52:95-118

Leung LR, Ghan SJ (1999) Pacific Northwest climate sensitivity simulated by a regional climate model driven by a GCM. Part I: Control simulations. J Climate 12:2010-2030

Leung LR, Wigmosta MS, Ghan SJ, Epstein DJ, Vail LW (1996) Application of a subgrid orographic precipitation surface hydrology scheme to a mountain watershed. J Geophys Res Atmosphere 101(D8):12803-12817

Levis S, Foley JA, Brovkin V, Pollard D (1999) On the stability of the high-latitude climate-vegetation system in a coupled atmosphere-biosphere model. Glob Ecol Biogeogr 8(6):489-500

Lhomme JP (1992) Energy balance of heterogeneous terrain: Averaging the controlling parameters. Agr Forest Meteorol 61:11-21

Liston GE (1999) Interrelationships among snow distribution, snowmelt, and snow cover depletion: Implications for atmospheric, hydrologic, and ecologic modeling. J Appl Meteorol 38(10):1474-1487

Liston GE, Pielke RA, Greene EM (1999) Improving first-order snow-related deficiencies in a regional climate model. J Geophys Res 104:19559-19567

Liu Y, Avissar R (1999a) A study of persistence in the land-atmosphere system using a General Circulation Model and observations. J Climate 12:2139-2153

Liu Y, Avissar R (1999b) A study of persistence in the land-atmosphere system using a fourth-order analytical model. J Climate 12:2154-2168

Liu C-H, Moncrieff MW, Grabowski WW (1999) A hierarchical approach to parameterizing organized convection. Proceedings of a Conference on the TOGA Coupled Ocean-Atmosphere Response Experiment (COARE), Boulder, Colorado, USA, 7-14 July 1998. WCRP-107, WMO/TD-No.940, pp357-358

Liu C-H, Moncrieff MW, Grabowski WW (2001a) Hierarchical modelling of tropical convective systems using explicit and parameterized approaches. Q J Roy Meteor Soc 127(572):493-515

Liu C-H, Moncrieff MW, Grabowski WW (2001b) Explicit and parameterized realizations of convective cloud systems in TOGA COARE. Mon Weather Rev 129(7):1689-1703

Lofgren BM (1995a) Sensitivity of land-ocean circulations, precipitation, and soil moisture to perturbed land surface albedo. J Climate 8(10):2521-2542

Lofgren BM (1995b) Surface albedo-climate feedback simulated using two-way coupling. J Climate 8(10):2543-2562

Loth B, Graf H-F, Oberhuber JM (1993) Snow cover model for global climate simulations. J Geophys Res 98:10451-10464

Lu L (1999) Implementation of a two-way interactive atmospheric and ecological model and its application to the central United States. Ph.D. Dissertation, Department of Atmospheric Science, Colorado State University, Fort Collins, Colorado, p134

Lu L, R Pielke A, Liston GE, Parton WJ, Ojima D, Hartman M (2000) Implementation of a two-way interactive atmospheric and ecological model and its application to the central United States. J Climate 14:900-919

Lynch-Stieglitz M (1994) The development and validation of a simple snow model for the GISS GCM. J Climate 7:1842-1855

Lyons SW (1990) Spatial and temporal variability of monthly precipitation in Texas. Mon Weather Rev 118:2634- 2648

Lyons WA (1999) Lightning. In: Pielke RA Sr, Pielke RA Jr (co-eds)Storms. Hazard and Disaster Series, Routledge Press, UK, pp60-79

Malhi Y, Pegoraro E, Nobre AD, Pereira MGP, Grace J, Culf AD, Clement R (2002) Energy and water dynamics of a central Amazonian rain forest. J Geophys Res - Atmosphere 107(D20):Art. No.8061

Martin PH, Valentini R, Jacques M, Fabbri K, Galati D, Quaratino R, Moncrieff JB, Jarvis P, Jensen NO, Lindroth A, Grelle A, Aubinet M, Ceulemans R, Kowalski AS, Vesala T, Keronen P, Matteucci G, Grainer A, Berbingier P, Loustau D, Grelle A, Schulze ED, Tenhunen J, Rebman C, Dolman AJ, Elbers JE, Bernhofer C, Grunwald T, Thorgeirsson H, Kennedy P, Folving S (1998) New estimate of the carbon sink strength of EU forests integrating flux measurements, field surveys, and space observations: 0.17-0.35 Gt(C). Ambio 27(7):582-584

Mason PJ (1988) The formation of areally-averaged roughness lengths. Q J Roy Meteor Soc 114:399-420

McBean G, Hantel M (eds) (1993) Interactions between global climate subsystems. AGU-Geophysical Monograph Series 75; co-published with IUGG 15; p155

McGregor JJ (1997) Regional climate modelling. Meteorol Atmos Phys 63:105-117

McNaughton KG (1994) Effective stomatal and boundary layer resistances of heterogeneous surfaces. Plant Cell Environ 17:1061-1068

Miguez-Macho G, Stenchikov GL, Robock A (2004) Spectral nudging to improve downscaling over North America using the Regional Atmospheric Modeling System (RAMS). submitted

Milly PCD, Dunne KA (1994) Sensitivity of global water cycle to the water-holding capacity of land. J Climate 7:506-626

Mölders N, Raabe A (1996) Numerical investigations on the influence of subgrid-scale surface heterogeneity on evapotranspiration and cloud processes. J Appl Meteorol 35:782-795

Mölders N, Raabe A, Tetzlaff G (1996) A comparison of two strategies on land surface heterogeneity used in a mesoscale-β meteorological model. Tellus 48A:733-749

Nabuurs GJ, Paivinen R, Sikkema R, Mohren GMJ (1997) The role of European forests in the global carbon cycle - a review. Biomass Energy 13(6):345-358

Nash JE, Sutcliffe JV (1970) River flow forecasting through conceptual models. Part I: A discussion of principle. J Hydrol 10:282-290

Nepstad DC, de Carvalho CR, Davidson EA, Jipp PH, Lefebvre PA, Negreiros GH, da Silva ED, Stone TA, Trumbore SE, Vieira S (1994) The role of deep roots in the hydrological and carbon cycles of Amazonian forests and pastures. Nature 372:666-669

Niehoff D (2002) Modellierung des Einflusses der Landnutzung auf die Hochwasserentstehung in der Mesoskala. Brandenburgische Umweltberichte 11, University of Potsdam p144

Noilhan J, Lacarrere P (1995) GCM grid scale evaporation from mesoscale modelling. J Climate 8:206-223

O'Brien KL (2000) Upscaling tropical deforestation: Implications for climate change. Climatic Change 44:311-329

Oglesby RJ (1991) Springtime soil moisture, natural climate variability and North American drought as simulated by the NCAR Community Model 1. J Climate 4(9):890-897

Pachur H-J, Wünnemann, B (1996) Reconstruction of the palaeoclimate along 30°E in the eastern Sahara during the Pleistocene/Holocene transition. In: Heine K (ed) Palaeoecology of Africa and the surrounding islands. AA Balkema, pp1-32

Paegle J, Mo KC, Nogues-Paegle J (1996) Dependence of simulated precipitation on surface evaporation during the 1993 United States summer floods. Mon Weather Rev 124:345-361

Phillips OL, Malhi Y, Higuchi N, Lawrance WF, Nuñez PV, Vasquez RM, Laurance SG, Ferreira LV, Stern M, Brown S, Grace J (1998) Changes in the carbon balance of tropical forests: Evidence from long-term plots. Science 282:439-442

Pielke RA (2001) Influence of the spatial distribution of vegetation and soils on the prediction of cumulus convective rainfall. Rev Geophys 39:151-177

Pielke RA, Avissar R, Raupach M, Dolman AJ, Zeng X, Denning AS (1998) Interactions between the atmosphere and terrestrial ecosystems: Influence on weather and climate. Global Change Biol 4:461-475

Pielke RA, Walko RL, Steyaert LT, Vidale PL, Liston GE, Lyons WA, Chase TN (1999) The influence of anthropogenic landscape changes on weather in South Florida. Am Meteorol Soc 127:1663-1672

Pielke RA Sr, Liston GE, Robock A (2000) Insolation-weighted assessment of Northern Hemisphere snow-cover and sea-ice variability. Geophys Res Lett 27:3061-3064

Rahmstorf S (1995) Bifurcations of the Atlantic thermohaline circulation in response to changes in the hydrological cycle. Nature 378:145-149

Randall DA, Dazlich DA, Zhang C, Denning AS, Sellers PJ, Tucker CJ, Bounoua L, Los SO, Justice CO, Fung I (1996) A revised land surface parameterization (SiB2) for GCMs. Part III: The greening of the Colorado State University general circulation model. J Climate 9(4):738-763

Raupach MR (1995) Vegetation-atmosphere interaction and surface conductance at leaf, canopy, and regional scales. Agr Forest Meteorol 73:151-179

Raupach MR, Finnigan JJ (1995) Scale issues in boundary-layer meteorology: Surface energy balances in heterogeneous terrain. Hydrol Process 9:589-612

Raupach MR, Finnigan JJ (1997) The influence of topography on meteorological variables and surface-atmosphere interactions. J Hydrol 190(3-4):182-213

Reale O, Dirmeyer P (2000) Modeling the effects of vegetation on Mediterranean climate during the Roman classical period. Part I: History and model sensitivity. Global Planet Change 25:163-184

Roeckner E, Arpe K, Bengtsson L, Brinkop S, Dümenil L, Esch M, Kirk E, Lunkeit F, Ponater M, Rockel B, Sausen R, Schleese U, Schubert S, Windelband M (1992) Simulation of the present-day climate with the ECHAM model: Impact of model physics and resolution. Max Planck Institute for Meteorology, Report No. 93, Hamburg

Running SW, Thornton PE (1996) Generating daily surfaces of temperature and precipitation over complex topography. In: Goodchild MF, Steyaert LT, Parks BO, Johnston C, Maidment D, Crane M, Glendinning S (eds) GIS and environmental modeling: Progress and research issues. GIS World, Fort Collins, USA, pp93-98

Schimel D, Melillo J, Tian HQ, McGuire AD, Kicklighter D, Kittel T, Rosenbloom N, Running S, Thornton P, Ojima D, Parton W, Kelly R, Sykes M, Neilson R, Rizzo B (2000) Contribution of increasing CO_2 and climate to carbon storage by ecosystems in the United States. Science 287(5460):2004-2006

Schlosser CA, Slater AG, Robock A, Pitman AJ, Vinnikov KY, Henderson-Sellers A, Speranskaya NA, Mitchell K (2000) Simulations of a boreal grassland hydrology at Valdai, Russia: PILPS phase 2(d). Mon Weather Rev 128(2):301-321

Schultz GA, Hornbogen M, Viterbo P, Noilhan J (1994) Coupling large-scale hydrological and atmospheric models. IAHS Special Publication No. 3, IAHS Press, Institute of Hydrology, Wallingford, UK

Scott R, Entekhabi D, Koster R, Suarez M (1997:) Time scales of land surface evapotranspiration response. J Climate 10:559-566

Sellers PJ, Mintz Y, Sud YC, Dalcher A (1986) A simple biosphere atmosphere model (SiB) for use within general circulation models. J Atmos Sci 43:505-531

Sellers PJ, Bounoua L, Collatz GJ, Randall DA, Dazlich DA, Los SO, Berry JA, Fung I, Tucker CJ, Field CB, Jensen TG (1996) Comparison of radiative and physiological effects of doubled atmospheric CO_2 on climate. Science 271(5254):1402-1406

Seth A, Giorgi F, Dickinson RE (1994) Simulating fluxes from heterogeneous land surfaces: Explicit subgrid method employing the biosphere-atmosphere transfer scheme (BATS). J Geophys Res 99D:18651-18667

Shao Y, Sogalla M, Kerschgens M, Brücher W (2001) Effects of land-surface heterogeneity upon surface fluxes and turbulent conditions. Meteorol Atmos Phys 78:157-181

Shuttleworth WJ (1991) The Modellion concept. Rev Geophys 29(4):585-606

Slater AG, Schlosser CA, Desborough CE, Pitman AJ, Henderson-Sellers A, Robock A, Vinnikov KY, Mitchell K, Boone A, Braden H, Chen F, Cox PM, de Rosnay P, Dickinson RE, Dai YJ, Duan Q, Entin J, Etchevers P, Gedney N, Gusev YM, Habets F, Kim J, Koren V, Kowalczyk EA, Nasonova ON, Noilhan J, Schaake S, Shmakin AB, Smirnova TG, Verseghy D, Wetzel P, Yue X, Yang ZL, Zeng Q (2001) The representation of snow in land-surface schemes: Results from PILPS 2(d). J Hydrometeorol 2(1):7-25

Solomon AM, Prentice IC, Leemans R, Cramer W (1993) The interaction of climate and land-use in future terrestrial carbon storage and release. Water Air Soil Poll 70(1-4):595-614

Steffen W, Noble I, Canadell J, Apps M, Schulze ED, Jarvis PG, Baldocchi D, Ciais P, Cramer W, Ehleringer J, Farquhar G, Field CB, Ghazi A, Gifford R, Heimann M, Houghton R, Kabat P, Korner C, Lambin E, Linder S, Mooney HA. Murdiyarso D, Post WM, Prentice IC, Raupach MR, Schimel DS, Shvidenko A, Valentini R (1998) The terrestrial carbon cycle: Implications for the Kyoto Protocol. Science 380(5368):1393-1394

Stewart JB, Engman ET, Feddes RA, Ker Y (1996) scaling up in hydrology using remote sensing. John Wiley, p255

Takle ES, Gutowski WJ, Arritt RW, Pan ZT, Anderson CJ, da Silva RR, Caya D, Chen SC, Giorgi F, Christensen JH, Hong SY, Juang HMH, Katzfey J, Lapenta WM, Laprise R, Liston GE, Lopez P, McGregor J, Pielke RA, Roads JO (1999) Project to Intercompare Regional Climate Simulations (PIRCS): Description and initial results. J Geophys Res 104:19443-19461

Tans PP, Wallace DWR (1999) Carbon cycle research after Kyoto. Tellus B 51(2): 562-571

Taylor CM, Lebel T (1998) Observational evidence of persistent convective-scale rainfall patterns. Mon Weather Rev 126:1597-1607

Taylor CM, Said F, Lebel T (1997) Interactions between the land surface and mesoscale rainfall variability during HAPEX-Sahel. Mon Weather Rev 125: 2211-2227

Taylor CM, Clark DB, Cox PM (2000) Land surface controls on west African rainfall in a GCM. In: World Meteorological Organisation (2000) Proceedings of the workshop on west African monsoon variability and predictability (WAMAP). Geneva, WMO Tropical Meteorology Research Programme, Report Series 63

Trenberth KE (1999) Atmospheric moisture recycling: Role of advection and local evaporation. Am Meteorol Soc 12:1368-1381

Tsvetsinskaya E (1999) Investigating the effect of seasonal plant growth and development in 3dimensional atmospheric simulations. Ph.D. Dissertation, University of Nebraska, Lincoln, Nebraska, p147

Turner CL, Seastedt TR, Dyer MI, Kittel TGG, Schimel DS (1992) Effects of management and topography on the radiometric response of a tall grass prairie. J Geophys Res 97D:18855-18866

UNESCO (1978:) World water balance and water resources of the earth. Studies and Reports in Hydrology 25, Prepared by the USSR Committee for the International Hydrological Decade, Paris

Valentini R, Dolman H, Ciais P, Schulze E, Freibauer A, Schimel D, Heimann M (2000) Accounting for carbon sinks in the biosphere, European perspective, scientific note to articles 3.3 and 3.4 of the Kyoto Protocol. CarboEuropean Office, Max Planck Institute for Biogeochemistry, Jena, Germany

Vinnikov KYa, Robock A, Speranskaya NA, Schlosser CA (1996) Scales of temporal and spatial variability of midlatitude soil moisture. J Geophys Res 101: 7163-7174

von Storch H, Zorita E, Cubasch U (1993) Downscaling of global climate change estimates to regional scales: An application to Iberian rainfall in wintertime. J Climate 6:1161-1171 von Storch H, Landenberg H, Feser F (2000:) A spectral nudging technique for dynamical downscaling purposes. Mon Weather Rev 128:3664-3673

Wahba G, Wendelberger J (1980) Some new mathematical methods for variational objective analysis using splines and cross validation. Mon Weather Rev 108: 1122-1143

Walko RL, Band LE, Baron J, Kittel TGF, Lammers R, Lee TJ, Ojima DS, Pielke RA, Taylor C, Tague C, Tremback CJ, Vidale PL (2000) Coupled atmosphere-biophysics-hydrology models for environmental modeling. J Appl Meteorol 39:931-944

Walsh JE, Richman MB, Allen DW (1982) Spatial coherence of monthly precipitation in the United States. Mon Weather Rev 110:272-286.

Wang G, Eltahir EAB (1999) The role of vegetation dynamics in the climate of West Africa. Department of Civil and Environmental Engineering Report No. 344, MIT, Cambridge, MA, p224

Wang G, Eltahir EAB (2000a) Ecosystem dynamics and the Sahel drought. Geophys Res Lett 27:795-798

Wang G, Eltahir EAB (2000b) Biosphere-atmosphere interactions over West Africa. I: Development and validation of a coupled dynamic model. Q J Roy Meteor Soc 126:1239-1260

Watson AJ, Lovelock JE (1983) Biological homeostasis of the global environment: The parable of Daisyworld. Tellus 35B:284-289

Wetzel PJ, Chang J-T (1988) Evapotranspiration from nonuniform surfaces: A first approach for short-term numerical weather prediction. Mon Weather Rev 116, 600-621

Wigley TML, Jones PD, Briffa KR, Smith G (1990) Obtaining sub-grid scale information from coarse resolution general circulation model output. J Geophys Res 95:1943-1953

Wilby RL (1997) Non-stationarity in daily precipitation series: Implications for GCM down-scaling using atmospheric circulation indices. Int J Climatol 17(4):439-454

Wilby RL, Dettinger MD (2000) Streamflow changes in the Sierra Nevada, CA simulated using statistically downscaled General Circulation Model output. In: McLaren S, Kniveton D (eds) Linking climate change to land surface change. Kluwer Academic Publishers, pp99–121

Wilby RL, Wigley TML (1997) Downscaling General Circulation Model output: A review of methods and limitations. Prog Phys Geog 21(4):530-548

Wilby RL, Hassan H, Hanaki K (1998a) Statistical downscaling of hydro-meteorological variables using General Circulation Model output. J Hydrol 205:1-19

Wilby RL, Wigley TML, Conway D, Jones PD, Hewitson BC, Main J, Wilks DS (1998b) Statistical downscaling of General Circulation Model output: A comparison of methods. Water Resourc Res 34:2995-3008

Wilby RL, Hay LE, Leavesley GH (1999) A comparison of downscaled and raw GCM output: Implications for climate change scenarios in the San Juan River basin, Colorado. J Hydrol 225:67-91

Wilby RL, Hay LE, Gutowski WJ, Arritt RW, Takle ES, Pan Z, Leavesley GH, Clark MP (2000) Hydrological responses to dynamically and statistically downscaled climate model output. Geophys Res Lett 27(8):1199-1202

Wilcox BP, Rawls WJ, Brakensiek DL, Wight JR (1990) Predicting runoff from rangeland catchments: A comparison of two models. Water Resour Res 26: 2401-2410

Williams PJ, Smith MW (1989) The frozen Earth. Cambridge University Press, New York

Winkler JA, Palutikof JP, Andresen JA, Goodess CM (1997) The simulation of daily temperature series from GCM output. Part II: Sensitivity analysis of an empirical transfer function methodology. J Climate 10:2514-2532

Wood EF, Lettenmaier DP, Zartarian V (1992) A land-surface hydrology parameterization with subgrid variability for general circulation models. J Geophys Res 97:2717-2728

Woodwell GM, Hobbie JE, Houghton RA, Melillo JM, Moore B, Peterson BJ, Shaver GR (1983) Global deforestation - contribution to atmospheric carbon-dioxide. Science 222(4628):1081-1086

Wu XQ, Hall WD, Grabowski WW, Moncrieff MW, Collins WD, Kiehl JT (1999) Long-term behavior of cloud systems in TOGA COARE and their interaction with radiative and surface processes. Part II: Effects of ice microphysics on cloud-radiation interaction. J Atmos Sci 56(18):3177-3195

Xu CY (1999) Climate change and hydrologic models: A review of existing gaps and recent research developments. Water Resour Manage 13(5):369-382

Xue Y, Shukla J (1993) The influence of land surface properties on Sahel climate: Part I. Desertification. J Climate 6(12):2232-2345

Yang Z-L, Dickinson RE, Robock A, Vinnikov K Ya (1997) Validation of the snow sub-model of the Biosphere-Atmosphere Transfer Scheme with Russian snow cover and meteorological observation data. J Climate 10:353-373

Yang Z-L, Dickinson RE, Hahmann AN, Niu G-Y, Shaikh M, Gao X, Bales RC, Sorooshian S, Jin J (1999) Simulation of snow mass and extent in General Circulation Models. Hydrol Process 13:2097-2113

Yershov E (1998) General geocryology. Cambridge University Press, New York.

Zhang T, Barry RG, Knowles K, Higginbottom A, Brown J (1999) Statistics and characteristics of permafrost and ground ice distribution in the Northern Hemisphere. Proceedings of American Meteorological Society 5th Conference on Polar Meteorology and Oceanography, Dallas Texas

Ziegler CL, Lee TJ, Pielke RA (1997) Convective initiation at the dryline: A modeling study. Mon Weather Rev 125:1001-1026

CHAPTER 2

Systems Approach: the Nature of Coupled Models

2.0 Introduction

Jesus Carrera

The early development of systems theory was closely linked to the formalisation of water resources systems. These were originally visualised as relatively complex networks consisting of nodes and connections that could be associated with physical entities and which had relatively simple individual behaviour. Thus, a basin could be viewed as a set of stores (reservoirs, aquifers, soil, etc.) and fluxes (through rivers, pipelines, etc.). Each element of the system obeyed well defined rules (mass balance, Darcy's law, Dalton's law, and the like). All this amounted to fertile land for formalisation, development and application of a robust water resources system. The books of Dooge (1973) and Eagleson (1970) can be viewed as pillars of such developments. Even though more than thirty years have passed since their publication, they are still valid and a good understanding of hydrology can be gained from studying them. This was probably the motivation behind the decision of the European Geophysical Union to reprint them (Dooge 2003; Eagleson 2003).

Water resources systems analysis placed a lot of emphasis on quantification for decision making, probably because it was viewed as a branch of civil engineering. Efforts concentrated on accurate, yet robust, representation of basins, which allowed optimisation. Maturity of this line of development has led to a high level of refinement and to the establishment of hydrology as a science (Gupta 1989, 2001; Basson et al. 1994; Loucks et al. 1981; etc.). However, emphasis is shifting from quantification to understanding, probably reflecting the multidisciplinary nature of hydrology, which has opened the field to many people with a scientific background. Moreover, hydrological models have grown in complexity, in the sense that they contain a continuously increasing number of parts and that processes of each part are described in increasing detail. As a result, they often become cumbersome, while their predictive capabilities remain doubtful at best. It is fair, therefore, to question the wisdom of such growth and whether simpler models would work just as well.

In parallel with the growth of hydrological models, the last two decades have witnessed the maturity in systems theory and engineering. Terms like chaos, games theory, operations research and stochastic analysis are often used without a well defined (or, rather, a well understood) meaning. This chapter is aimed at bridging the gap between these two branches of knowledge. It must be kept in mind that model development will always have a chaotic component (i.e. it cannot be fully subjected to the "optimality" purported by systems analysis). Inversely, it would be ludicrous to pretend to convey all the formalities of systems theory in a single chapter. The relevant topics span a range too broad: from statistics to numerical methods; from spectral analysis to topology. Still, we feel that introducing some formalisation may be appropriate.

In this context, the objective of this chapter is not to supersede traditional water resources systems books, which are and will be needed in the future, but rather to complement them. The chapter starts by revising the concepts of fluxes and compartments (stores) and how they interact. These interactions are formalised into a matrix that helps in identifying the nature of links and feedbacks. A second section is devoted to revising the concept of non-linearity, its definition and types and why it can lead to unusual behaviours, such as oscillations and multiple equilibria. This second section is also devoted to presenting a broad framework for the solution of non-linear coupled problems. The third, final, section aims at formalising the modelling process itself. Therefore, it is somewhat abstract. Yet it introduces concepts such as calibration, uncertainty, vulnerability, risk analysis, stochastic versus deterministic modelling and others that appear to mark the future of hydrological modeling developments.

2.1 Fluxes, Compartments and Ordering of Feedbacks

Ekkehard Holzbecher, Mike Bonell, Axel Bronstert, Oleg F. Vasiliev

Several different approaches are possible to systemise the interaction of hydrological and related processes. Here, one approach is presented in more detail which utilises the concept of compartments and fluxes. Very often modelling concepts are illustrated in box-and-arrow diagrams or flow charts, where boxes, representing compartments, are connected by arrows, representing links, of which fluxes are a subset. In such a network of compartments, terms as coupling, linkage, feedback can be defined in a systematic way. The basic distinction between inter-compartmental and intra-compartmental coupling is introduced.

The term "compartment" is used with different meanings in the literature and what is said here will surely not put an end to the discussion. While the framework presented is applicable and useful, when specific problems are tackled there may be cases in which alternative systematisations work as well or better. Some examples are given in Chapters 3 and 4. Mathematical formulation of the problem will be presented in Sect. 2.2.

2.1.1 Introduction and Definitions

Conceptual models, which are used to structure scientific expertise and research and which are the basis of any computer model, are often built as a network of compartments. Such modelling approaches can be found quite often, particularly in hydrology. What follows is based on the idea that the structure of a model is already reflected in the modelling concept, which often can be represented as a network model. Basic properties concerning coupling can thus be investigated in the network model, which is an easy task compared to an investigation of the computer model with all its codes, modules and data.

A network model is defined by the linkage of different compartments and/or sub-compartments of the atmosphere, the biosphere and/or the pedosphere. The linkages are usually determined by energy, by water and/or by biogeochemical fluxes or by interacting processes. As an example, Fig. 2.1-1 displays a pictorial version of a coupled landscape model where pools of matter are in bold face type, the fluxes between pools (bold italic type) are shown by arrows (double headed arrows indicate two way transfers), state properties which limit fluxes (such as soil hydraulic properties) are shown in italic, and trigger mechanisms such as frozen soil are shown in regular type. This model conserves mass and energy, and uses energy to move mass from one pool to another. The model also contains exogenous inputs of energy through atmospheric variation and through potential land-use and land-cover changes.

Generally, compartments are pools, in which mass and energy can be stored or from which mass and/or energy can be removed. Reviewing the scientific literature, various terms can be found, which are used synonymously with the term *pool*, such as stores, reservoirs, spheres, phases, etc. In the following discussion the term *compartment* is preferred. In hydrological sciences, it is water that, in the first instance, is stored in or removed from a pool. When the migration of chemical components through the network of compartments is studied, the storage or removal of component mass is considered.

Water pools of our planet, which are commonly used as compartments in studies with the hydrological cycle, are:

1. Atmospheric water
2. Soil moisture
3. Lakes and reservoirs
4. Rivers
5. Groundwater
6. Frozen water
7. Oceans

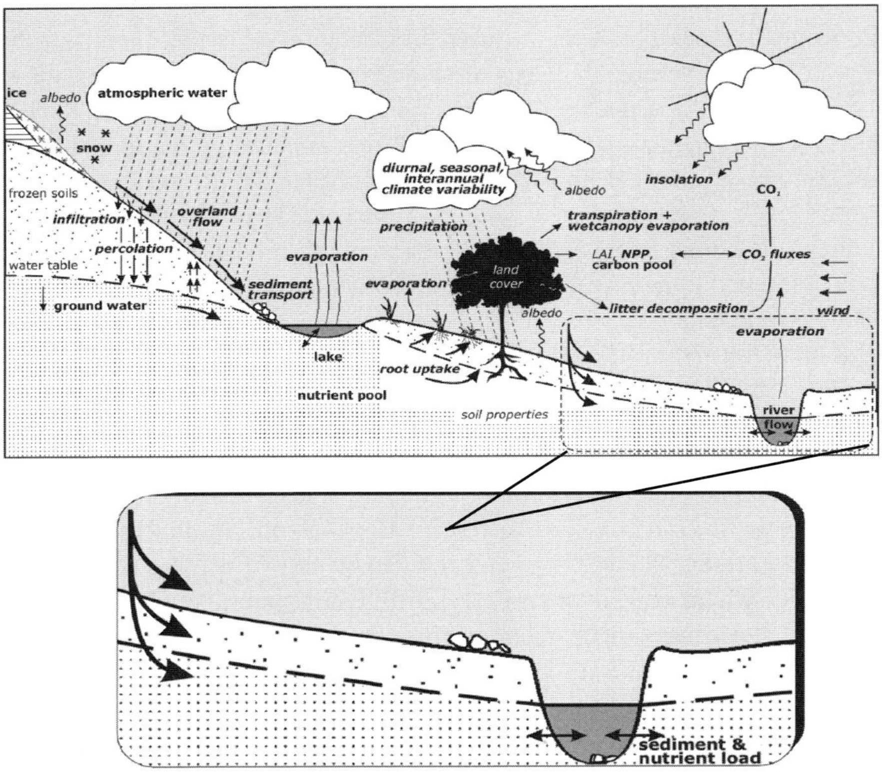

Fig. 2.1-1 Landscape model representing part of the hydrological cycle

Most models concerning the hydrological cycle contain several of these compartments. Processes, as listed in the matrix given in Sect. 3.1, act within one of these compartments or describe an interaction between two compartments. Depending on the specific study, one or more of the compartments mentioned may be omitted, others may have to be divided into (sub)-compartments.

Several sub-compartments can also be distinguished in hydrological studies:

1. Surface water bodies
    1.1. Ocean
    1.2. Rivers
    1.3. Lakes and river impoundments
    1.4. Canals
    1.5. Reservoirs
2. Soil moisture
    2.1. Root zone
    2.2. Vadose zone
    2.3. Capillary fringe
3. Groundwater
    3.1. Phreatic aquifer
    3.2. Deep aquifers
4. Frozen water
    4.1. Snow
    4.2. Ice
    4.3. Frozen soil water and permafrost

Compartments are limited, and defined, by separating boundaries. The ground surface, for example, is the boundary between pedosphere and atmosphere. The water surface is the boundary between surface water bodies and the atmosphere. The bottom of surface water bodies is the interface between free water and a porous medium.

Special boundaries may have to be considered, depending on the problem to be tackled. In subsurface regions, the water table constitutes the interface between the saturated and the unsaturated zone. Some problems require a closer look and it may be that the boundary concept chosen a priori may not be sufficient. A division of the unsaturated zone into root zone, vadose zone and capillary fringe may be adequate but a further distinction between different soil horizons may be necessary. The atmosphere is usually divided into several layers. As an example from limnology, simple models of stratified lakes distinguish between one compartment above and one below the pycnocline.

Other terms are used synonymously with the term "compartment"; using the term *sphere* and distinguish hydrosphere, biosphere, atmosphere, pedosphere and geosphere. Seigneur (1993) speaks of environmental media and classifies atmosphere, surface water, soil (including groundwater) and biota. The term *phase* can frequently be found, which is mostly referring to the physical state, i.e. gaseous, liquid or solid. Further distinctions are necessary in certain applications, e.g. certain geo-systems require partition into several liquid phases such as water and oil. Lastly, note that the term *legs* appears in some publications on environmental modelling.

Seigneur (1993) proposes to distinguish several phases within one environmental medium. In some media there is clearly one dominant phase, for example, the gaseous phase in the atmosphere, the liquid phase in surface water bodies. Nevertheless, other phases are present, as for example clouds, rain and fog represent the liquid phase in the atmosphere. Even the solid phase, in the form of snow flakes and ice-crystals, may be relevant for certain processes in the atmosphere. In surface water bodies, suspended load and sediment load represent the solid phase and are extremely relevant for the migration of those chemical components which interact with solid particles.

In the subsurface, there is no dominant phase as in the above examples. The solid phase dominates, in the saturated zone, below the groundwater table, but the fluid phase is also present, with a volumetric share mostly in the range of 20 to 30%, while the gaseous phase is mostly negligible. The fluid phase dominates in the benthic zone at the bottom of surface water bodies, but the solid phase is also present with a relevant share. The gas phase has to be added as the third major contributor in the unsaturated zone.

Concerning the behaviour of substances in the subsurface, a different setting for compartments, media and/or phases may be suitable. Most chemical components interact in one way or the other with the porous medium that surrounds the pore water. The solid phase plays an important role in the fate of these substances.

The separating interface can be identified directly in the above examples, but a more complex boundary concept becomes necessary for some problems. The saturated subsurface compartment sometimes needs to be thought of as a combination of a fluid and a solid phase which, in a model, are taken as two different compartments. The unsaturated zone can be understood as being combined from fluid, solid and gaseous phases (three compartments!). For modelling transport in the subsurface, one sometimes has to distinguish further, for example, an interfacial or an organic phase. In multiple porosity approaches even the fluid phase is divided into several sub-compartments. These examples show that the compartment concept does not only distinguish different spatial regions but can also contain several compartments in a region which is specified within well-defined boundaries.

Biota are present in all the three other environmental phases; biota also incorporate all three physical state phases. Thus it is difficult to distinguish media and phases clearly. Seigneur (1993) finds it essential to separate biota from the media, "because it can play a major role in the exposure to chemicals and it cannot be treated by the same models as those used to simulate the transport and fate of chemicals in the host medium". To avoid the conflict between the terms as described above, the term *compartments* will be used throughout as a "neutral" term. Compartment is used here from a general perspective as a class of objects with no particular physical characteristics.

Compartments in a network are connected by links, so that a link network emerges which can be represented by a flow chart. More details and examples are given below. In a general form, the link may just stand for influence. If there is an influence from compartment A to compartment B, this is represented by a link from A to B.

Fig. 2.1-2 Uni-directional link

A set of linked compartments is a network, or network model. Fig. 2.1-3 depicts a very simple example (think of compartments A: atmospheric moisture, B: surface water bodies, C: unsaturated zone, D: groundwater)

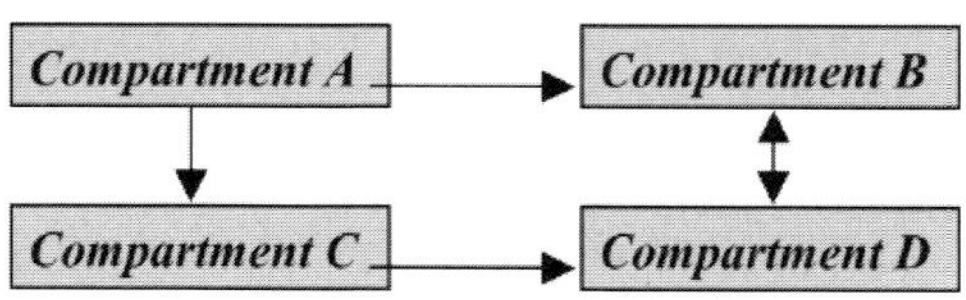

Fig. 2.1-3 Example of a compartment network

A network model accounts for pools of water and/or chemical/biochemical species in each of the pools, and tracks the available mass and/or energy and how that energy and/or mass flows through the system. A network of compartments is usually set up to study and model certain environmental processes, the flow of water and the migration of components and/or heat through this network. Depending on the type of processes (flow, transport, sorption, etc.) and the problem to be solved, the network of compartments may be very different.

A particular form of link is given when compartments are connected by fluxes. Through fluxes, one store losses and the other gains some kind of storable variable, mostly of mass or energy. Fig 2.1-4 shows a part of the hydrological cycle represented in a network model, in which links denote fluxes. Examples of mass and energy fluxes are given later (see Sect. 3.2).

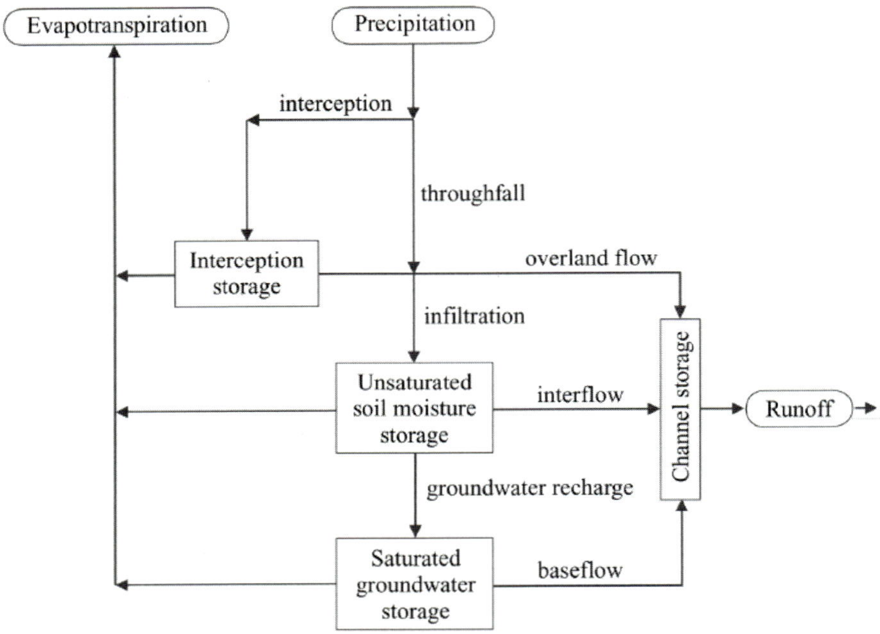

Fig. 2.1-4 Network model of part of the hydrological cycle after Freeze and Cherry 1979

Thus far, it can be understood quite well what the term compartment means in hydrology. It is more difficult to understand what is meant when a system of compartments is coupled. The term *coupling* is conceived differently in the scientific literature on computer modelling. The following is an attempt to give an overview and systematic view on the subject, showing that several types of coupling can be distinguished.

One major classification distinguishes between *inter-* and *intra-compartmental* coupling, in order to make a difference for couplings between compartments or within a compartment. Complex models, in which the atmosphere, ocean and/or land surface and subsurface are considered, include both types of couplings.

2.1.2 Inter-compartmental Coupling

Fig 2.1-5 shows a system of three linked compartments of the hydrological cycle: (1) soil moisture, (2) groundwater and (3) surface water. When rainfall reaches the ground surface, part returns to the atmosphere by evapotranspiration. Another part percolates into the subsurface and returns to the surface when plants take the water in the root-zone. All these processes in the vadose or unsaturated zone are mostly treated as being one-dimensional because the vertical movements are dominant. Finally, some water seeps down to the groundwater table, recharging the groundwater reservoir. In the figure, groundwater recharge is represented by the flux across the interface between saturated and unsaturated zones.

In contrast to the unsaturated zone, the main flow direction in the saturated subsurface is horizontal. There are vertical components but these can often be neglected. Groundwater hydrologists often refer to the idea of horizontal flow as Dupuit assumption (see for example, Anderson and Woessner 1991). In phreatic aquifers near the ground surface, groundwater flow is determined by relatively strong horizontal gradients at the water table , while these tend to be weaker in deep aquifers. Aquifers can be connected with surface water bodies, which may be recharged by groundwater (groundwater discharge).

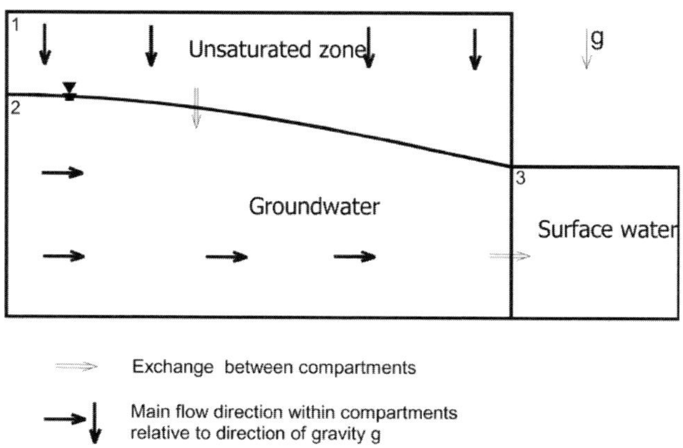

Fig. 2.1-5 Schematic view of a one-way coupling of the unsaturated zone, groundwater and surface water; g=gravity

In the network of compartments represented in Fig 2.1-5, flow is unidirectional (or one-way). The compartments are passed through in the order (1), (2) and (3). Within this system there is no *feedback* from compartment (2) to (1), nor from (3) to (2), nor from (3) to (1). This characteristic is important to understand by everyone studying the system, because it allows a straightforward treatment. First, compartment (1) has to be examined, which determines the flux-term between compartments (1) and (2) (groundwater recharge). Compartment (2) is next, determining groundwater discharge. Finally comes compartment (3). Computer algorithms make use of this feature when solving models.

Such a simple straightforward procedure is not possible in all network models. In fact it will be shown that it is the feedback property that determines if the simple strategy is possible. To illustrate the point, Fig. 2.1-6 shows the same set of compartments with two-way fluxes; the groundwater table moving in the vertical direction may affect soil moisture and water fluxes in the unsaturated zone, which was neglected in Fig. 2.1-5.

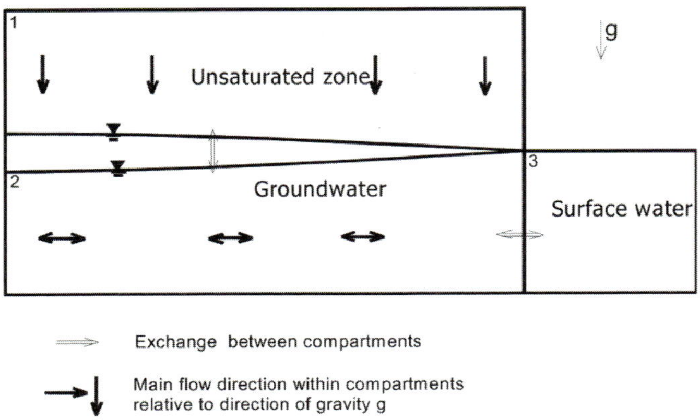

Fig. 2.1-6 Schematic view of a two-way coupling of the unsaturated zone, groundwater and surface water

In the situation depicted in Fig. 2.1-6, the groundwater table may fall below the surface water level. Under these circumstances there is no groundwater discharge into the river but surface water recharges the aquifer. While the flux-direction at any point at the interface and at any time is easy to understand, actual computation depends on both compartments. Thus groundwater and surface water cannot be studied or computed without taking the other compartment into account. There is a feedback in the system: groundwater influences surface water, but surface water influences groundwater as well.

The same situation is given at the interface between compartments (1) and (2). There is indeed an influence from soil moisture on groundwater through recharge, but there is an influence in the other direction as well, when the moving water table has to be taken into account. The position of the water table marks the lower boundary of the soil compartment. Here the modeller faces a much more complex situation, as a straightforward run through the compartments is no longer possible.

As the property of feedback is so important for the design and performance of models, it will be examined further. Let us first note, however, that the main features of a compartment network can be represented and observed already in a link chart, as shown below.

Fig. 2.1-7 Three compartment system without and with feedback

Coupling properties in a network can be represented by a matrix. Each row and each column represent a compartment. Influencing compartments are listed in rows, while influenced compartments are listed in columns. When the compartment represented by i has a link to the compartment represented by column j, the matrix has an entry in element (i,j) of the matrix[1]. In the following, matrix entries are shown as "x". This form will be used in Sect. 3.1, to describe links of the hydrological cycle.

Fig. 2.1-8 shows the two forms (network and matrix) of link representation for one single link between two compartments A and B.

Fig. 2.1-8 Example of link representation in network form and matrix form

Each network model can be represented by a matrix and vice versa. Every entry in the matrix corresponds to one arrow in the network representation. Double-sided arrows, two-way coupling, appear in the matrix as two entries in corresponding fields on both sides of the main diagonal of the matrix.

In the following figures, three examples of a three compartment network are given. Fig. 2.1-9 shows a simple connection by one-way

[1] Some authors prefer to list influenced compartments in a column and influencing compartments in a row (Patten et al. 1990); this is not of any importance for the points made in this contribution.

links. Compartment A has a direct link on compartment B and an indirect influence on compartment C. There is no influence from compartments B or C on A. There is no influence from C to B. Thus there is no feedback in the entire network. Compartments can be treated in a simulation in the order A, B and C, with no further complications.

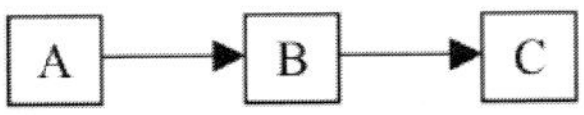

	A	B	C
A		x	
B			x
C			

Fig. 2.1-9 One-way connection

Fig. 2.1-10 depicts a more complex situation in which compartments are connected by two-way couplings. Here the development in each of the compartments is influenced by developments in the other compartments. There are several feedbacks in the network. A directly influences B and is directly influenced by B. There are indirect influences from A to C and from C to A.

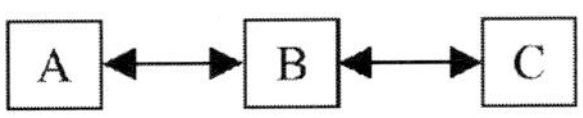

	A	B	C
A		x	
B	x		x
C		x	

Fig. 2.1-10 Feedback through two-way coupling

In a model, the compartments cannot be treated in the same manner as in the network linked by one-way coupling. Usually, an iterative approach is used to capture the feedbacks. For example, separate models for the compartments can be applied in the order A, B, C, A, B, C... until it converges to a solution which fulfils all requirements. The Newton or Newton-Raphson method, another numerical approach, treats several coupled compartments together in one iteration step. Solution strategies for these cases are discussed in Sect. 2.2.

Two-way couplings are not a prerequisite for feedback. As shown in Fig 2.1-11, feedbacks can occur in a network coupled by one-way links. There is an indirect influence from compartment A to compartment C, and a direct link from C to A. In a similar manner, compartments B and A are coupled, as are C and B too. Obviously there is feedback in the network, which has to be accounted for in any study on the dynamic behaviour of the system and particularly in numerical approaches. The situation shown may also be termed a cycle or a loop.

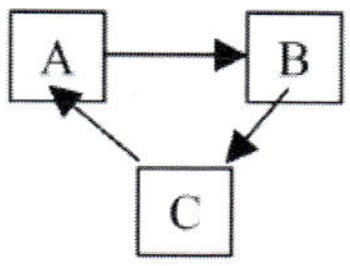

	A	B	C
A		x	
B			x
C	x		

Fig. 2.1-11 Feedback through one-way connections

It is clear that the case with feedback is the more complex, not only from the physical point of view but also for numerical reasons. As an iteration becomes necessary, the same steps have to be computed again and again, until a certain accuracy criterion is fulfilled. Often, hundreds of iterations are required. Each iteration needs approximately the same amount of computer resources as the entire solution for the system with no feedback. It is therefore important to distinguish between systems with and without feedback.

As an example, take the flow network given by Freeze and Cherry (1979), which was shown above in Fig. 2.1-4. As a model for a part of the hydrological cycle, four compartments are available for water storage: interception, soil moisture, groundwater and surface water. These are connected by overland flow, interflow, groundwater recharge and base flow, which are represented by four x-entries in the matrix built from the first four lines and columns in Fig 2.1-12.

Storage	Interception	Soil Moisture	Ground-water	Channel	Atmosphere
Interception				x	x
Soil Moisture			x	x	x
Groundwater				x	
Channel			o		o
Atmosphere	x	x		x	

Fig. 2.1-12 Matrix for a part of the hydrological cycle (Fig. 2.1-4), as given by Freeze and Cherry (1979) (x-entries); extended version (o-entry), for details see text

Note that this simple network has no feedback. An implemention of the conceptual model was presented by Blackie and Eeles (1985). In the model design the compartments should be treated in the given order to avoid unknown flux terms. It should also be noted that, depending on the particular situation in the case studies, the network given may be inappropriate. In other published conceptual models (for example see Imbe et al. 2000), there is a link from surface water to groundwater which is missing in the example above. This gives an additional entry in the matrix for the four compartments, indicated by one o-entry in Fig. 2.1-12. With this link the network undergoes a qualitative change concerning feedback. While there is no feedback without that link, the addition introduces one cycle of length two (with two-way feedback) as there is a mutual influence between the groundwater and the channel compartment. The system becomes much more complex and that is reflected when the system is simulated in a computer model.

The given network of four compartments, shown in Fig 2.1-4, can be extended to incorporate a greater part of the hydrological cycle. When the atmosphere is added as a fifth compartment, several flow entries have to be added, as shown in Fig 2.1-12. The x-entries in the last row represent precipitation, divided into interception, infiltration and overland flow. The entries in the last column represent evaporation and evapotranspiration. The x-entries were mentioned by Freeze and Cherry (1979); the o-entry in the last column is essential when surface water bodies with large surfaces, like lakes, are included in the model. Note that by adding the air-compartment, several feedbacks are introduced in the network.

Comparison of the last row and column shows that three two-way couplings appear when the atmosphere is added as a compartment. A feature that is not so easy to identify, is that several longer cycles emerge as well. There is a three compartment cycle, like the one shown in Fig 2.1-11, including compartments for the atmosphere - interception store - channel storage. Moreover, there is a four compartment cycle: atmosphere - soil moisture - groundwater - channel storage. An analytical tool to identify such cycles in complex model networks is described below.

In their textbook Ward and Robinson (1990) show a more complex compartment network (Fig. 2.1-13). The same part of the hydrological cycle is represented as was given in Fig 2.1-4. Two other compartments appear, vegetation and surface, replacing the now missing interception storage. Besides the main flux components, which coincide with the aforementioned concept of Freeze and Cherry (1979), some secondary fluxes are indicated[2]. Floods introduce a flux from the channel store to the surface store. Through

[2] Secondary fluxes are understood here as fluxes of minor importance in relation to the major fluxes in the network. In the same manner secondary links are understood as relatively irrelevant compared to primary links. Note that a secondary link is not identical to an indirect link (Fath and Patten 1999).

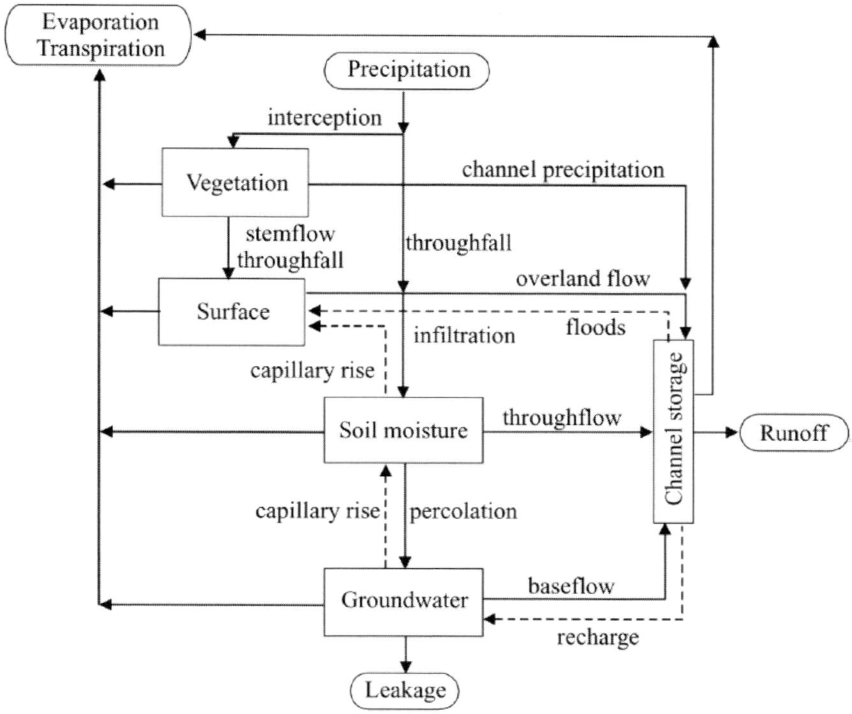

Fig. 2.1-13 Network model of a part of the hydrological cycle containing secondary fluxes (after Ward and Robinson 1990)

capillary rise there are water fluxes from groundwater to soil moisture and from soil moisture to the surface.

In the following, the general case of a network including N compartments is treated. At first, it may be very difficult to decide whether there is feedback in the system or not. Nevertheless there are simple cases. When the matrix, which represents the compartment network, has a triangular form, the computation of the compartments can follow the procedure described in the example above for the three compartment system with no-feedback (Fig. 2.1-9). Matrices of lower and upper triangular form are shown in Fig 2.1-14.

In the upper-triangular matrix case, the algorithm starts with the compartment represented in the last row and column. Then, one after the other, all the compartments are solved with back-reference only to compartments which are already solved. After N (=number of compartments) steps, the procedure stops. No iteration is necessary, as it is in systems with feedback.

Fig. 2.1-14 Upper (left) and lower (right) triangular matrices; elements in region, marked by 0, are all zero; elements, marked by *, can be non-zero; elements in diagonal are non-zero

Networks represented by triangular matrices display no feedbacks. As re-ordering of compartments does not change the feedback-property, it is possible to classify matrices, based on whether the transformation into triangular form is possible or not. A necessary condition for no-feedback can be formulated in terms of matrix operations. A matrix which can be transformed into triangular form by simple line and column exchange operations has no feedback: vice versa, feedback within a system is given when the matrix cannot be transformed into an upper or a lower triangular form.

Obviously the feedback property of a network is connected with the existence of cycles within the network, i.e. of paths having the same start and end point. In a network with N compartments (in graph theory terminology, a graph of order *N*) the maximum cycle length is *N*. The minimum cycle length is always 2. When there is a cycle in the network, there is feedback and vice versa, the opposite holds: in the case of feedback al least one cycle can be identified. Cycles play an important role in all environmental systems; Patten et al. (1990) show that mass and heat transport are enhanced within cycles.

Criteria can be formulated making use of the *adjacency matrix* **A**. The adjacency matrix emerges from listing of compartments in rows and columns. There are only 0 and 1 entries: the element in row *i* and column *j* gets a 1 if there is a link from compartment *i* to compartment *j*. Some publications use binary entries "true" and "false". The diagonal terms of the adjacency matrix are set to 0 by convention.

	A	B	C
A		x	
B			x
C	x		

0 1 0
0 0 1
1 0 0

Fig. 2.1-15 An example of a network matrix and adjacency matrix for determinant calculation

If, additionally, the main diagonal of the matrix is filled by 1s (each compartment is linked with itself), one may formally write **I+A**. It its easy to see that for all no-feedback systems the determinant of the matrix **I+A** is 1. This holds, because permutations of lines and columns do not change the determinant and triangular matrices which emerge according to the procedure described above have the determinant 1.

Unfortunately the determinant criterion is not sufficient. There are systems with feedback which have the determinant 1 as well. An example is shown in the following.

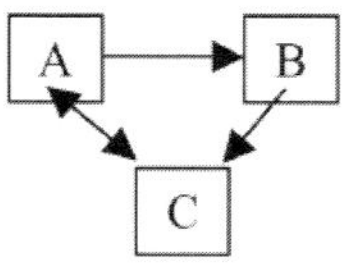

	A	B	C
A		x	X
B			x
C	X		

Fig. 2.1-16 A network with feedback and determinant 1

The adjacency matrix is connected with cycles in the network. It can be shown that the (i,j) element of the n-th power of the matrix $\mathbf{A}^n$ gives the number of different paths of length n from compartment i to compartment j (Chartrand 1977). Thus the existence of cycles can be read from the diagonals of the matrices $\mathbf{A}^i$. In the following example (compare Fig. 2.1-11) there is one 3-cycle, connecting all three compartments of the network :

$$\mathbf{A}=\begin{pmatrix}0&1&0\\0&0&1\\1&0&0\end{pmatrix},\quad \mathbf{A}^2=\begin{pmatrix}0&0&1\\1&0&0\\0&1&0\end{pmatrix},\quad \mathbf{A}^3=\begin{pmatrix}1&0&0\\0&1&0\\0&0&1\end{pmatrix}$$

The matrices $\mathbf{A}^2$ and $\mathbf{A}^3$ show that there is no 2-cycle, but a 3-cycle for all compartments. The example of Fig. 2.1-16

$$\mathbf{A}=\begin{pmatrix}0&1&1\\0&0&1\\1&0&0\end{pmatrix},\quad \mathbf{A}^2=\begin{pmatrix}1&0&1\\1&0&0\\0&1&1\end{pmatrix},\quad \mathbf{A}^3=\begin{pmatrix}1&1&1\\0&1&1\\1&0&1\end{pmatrix}$$

has one 2-cycle (represented by two 1-entries in the main diagonal of $\mathbf{A}^2$) and a 3-cycle.

Using this property and the fact that the existence of feedback is identical to the existence of cycles, one may formulate a necessary and sufficient criterion as follows: *feedback is equivalent with at least one non-zero entry in the diagonal of matrices* $\mathbf{A}^2$, $\mathbf{A}^3$,....$\mathbf{A}^n$. Another formulation of the feedback criterion is given by:

$$spur\sum_{i=2}^{n}\mathbf{A}^i \neq 0 \qquad (2.1\text{-}1)$$

where *spur* denotes the sum of all diagonal elements. This condition is referred to as *criterion for inter-compartmental feedback.*

Another problem is the convenient queuing of the compartments in a simulation run in the case of no-feedback, in order to avoid iterations. The answer can be derived analytically: as the transformation in a triangular matrix is possible if there is no feedback, the entire system can be solved by treating the compartments subsequently in the order given in the triangular matrix (Holzbecher 2003).

2.1.3 Intra-compartmental Coupling

While inter-compartmental coupling is a characteristic of the network, complex model approaches within compartments may contain complex couplings of variables. The result is usually a system of coupled partial differential equations. The coupling of processes is reflected by the coupling of variables in the resulting set of differential equations. The situation can be explained best by an example. Flow modelling in most cases is based on mass conservation described by a modified version of the continuity equation:

$$\frac{\partial \rho}{\partial t} = -\nabla \cdot \rho \mathbf{v} \qquad (2.1\text{-}2)$$

including density ρ and velocity vector $\mathbf{v}$. Density ρ may vary through spatial and/or temporal temperature changes. On the other hand, the temperature distribution is computed mostly using the heat transport equation, resulting from the energy conservation principle. The transport equation in the advective term contains the velocity $\mathbf{v}$. Thus in the set of differential equations there is a dependence of one variable (ρ and $\mathbf{v}$) from the other. Moreover, there is feedback similar to the inter-compartmental case discussed above.

Flow and transport links can be represented in a link matrix as well. The main difference remains that here the processes are linked and are not compartments as in the inter-compartmental case. Two matrices, representing the feedback and the no-feedback situations are shown in Fig 2.1-17 and Fig 2.1-18.

Process	Flow	Transport
Flow		x
Transport		

Fig. 2.1-17 Matrix representing intra-compartmental links for the no-feedback situation

Flow always has an influence on transport. The exceptions are those degenerate cases, when velocity becomes zero, or when component and mass gradients vanish. This influence is represented by the one entry in the matrix above. Application studies for this situation utilise the fact that there is only this one entry. First, the flow field is computed and, based on that, the transport solution is obtained in a second step. This straightforward procedure is not possible, when flow is affected by transport.

Transport has an effect on flow when one of the relevant flow parameters is influenced by the distribution of a transported component, i.e. heat or salt or some chemical species. Basic flow parameters, which may be influenced that way, are fluid viscosity and/or fluid density, but there are others as well: in porous media, porosity and/or conductivity may change in response to transport. The link matrix is shown below.

Process	Flow	Transport
Flow		X
Transport	X	

Fig. 2.1-18 Matrix representing intra-compartmental links for the situation with feedback

When there is a mutual influence between flow and transport, as shown in the matrix above, the variables depend on each other. There is a feedback, which can be identified from the link matrices in the very same way as described above for the compartment networks. Models of the gaseous phase almost always need to take this coupling into account as, for example, in models of the atmosphere (Petoukhov et al. 2000).

In the case of feedback, a simple sequential procedure, in which flow and transport are solved completely one after the other, generally does not deliver results of a required accuracy. There are exceptions, as where a sequential approach is possible and there are papers in which the sequential approach is applied. Nevertheless, general problem solvers need to take the mutual influence into account and do so mostly by an iterative procedure (see below).

The following Table 2.1-1 gives some examples of phenomena for which intra-compartmental coupling between variables may be necessary. All entries in the table should be understood as one-way or two-way links between the processes under consideration. There is a coupling, i.e. a feedback from one process to another, in all cases shown.

Table 2.1-1 Examples of phenomena for which intra-compartmental model approaches are necessary

Phenomenon	**Flow**	**Transport**		
		Heat	**Salt**	**Components**
Thermal convection	x	x		
Saltwater intrusion	x		x	
Double diffusive convection	x	x	x	
Thermodynamics		x		x
Reactive transport			x	x
Reactive flow and transport	x		x	x
Multispecies thermodynamics	x	x	x	x

Formally, one may represent intra-compartmental feedback in a diagonal matrix $\mathbf{J}$. A 1-entry in the i-th element of the diagonal vector denotes internal feedback in the i-th compartment. No intra-compartmental feedback is represented by a 0-entry. There is intra-compartmental feedback in a network when at least one compartment has this property. By analogy to the condition noted above for inter-compartmental feedback, one may write

$$spur(\mathbf{J}) \neq 0 \tag{2.1-3}$$

as the *criterion for intra-compartmental feedback*.

Both criteria, for inter- and intra-compartmental feedback, can be incorporated in a general *feedback criterion* for network models. The validity of the inequality

$$spur\left(\mathbf{J}+\sum_{i=1}^{n}\mathbf{A}^{i}\right)\neq 0 \qquad (2.1\text{-}4)$$

is equivalent with feedback. Note that it does not matter practically if the starting index in the sum is 1 or 2, as the diagonal elements of the adjacency matrix **A** are all zero by definition.

2.1.4 Final Remarks

A network matrix will be introduced in Chapter 3 to display the complex interaction of processes affecting the hydrological cycle. There are 22 different processes listed, mainly hydrological, but including atmospheric, land-cover and biogeochemical fluxes as well. Two-way couplings already leave no doubt that there are several feedbacks in the system. Note that in contrast to the frequent use of network models in hydrology, it is not compartments which are listed in rows and columns but processes. Similarly structured matrices were introduced in the section on intra-compartmental coupling in 2.1.3.

It is almost impossible to simulate the entire system represented in the matrix. For specific case studies and sites there are considerably less compartments and links. Conceptual models, which include only parts of the hydrological cycle, are definitely simpler and sometimes can be structured without feedback. In all cases it is important to identify the relevant feedbacks within a system first.

Box 2.1-1 ***Coupled Flow and Transport Modelling a Convection Cell***

Ekkehard Holzbecher

One of the classical examples for the complex interaction of variables is density-driven flow. In his classic paper Lord Rayleigh (1916) explained convective flow patterns as interaction of flow and transport. In his approach heat transport determines the spatial and temporal distribution of temperature and temperature determines density. Lapwood (1948) applied the same approach for flow in porous media. A broad discussion of convective motions in porous media is provided by Nield and Bejan (1992). Convective motions may also be induced by salt gradients. Saltwater intrusion is one of the most prominent phenomena, which can be explained by the influence of salt distribution on water density – Reilly and Goodman (1985) published a study about it from the historical perspective. An overview on double-diffusive convection, where influences from salt and heat on density are

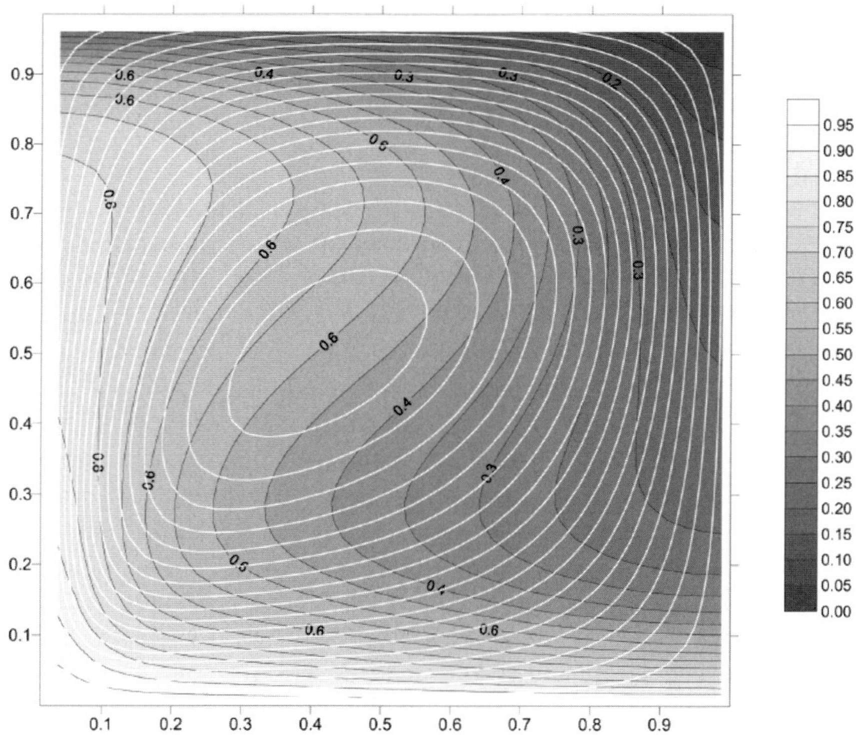

Fig. 2.1-19 Steady convection cell in unit square, taking into account variable density and viscosity of water in porous media; streamlines (white) and isotherms (black); simulated using FAST-C(2D) code (Holzbecher 1998)

both relevant, is given in the monograph by Brandt and Fernando (1995). Subramanian and Balakotaiah (1997) classify reaction-driven convective patterns in porous media.
An example for thermal convection is shown in Fig 2.1-19, calculated using FAST-C(2D) code (Holzbecher 1998). Clearly visible are closed streamlines as contours of the computed stream function. There is upward flow on the left vertical boundary and downward flow on the right boundary. Fill patterns in the background indicate temperature. The simulated temperature scale is from 20°C at the top to 280°C at the bottom. The temperature scale, shown in markers and scale bar, are normalised with 1 at the bottom and 0 at the top. Simulation was performed using the Picard iteration scheme, in which flow and transport are solved sequentially until both solutions fit.

2.2 Non-linearities

Jesus Carrera, Lawrence E. Band, Axel Bronstert, Pavel Kabat, Nicole Mölders

2.2.0 Overview of Non-linear Dynamics in Hydrology

The basic biophysical and biogeochemical processes and feedbacks characterising hydrological systems range from fully linear to highly non-linear. Non-linear dynamics can contribute to unstable and chaotic behaviour of these systems, making prediction difficult and sometimes impossible. In addition, much of our analytical frameworks in hydrology have been developed for linear systems, of which the best example is the unit hydrograph method but also include linear reservoirs and a number of linear diffusion processes. Hydrological phenomena that result from the interactions and feedbacks that are generally linear or weakly non-linear are often well characterised by these methods. However, application of these simple linear approaches to non-linear systems can often result in significant error, both in predictive mode and in assessment of system stability and vulnerability. Non-linearity can take a variety of forms and arise for a variety of reasons. Many hydrological processes characterising energy, water and momentum transfer are non-linear at short time scales, and feedbacks linking hydrological, carbon and nutrient cycles can be very non-linear at longer time scales. Some of these arise from the interactions between or within the atmospheric, land and water compartments in the Earth system and some are intrinsic properties of specific processes.

As an example, soil water characteristic functions, such as the van Genuchten equations, are strongly non-linear relationships between the hydraulic conductivity, moisture content and tension. Variation in the soil water state tends to enhance or diminish drainage and evaporation/ transpiration under wet and dry conditions, respectively. Typical field heterogeneity in the basic storage variable of soil moisture, even within a hypothetical uniform soil, produces potentially large variations in tension and conductivity. Moreover, the linear averaging of soil moisture as input to characteristic functions does not yield the average soil water conductivity or tension. Under these circumstances, specific methods need to be incorporated for quantifying and handling the effects of non-linearity in heterogeneous systems.

The interaction of multiple compartments in a land/atmosphere/water system can also produce non-linear feedbacks that tend to dampen or augment changes or perturbations to the system, making prediction considerably more difficult. In this case, the coupling of system compartments can produce fundamentally different system behaviour compared to simpler modelling approaches that may prescribe one or more compartments (e.g. prescription of meteorological conditions or surface states). These non-linear feedbacks, while occurring in the natural systems we are interested in, would not be captured if the modelling frameworks did not incorporate sufficient intercompartment feedbacks. The coupling between land surface and atmosphere effectively converts a set of land-surface processes from spatially independent to dependent processes due to atmospheric feedback. It is only with this coupling that the feedbacks, specifically non-linear feedbacks, can be reproduced. The presence of various forms of non-linear feedbacks can tend to produce single stable or multiple stable equilibrium states, or can produce inherent instability under given circumstances. Even if accurate prediction of system state is difficult or unattainable due to strong non-linearity, the relative stability and vulnerability of natural systems to perturbations is an important property to assess. This can only be done if appropriate feedbacks and non-linearity in system response are identified and incorporated.

In this section, different forms of non-linear hydrological processes and feedbacks that typically arise in natural coupled systems are described, with specific examples and case studies given. Computational methods and frameworks for dealing with this non-linearity are then discussed.

2.2.1 Definition and Scope

The concept of non-linearity is a negative one in that it is defined by opposition (negation) of linearity. A system is said to be non-linear if it is not linear. Obviously, this leads to a need for defining linearity.

A system is linear if it satisfies the following definition. Let Q_A be the system output to an input P_A, that is $Q_A=L(P_A)$. Similarly, Q_B would be the output to an input P_B. Then, the system is linear if the output to aP_A+bP_B *is* aQ_A+bQ_B. That is,

$$L(a\ P_A + b\ P_B) = a\ L(P_A) + b\ L(P_B) \qquad (2.2\text{-}1)$$

This equation is often termed principle of superposition and is indeed the most straightforward manner with which to test whether a system is linear. If the system does not satisfy the above equation, it is said to be non-linear (an actual application for the linearity of the effect of land use on atmospheric models is described in Box 2.2-1).

The terms L, Q and P in the above definition may refer to the whole system or to any of its parts (subsystems, compartments or pools). For example, assume that L represents the operations needed to compute the basin response (measured by river flow Q) to rainfall, P. Rainfall may be prescribed (e.g. measured and entered into a real-time flood predictor) or the output of a weather simulator (e.g. for a system that is trying to evaluate risks associated to climate change). The operations themselves can be very simple (e.g. the unit hydrograph described below) or complex. Examples of the latter are models (e.g. the Sacramento model) that consist of many compartments to represent interception, soil water, runoff generation, river flow routing, etc. The variables representing these compartments (e.g. water content at the canopy or the soil) are called state variables. These are needed for internal computations, but not necessarily for the system output.

It is difficult to define whether a system is linear (or nearly so) from its compartments. If all of them are internally linear (that is, if all state variables and fluxes are linear functions of the inputs), then the overall system will also be linear. However, when any compartment is non-linear, then the overall system may become highly non-linear, if those remaining amplify such non-linearity. On the other hand, the fact that all compartments are highly non-linear does not necessarily imply a highly non-linear behaviour for the whole system.

The definition in Eq. 2.2-1 should be understood as an abstract one, in the sense that both the input, P, and the output, Q, may consist of many components. In the rainfall-runoff model example above, P may include rainfall, temperature, air humidity, etc at different locations. The output, Q, may include river flow, evaporation (which may be a feedback to the weather simulator), albedo, etc.

The most traditional example of linear superposition in hydrology is the unit hydrograph, which is the outflow caused by a unit rainfall during a unit time (say 1 hour). In this case, the time dimension must be taken into account, so that the unit hydrograph is a function $Q_u(t)$. If the rainfall is R_1 during the first hour and R_2 during the second hour, superposition would yield (see Fig. 2.2-1):

$$Q(t) = R_1 Q_u(t) + R_2 Q_u(t-1) \quad (2.2\text{-}2)$$

While most hydrologists know that basins rarely (if ever) behave linearly, this type of superposition is still used as a reference and to test the degree and kind of non-linearity (again, see Box 2.2-1). Yet, they cannot help feeling somewhat surprised when the actual outcome fails to be close to linear.

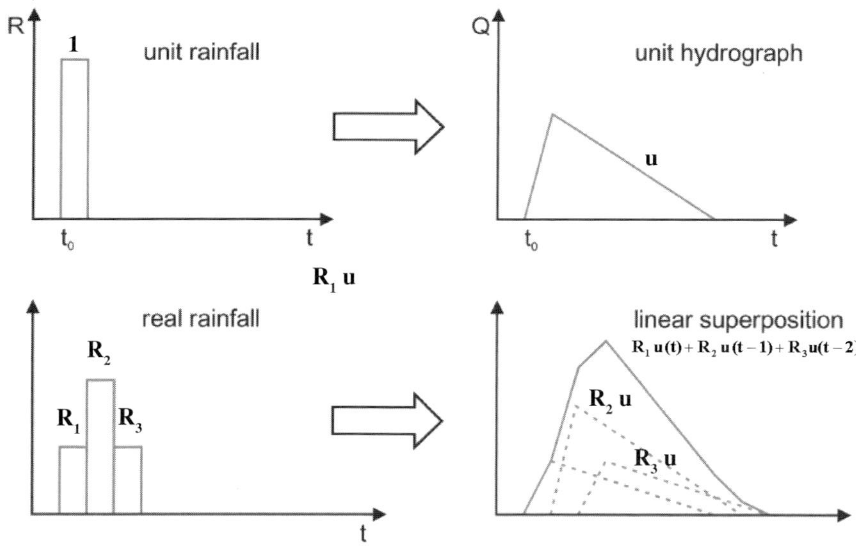

Fig. 2.2-1 Unit hydrograph (above) is the runoff response to a unit rainfall during a unit time. If (when) the basin behaves linearly, the response to any rainfall can be obtained by superimposing the response to elementary "unit" rainfalls

It is precisely this unpredictability notion that makes non-linearity a source of concern. In fact, non-linear systems can lead to unstable (chaotic) predictions. Moreover, non-linearity appears to be ubiquitous. All systems contain parts that behave non-linearly. Unpredictability and ubiquity make coupling difficult. This is why non-linearity is a specific source of concern for coupled models. Nature appears to behave non-linearly. Yet, scientists only know how to deal with linear processes. In fact, as we shall see below, the most effective way of dealing with non-linearity is by linearisation (when it works!).

2.2.2 Types and Effects of Non-linearity

Coupling and multi-dimensions make it difficult to systemise non-linearities. For the sake of discussion, we classify non-linearities in three classes: affine, continuous and thresholds (see Fig. 2.2-2).

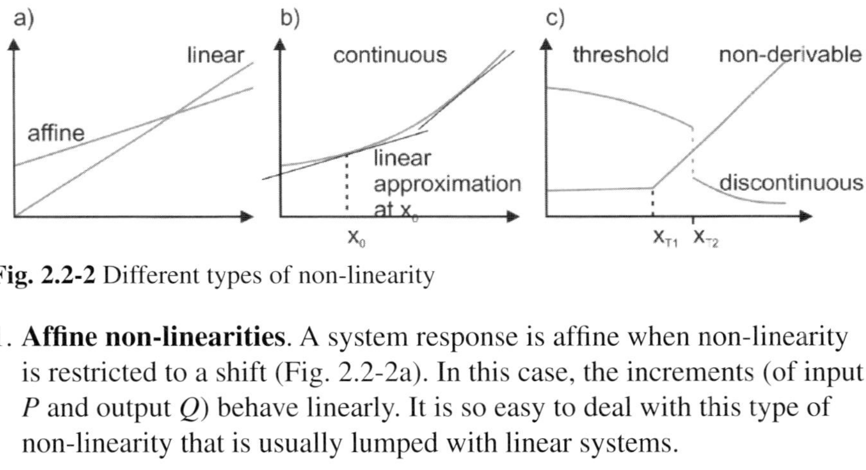

Fig. 2.2-2 Different types of non-linearity

1. **Affine non-linearities**. A system response is affine when non-linearity is restricted to a shift (Fig. 2.2-2a). In this case, the increments (of input *P* and output *Q*) behave linearly. It is so easy to deal with this type of non-linearity that is usually lumped with linear systems.

2. **Continuously differentiable non-linearities**. The continuous non-linearity refers to the case in which one can ensure that small changes in *P* will lead to small changes both in *Q* and its derivative (rate of change) such as shown in Fig. 2.2-2b. In this case, one can linearise the response as:

$$Q - Q_1 \approx \frac{\partial Q}{\partial P}(P - P_1) \qquad (2.2\text{-}3)$$

This approximation is linear (actually, affine) and good for small perturbation (for small $P\text{-}P_1$). In fact, one way of evaluating the degree of non-linearity is by analysing the validity of the linearisation with respect to typical perturbations. If the above equation is approximately valid, then the system is said to be "weakly" non-linear. Otherwise, it is "highly" non-linear (hydrologists often feel their system is the most highly non-linear in the world when agonising about why their code does not work).

This is the simplest, yet most frequent, case of non-linearity. Threshold non-linearities can be viewed as extreme cases.

One particular case of interest refers to time-evolving (feedback) systems, in which the future evolution depends on the current state (among other things). Ignoring external impacts, one can write:

$$x_{k+1} = f(x_k) \tag{2.2-4}$$

where x_k is the state of the system at time k and f represents the system response function. An important question about this system is whether its evolution will lead to a stable equilibrium, whether it will oscillate around an equilibrium state, whether its value becomes unbounded, etc. Much of the literature on chaotic systems is related to this topic and is all related to the non-linear nature of f (see Fig. 2.2-3).

Fluctuating or monotonous evolution. The fluctuating or monotonic evolution of x depends on the sign of $\partial f / \partial x$. A positive sign (positive feedback) yields a monotonic evolution (Fig. 2.2-3b) while a negative slope of the system response (negative feedback) tends to yield fluctuating evolutions (Fig. 2.2-3a). Obviously, we are only referring here to the first order feedback (by the own state). Other inputs and coupling to other systems may lead to significantly more complex behaviours. In general, there will be many inputs, **x**, (bold-face meaning a vector of inputs) and outputs, **f**. In such cases, one needs to compute the Jacobian matrix $\partial f / \partial x$ to study the response of each output to each factor.

Stable or unstable equilibria. An equilibrium state is stable if perturbations around it are smoothed out and the system returns to its equilibrium state (Fig. 2.2-3b). On the other hand, it is unstable if small perturbations cause the system to evolve unboundedly (Fig. 2.2-3a). As it turns out, local stability solely depends on the absolute value of df / dx at the equilibrium. If this derivative is larger than 1, the system is unstable (Fig. 2.2-3b). If it is smaller than 1, the system is stable. In multivariate problems, the eigenvalues of the Jacobian matrix determine stability of the system.

The above classification refers to the local behaviour of the system response function f. Things can get more complex when analysing the global behaviour of f. For example, if the derivative grows for small values of x (concave function) and decreases for large values of x (convex function) chances are that there will be **multiple equilibria**. In the example of Fig. 2.2-3c, there are two stable equilibrium points, x_{F1} and x_{F2}, and an unstable equilibrium point x_T. Because of small external perturba-

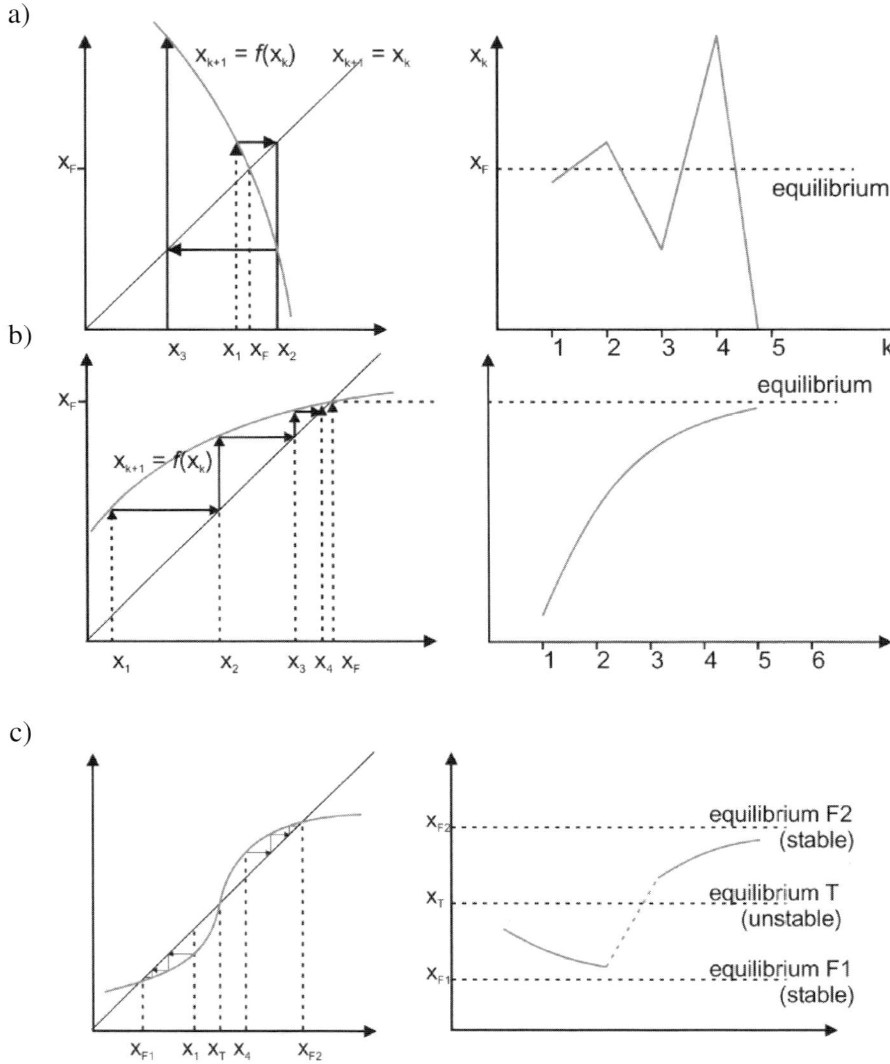

Fig. 2.2-3 Non-linear response function and system evolution

a) Evolution is fluctuating ($df/dx<0$) and unstable ($|df/dx| >1$)

b) Evolution is monotonic ($df/dx>0$) and remain stable ($|df/dx| <1$) around the equilibrium point x_F

c) Because of the changes in df/dx, multiple equilibria do exist. Below x_T, the system will evolve to x_{F1}. However, if perturbed above x_T, it will evolve towards x_{F2}

tions, the system may be fluctuating around a stable point x_{F1}. However, if a large impact brings it above x_T, it will converge to a new equilibrium point x_{F2}. The topic of multiple equilibria is very important to global cycles and coupled models and has been extensively discussed in Sect. 1.3.

3. **Thresholds and discontinuities**. These types of non-linearity are illustrated in Fig. 2.2-2. Thresholds and discontinuities refer to sudden changes in either the value of the system response function, such as x_{T2} in Fig. 2.2-2c (discontinuities), or in its derivative (non-derivable response), such as x_{T1} in Fig. 2.2-2c. The system response (function Q in 2.2-1 or 2.2-3, or function *f* in 2.2-4) is then said to be **discontinuous** or **non-differentiable** (or non derivable) and the value of x for which the sudden change occurs is called a **threshold** (x_{T1} or x_{T2} in Fig. 2.2-2). In either case, the response may be markedly different above and below the threshold value. When this threshold is explicitly known, dealing with it is not difficult, in the sense that the modeller knows which equations to use when. A more difficult, often subtle case, arises in coupled systems when no discontinuity is apparent anywhere and the change in behaviour is due to non-linearity such as the one in Fig. 2.2-3c. In this type of problem, linearisation is of little use and numerical methods get tricky (enumeration, random search and the like are required).

2.2.3 Solution of Non-linear Problems

We must say from the outset that there is no general method for solving all kinds of non-linear problems. In fact, as we shall see, the most effective methods are based on linearising the problem. Because of the difficulties associated with solving non-linear problems, many researchers have devoted significant efforts to devising solution methods. Thorough descriptions of them are presented in many textbooks (e.g. Argyros and Szidarovszky 1993) and many hydrology books devote large portions to solution methods. A full description of such methods falls well beyond the scope of this book and, therefore, this section is restricted to simple outlines, with special emphasis on the coupled nature of many hydrological problems. We start by describing the mathematical formulation of the problem. Second, we describe the Picard family of methods (sequential iterations). Third, we outline the Newton family (direct substitution). We conclude with other methods.

Formulation of the Problem

One-dimensional problems (lumped compartments)

Let us start by considering one compartment with a single state-variable, x. In the example of Fig. 2.1-4, x may represent soil water content if the soil is treated as a lumped compartment which, at a given time, is given by:

$$V(x - x_{old}) - \Delta t \left[I - E(x) - R(x) - Q(x) \right] \tag{2.2-5}$$

where V is the volume of soil; x is the volumetric water content we are seeking; x_{old} is the initial water content (x and x_{old} were denoted x_{k+1} and x_k, respectively, in Eq. 2.2-4; we are dropping these sub-indices for simplicity); Δt is the time increment; I is the infiltration; E the evaporation from the soil; R the groundwater recharge; and $Q(x)$ the interflow. This equation represents the water mass balance in the soil. Since compartment equations often represent some conservation law (mass, energy, momentum, etc.), Eq. 2.2-5 is typical of most lumped compartments. It simply expresses the notion that the variation in volume of water stored in the compartment is equal to the volume of water exchanged with adjacent compartments. In fact, if we add up all fluxes, we can write:

$$V(x - x_{old}) - \Delta t\, q(x) = 0 \tag{2.2-6}$$

where $q(x) = I - E(x) - R(x) - Q(x)$. For our purposes here, it is convenient to simply write

$$f(x) = 0 \tag{2.2-7}$$

where f represents the whole algorithm to compute x given x_{old} (previous state) and I. If the links $E(x)$, $R(x)$ or $Q(x)$ are difficult to compute, then solving Eq. 2.2-7 will also be difficult.

Some of the methods we will describe later require writing x explicitly. This can be done in many different ways. For example, Eq. 2.2-6 can be rewritten as

$$x = x_{old} + \frac{\Delta t}{V} q(x) \tag{2.2-8}$$

We will rewrite it formally as

$$x = g(x) \tag{2.2-9}$$

In summary, lumped compartments can be represented by equations like Eqs. 2.2-7 or 2.2-9. Solving any of these yields x, which can be used to evaluate the links with other compartments ($E(x)$, $R(x)$, etc.). On occasions, however, it may not be appropriate to lump the behaviour of the whole compartment with a single state variable. In the above example, evaporation may be dependent on soil moisture at the upper soil layers (root zone) while, depending on the structure of the soil, interflow and recharge may depend on soil moisture in the deep soil layers. In such cases, it may be more appropriate to work with a distributed compartment, consisting of cells (sub-compartments) that are linked among themselves. Another difficulty arises when fluxes are affected by the state of the adjacent compartments. Both cases are presented below, prior to discussing solution methods.

As discussed in Sect. 2.1.3, modelling complex compartments may lead to partial differential equations (this will be also addressed in Sect. 2.3, when dealing with model building). Solving these equations requires dividing (discretising) the spatial domain in a number of cells (sub-compartments). Each cell is characterised by one or several state variables. Let us denote such state variable as x_i (in the example of Eq. 2.2-5, x_i would be the water content in the *i-th* cell). Then, a water balance analogous to Eq. 2.2-5 leads to

$$V_i(x_i - x_{old\,i}) + \Delta t\left[\sum_j A_{ij}x_j - q_i(x_1, x_2 ...)\right] = 0 \qquad (2.2\text{-}10)$$

where the term $A_{ij}x_j$ expresses the exchanges of water with adjacent cells. It should be noticed that A_{ij} involves geometrical terms (distances between centres of cells, contact areas, and the like), but also physical parameters (e.g. permeability). These may vary with the state, so that A_{ij} may end up being also a function of $x_1, \ldots x_N$.

Since the notation starts becoming cumbersome, we define the vectors of state variables, the fluxes and the conductance and volume matrices:

$$\mathbf{x} = \begin{Bmatrix} x_1 \\ \vdots \\ x_N \end{Bmatrix}$$

$$\mathbf{q} = \begin{Bmatrix} q_1 \\ \vdots \\ q_N \end{Bmatrix}$$

$$\mathbf{A} = \begin{pmatrix} A_{11} \cdots A_{1N} \\ \vdots \quad\quad \vdots \\ A_{N1} \cdots A_{NN} \end{pmatrix}$$

$$\mathbf{D} = \begin{pmatrix} V_1 & \cdots & 0 \\ & \ddots & \vdots \\ \vdots & & \ddots \vdots \\ 0 & \cdots & V_N \end{pmatrix}$$

Then, Eq. 2.2-10 can be written as:

$$\mathbf{f(x)} = \mathbf{D}\,(\mathbf{x} - \mathbf{x}_{old}) + \Delta t\,(\mathbf{Ax} - \mathbf{q}) = \mathbf{0} \qquad (2.2\text{-}11)$$

Notice that this equation is formally analogous to its one-dimensional counterpart Eq. 2.2-7. The fact that f and x are now written in bold-face reflects the fact that Eq. 2.2-11 represents a vector of equations (N equations, one for each cell) with a vector of unknowns.

By solving for **x**, we can also write Eq. 2.2-11 in a form analogous to Eq. 2.2-9:

$$\mathbf{x} = (\mathbf{D} + \Delta t\mathbf{A})^{-1}(\mathbf{Dx}_{old} + \mathbf{q}\Delta t) = \mathbf{g(x)} \qquad (2.2\text{-}12)$$

An alternative way of writing this equation is presented in Sect. 2.3. In summary, however, equations such as Eqs. 2.2-7 and 2.2-9 can also be written for the case of distributed variables compartments. A considerable volume of research is devoted to writing equations analogous to Eq. 2.2-10 in such a form that they represent accurately the behaviour of the compartment (see, e.g. Huyakorn and Pinder 1983, for groundwater models). However, these can be usually reduced to forms such as Eqs. 2.2-11 or 2.2-12.

Multi-compartmental problems

The motivation behind this book is coupled models. As discussed extensively in Sect. 2.1 this notion arises from the belief that one can solve each compartment provided that the links between compartments are known. This is formalised in Fig. 2.2-4, which is the mathematical equivalent of Fig. 2.1-3, and not fundamentally different from Figs. 2.1-4, 2.1-6 or 2.1-12.

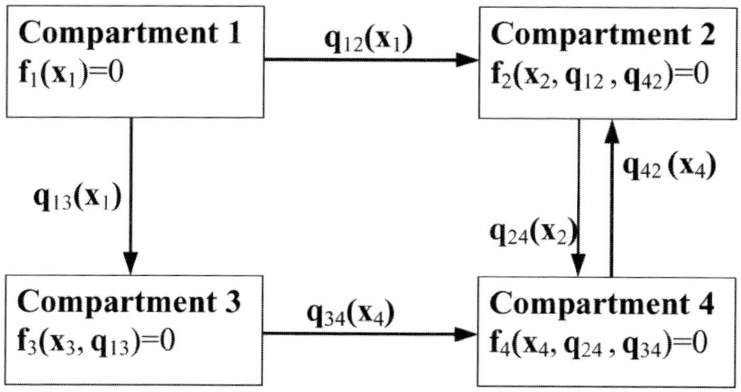

Fig. 2.2-4 Schematic representation of a multi-compartment system

In essence, the behaviour of each compartment is characterised by an equation similar to Eqs. 2.2-7 or 2.2-11. Influences (links) between compartments are characterised by $q_{ij}(x_j)$,which represents the impact of compartment i on compartment j. In one-way interactions, there is only one arrow pointing from compartment i to compartment j. However, in the case of two-way links (direct feedbacks), a second arrowhead appears and compartment j influences compartment i through a term $q_{ji}(x_i)$. Sometimes, these two-way feedbacks are written as $q_{ij}(x_i, x_j)$ to explicitly stress the reciprocal influence of the state variables. While this may be more appropriate to explain the physical dependence of terms, we will stick to the notation of Fig. 2.2-4 for simplicity.

Inter-compartmental influences q_{ij} are not restricted to fluxes (q in Eq. 2.2-11), but include any kind of influence. In the example of Box 2.1-1, heat transport depends on water flow through water fluxes, but flow depends on temperature through the densities.

With this notation, the coupled problem of Fig. 2.2-4 can be written as:

$$\mathbf{f_1(x_1) = 0} \tag{2.2-13a}$$

$$\mathbf{f_2[x_2, q_{12}(x_1), q_{42}(x_4)] = 0} \tag{2.2-13b}$$

$$\mathbf{f_3[x_3, q_{13}(x_1)] = 0} \tag{2.2-13c}$$

$$\mathbf{f_4[x_4, q_{24}(x_2), q_{34}(x_3)] = 0} \tag{2.2-13d}$$

Some of the features of inter-compartmental coupling discussed in section 2.1.2 are apparent in this simple example. For instance, compartment 1 is indirectly linked to compartment 4 though compartments 2 and 3. Indeed, $\mathbf{x}_1$ will affect $\mathbf{x}_2$ and $\mathbf{x}_3$ through $\mathbf{q}_{12}$ ($\mathbf{x}_1$) and $\mathbf{q}_{13}$ ($\mathbf{x}_3$), respectively. In turn, they will affect $\mathbf{x}_4$ through $\mathbf{q}_{24}$ ($\mathbf{x}_2$) and $\mathbf{q}_{34}$ ($\mathbf{x}_3$).

Picard Family of Solution Methods

One and multidimensional problems

In these methods, one should first write the equations in a recursive form, such as Eqs. 2.2-9 or 2.2-12. Then, the solution can be written iteratively as:

$$\mathbf{x}^{m+1} = \mathbf{g}(\mathbf{x}^m) \qquad (2.2\text{-}14)$$

where the super-index denotes iteration number. Then the algorithm can be written as:

Step 1: Initialisation. Set $m = 0$. Define $\mathbf{x}^0$

Step 2: Compute $\mathbf{x}^{m+1} = \mathbf{g}(\mathbf{x}^m)$

Step 3: $m = m+1$

Step 4: Convergence check. If $\mathbf{x}^m$ is very close to $\mathbf{x}^{m-1}$, stop. Otherwise, return to step 2.

The behaviour of this algorithm can be illustrated by Fig. 2.2-3. While this figure was originally designed to illustrate the time evolution of non-linear systems, it can also be used to describe the iterative evolution of recursive equations, such as Eq. 2.2-14. It is sufficient to substitute the sub-indices k and $k+1$, denoting time in Fig. 2.2-3, by m and $m+1$, respectively, which denote iteration number in equation Eq. 2.2-14.

As illustrated by Fig. 2.2-3 for the single-dimensional case, the method converges whenever $|dg/dx|$ is smaller than one. Otherwise, it might diverge. Convergence will be fluctuating if dg/dx is negative, or monotonous. In the case of divergence, one can apply a relaxation method, which consists of substituting Eq. 2.2-14 by

$$\mathbf{x}^{m+1} = \alpha\, \mathbf{g}(\mathbf{x}^m) + (1 - \alpha)\, \mathbf{x}^m \qquad (2.2\text{-}15)$$

If the slope of **g** is negative (Fig. 2.2-3a), α should be smaller than 1. Otherwise, α may be slightly larger than 1 (acceleration instead of relaxation). The problem is that convergence may become extremely slow if one chooses a value of α smaller than 1 in the case of Fig 2.2-3b. On the other hand, a value of α larger than 1 may cause divergence in the case of Fig. 2.2-3a. Hence, it is not surprising that researchers have developed adaptive algorithms in which the value of α is updated as iteration progresses.

The analysis becomes a bit more complex in the multivariate case. There, one needs to compute the Jacobian matrix. The absolute values of its eigenvalues at the solution point must be smaller than one absolute values to guarantee convergence.

Multi-compartmental problems

Solution of compartmental networks is a natural extension of the algorithm associated to Eq. 2.2-14. The procedure requires, first, ordering the compartments so as to minimise the number and importance of feedback links (that is with i greater than j, such as $\mathbf{q}_{42}$ in Fig. 2.2-4). This ordering has been discussed in Sect. 2.1 and will be revised in Chapter 3. In general, this ordering comes quite naturally. Then, the algorithm becomes:

Step 1: Initialisation. Set $m = 0$. Define $\mathbf{x}_i^o$ for i =1 through N

Step 2: Set $m = m + 1$
For i=1 through N solve steps 3 and 4

Step 3: Solve compartment $i = \mathbf{f}_i\left[\mathbf{x}_i^m, \mathbf{q}_{ij}(\mathbf{x}_j^n)\right]$, where $n = m$ for $j < 1$ and $n = m - 1$ for $j > 1$ (basically, use the most recent links $\mathbf{q}_{ij}$)

Step 4: Compute links starting at compartment $i\left[\mathbf{q}_{ij}(\mathbf{x}_i^m)\right]$

Step 5: Convergence check. If $\mathbf{x}_i^m$ is close to $\mathbf{x}_i^{m+1}$ for all i, stop. Otherwise, return to step 2.

In fact, this algorithm only requires iterating in the case of feedbacks. If all links are one-way and there are no closed loops (such as that in Fig. 2.1-11), then no iterations are required. This is the case depicted in Fig. 2.1-4. Hydrologists usually take rainfall as an input and then compute interception storage (compartment 1) to evaluate infiltration ($\mathbf{q}_{12}$) and overland flow ($\mathbf{q}_{14}$), as well as direct evaporation from the soil surface and vegetation ($\mathbf{q}_{15}$). The second compartment is the unsaturated soil moisture storage, which is solved by Eqs. such as 2.2-8, 2.2-12 or similar. Having computed soil moisture, the scheme of Fig. 2.1-4 allows one to evaluate recharge ($\mathbf{q}_{23}$), interflow ($\mathbf{q}_{24}$) and evapotranspiration from the unsaturated soil ($\mathbf{q}_{25}$). Recharge allows computing groundwater storage variations (compartment 3) and base flow ($\mathbf{q}_{34}$) into channel storage (compartment 4) as well as phreatophytic transpiration ($\mathbf{q}_{35}$). Floods can be computed from compartment 4 and total evaporation is the sum of ($\mathbf{q}_{15}$), ($\mathbf{q}_{25}$) and ($\mathbf{q}_{35}$). Since the scheme of Fig. 2.1-4 contains no feedbacks, this would end the computation for one time step. Absence of feedbacks can also be analysed by means of influence matrices (recall Fig. 2.1-12).

If the model contains feedbacks, then iterations may be needed. However, if these feedbacks are weak then the above algorithm will converge very fast. Therefore, modifications of the basic scheme of Fig. 2.1-4 to include secondary fluxes (Fig. 2.1-13) can be accommodated well by most hydrological models, because it only requires iterating.

Given the condition for convergence of Picard methods (the absolute values of the Jacobian eigenvalues must be smaller than one), it should come as no surprise that its behaviour for solving network models depends on the magnitude of $\partial \mathbf{g}_i / \partial \mathbf{x}_j$. In fact, coupling becomes negligible when, $\partial \mathbf{g}_i / \partial \mathbf{x}_j$ and thus its eigenvalues, approach zero.

The Picard method, or some of its variants, has traditionally been the standard method for solving hydrologic network models for three reasons: (1) it is relatively simple and intuitive, (2) it is rather general and versatile, and (3) it works well when feedbacks are weak. In fact, as shown by Fig. 2.1-13, strict hydrological models work quite well because, indeed, hydrological feedbacks are relatively weak (they can be highly non-linear, though). In fact, strong feedbacks have become apparent when coupling hydrology to other processes. The most singular case has been the coupling between hydrological and atmospheric circulation models. As shown by Fig. 2.1-12, strong links (feedbacks) do exist with the atmosphere. A good portion of the evaporation returns as rainfall within the catchments area. Therefore, rainfall cannot be considered as an external input for many purposes. Instead, it must be considered to be, at least partially, driven by evaporation within the basin. Such strong feedbacks cause Picard methods to fail. In fact, divergence may be fluctuating in the case of negative feedbacks (recall Fig. 2.2-3), so that when analysing model results, one is sometimes left wondering whether fluctuating responses are natural or simply reflect a poor solution algorithm. In any case, it is clear that more powerful methods may be needed, as discussed below.

Newton Family of Methods

One and multi-dimensions problems

Newton's methods (often referred to as Newton-Raphson methods in the multi-dimension case) are based on linearising the basic Eqs. 2.2-7 or 2.2-11. Linearisation can always be performed by means of a series expansion:

$$\mathbf{f}\,(\mathbf{x}^{m+1}) \approx \mathbf{f}\,(\mathbf{x}^{m}) + \frac{\partial \mathbf{f}}{\partial \mathbf{x}}\,(\mathbf{x}^{m+1} - \mathbf{x}^{m}) = \mathbf{0} \qquad (2.2\text{-}16)$$

where super-indices again denote iteration number and $\partial \mathbf{f} / \partial \mathbf{x}$ is the Jacobian matrix, $\mathbf{J}$, evaluated at $\mathbf{x}$. Then, one can solve Eq. 2.2-16 for $\mathbf{x}^{m+1}$ as:

$$\mathbf{x}^{m+1} = \mathbf{x}^{m} - \mathbf{J}^{-1}\,\mathbf{f}\,(\mathbf{x}^{m}) \qquad (2.2\text{-}17)$$

Actually, one does not need to compute the inverse of the Jacobian matrix, as implied by Eq. 2.2-17. Instead, it is sufficient to solve the linear system

$$\mathbf{J}(\mathbf{x}^{m+1} - \mathbf{x}^{m}) = -\,\mathbf{f}\,(\mathbf{x}^{m}) \qquad (2.2\text{-}18)$$

This leads to the conventional Newton-Raphson algorithm:

Step 1: Initialisation. Set $m = 0$. Define $\mathbf{x}^0$. Compute $\mathbf{f}\,(\mathbf{x}_0)$

Step 2: Compute $\mathbf{J} = \partial \mathbf{f} / \partial \mathbf{x}^m$.

Step 3: Solve Eq. 2.2-18 to compute $\mathbf{x}^{m+1}$

Step 4: Compute $\mathbf{f}\,(\mathbf{x}^m)$

Step 5: Convergence check. If $\mathbf{f}\,(\mathbf{x}^{m+1})$ is close to zero and/or $\mathbf{x}^{m+1}$ is close to $\mathbf{x}^m$, stop. Otherwise, set $m = m + 1$ and return to step 2.

It is worth noticing that convergence is checked in step 5 by ensuring that $\mathbf{f}\,(\mathbf{x}^{m+1})$ is close to zero. Often, this is a safer (and tougher!) test of convergence than ensuring that $\mathbf{x}^{m+1}$ is close to $\mathbf{x}^m$.

An illustration of the Newton algorithm for the one dimension case is displayed in Fig. 2.2-5. As one can see, when things work properly, Newton's method is extremely accurate and converges very fast. Unfortunately, it may also fail. For one thing, Eq. 2.2-18 may not have a solution. What is worse, it may fail to converge when $d^2\,\mathrm{f} / dx^2$ is of opposite sign to $\mathbf{f}(\mathbf{x})$. This is implicitly shown in Fig. 2.2-5. If the algorithm had started far to the right of

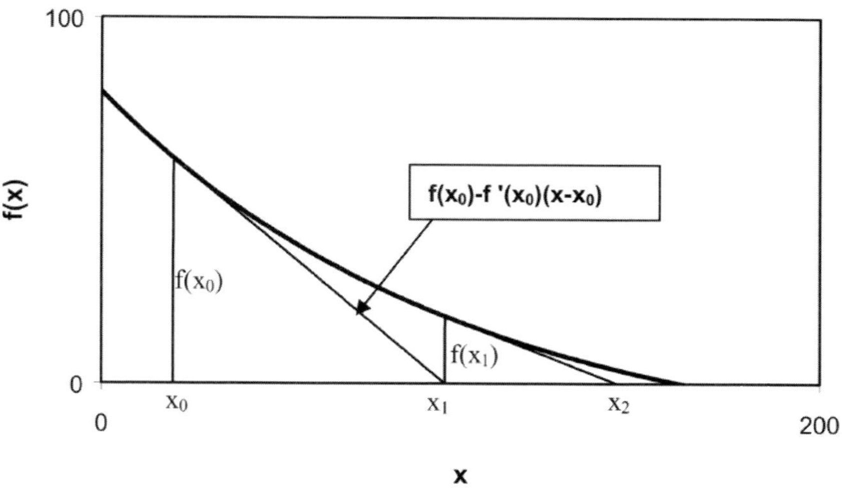

Fig. 2.2-5 Schematic description of Newton´s method. Starting at x_0, one computes sequentially $f(x_0)$ and $f\ (x_0)$, which leads to the tangent of f(x) at x_0. The point x_1 is the intersection of this tangent with the abscissas axis. The procedure is repeated to obtain x_2, and so on

the solution, it would not have converged. This problem can only be solved by a good choice of the initial point x_0. This is why initialisation is so important for Newton methods. However, the reason why this method has not become more popular is its relative difficulty. One needs to program not only the function **f(x)**, which may be hard enough, but also its derivative. Moreover, convergence of the method requires that **f(x)** is not only continuous but also continuously derivable (recall Sect. 2.2-2). Yet, when these difficulties are overcome, Newton's method tends to be extremely efficient.

Multi-compartmental problems

The Newton's method for solving network models is a natural extension of the algorithm described above. The only difficulty lies in the computation of the derivatives. One needs to compute the derivatives of not only the state equations but also the links. Specifically, the linearisation of the compartment equations is:

$$\mathbf{f}_i\left[\mathbf{x}_i^{m+1},\mathbf{q}_{ij}\left(\mathbf{x}_j^{m+1}\right)\right]=\mathbf{f}_i\left[\mathbf{x}_i^{m},\mathbf{q}_{ij}\left(\mathbf{x}_j^{m}\right)\right]+\sum_{j=1}^{N}\frac{\partial\mathbf{f}_i}{\partial\mathbf{x}_j}\left(\mathbf{x}_j^{m+1}-\mathbf{x}_j^{m}\right) \qquad (2.2\text{-}19)$$

where, for $i \neq j$,

$$\frac{\partial \mathbf{f}_i}{\partial \mathbf{x}_j} = \frac{\partial \mathbf{f}_i}{\partial \mathbf{q}_{ij}} \cdot \frac{\partial \mathbf{q}_{ij}}{\partial \mathbf{x}_j} \tag{2.2-20}$$

It is clear that programming all these derivatives is difficult and prone to errors. However, if one does manage to compute them, the full system becomes:

$$\begin{bmatrix} \frac{\partial \mathbf{f}_1}{\partial \mathbf{x}_1} & \frac{\partial \mathbf{f}_1}{\partial \mathbf{x}_2} \cdots & \frac{\partial \mathbf{f}_1}{\partial \mathbf{x}_N} \\ \vdots & \vdots & \vdots \\ \frac{\partial \mathbf{f}_N}{\partial \mathbf{x}_1} & \frac{\partial \mathbf{f}_N}{\partial \mathbf{x}_2} \cdots & \frac{\partial \mathbf{f}_N}{\partial \mathbf{x}_N} \end{bmatrix} \begin{bmatrix} \mathbf{x}_1^{m+1} - \mathbf{x}_1^m \\ \vdots \\ \vdots \\ \mathbf{x}_N^{m+1} - \mathbf{x}_N^m \end{bmatrix} = - \begin{bmatrix} \mathbf{f}_1(\mathbf{x}^m) \\ \vdots \\ \vdots \\ \mathbf{f}_N(\mathbf{x}^m) \end{bmatrix} \tag{2.2-21}$$

This equation is formally identical to Eq. 2.2-18, so that the iterative algorithm is also formally identical. We must stress, however, the potentially formidable size of Eq. 2.2-21 – just consider an atmospheric circulation model coupled to groundwater models, or distributed parameters surface water models, etc. The resulting system may well run into the billions of unknowns and may exceed current computing capabilities.

In summary, Newton's method may be extremely complex and computer demanding. However, when feasible, it is extremely efficient in the sense that it tends to converge faster than Picard methods (e.g., Saaltink et al. 2001).

Other methods

A large number of methods is available blending Picard and Newton's methods. Typically, these aim at maximising the advantages of both. However, it is difficult to demonstrate a priori that the opposite will not occur. As a result, their efficiency is problem-dependent.

Yet, when proven, they can be extremely good. Unfortunately, the experience in coupled models is limited.

Other methods that have been used in hydrology, such as neural networks or other enumerative methods (Beven 1993; Gupta et al. 1999) seem cumbersome for complex network models.

Box 2.2-1 ***Testing of Linearity by the Principle of Superposition***

Nicole Mölders

This example illustrates how the principle of superposition can be applied to examine whether the change in latent heat fluxes caused by various simultaneously occurring land-use changes (e.g. deforestation, urbanisation, afforestation and recultivation of open-pit mines, installation of open-pit mines, drainage) can be expressed by the sum of the changes caused by the single land-use modifications. The study was performed with a mesoscale-β atmospheric model prescribing the same "typical day" for the origin land use (Fig. 2.2-6a) and the modified land use (Fig. 2.2-6b). The land-use changes (between 1930 and 1980) took place in about 45% of the area. In the model, they were realised by replacing the parameter of the dominant land-use type whenever there was a change in dominant land use for the area represented by the 1x1 km² subgrid cell[1]. Simulations were carried out changing only one of the land-use types at a time.

a) b)

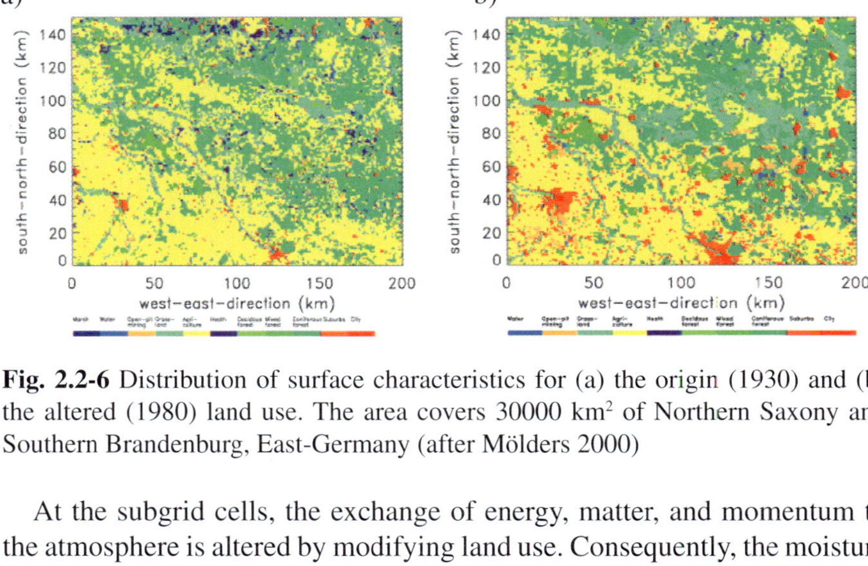

Fig. 2.2-6 Distribution of surface characteristics for (a) the origin (1930) and (b) the altered (1980) land use. The area covers 30000 km² of Northern Saxony and Southern Brandenburg, East-Germany (after Mölders 2000)

At the subgrid cells, the exchange of energy, matter, and momentum to the atmosphere is altered by modifying land use. Consequently, the moisture and temperature states of the atmosphere differ in the altered land-use conditions. Thus, advection is modified and affects the latent heat fluxes also at the unchanged subgrid cells. At a cell without land-use changes, upwind modifications still lead to changes in wind speed, air humidity and moisture, cloudiness as well as thermal stability.

[1] For further details on the explicit subgrid scheme see Seth et al. (1994), Mölders et al. (1996).

Each cell, j, has to be examined to see whether the land-use changes that took place at some subgrid cells only led to a linear or non-linear change in latent heat fluxes. To this end, let k denote the origin landscape simulation and i (i=1,…,n), each of the modifications. Then, linearity is tested by comparing the cumulative change, $\sum_{i=1}^{n}(x_j^k - x_j^i)$, and the change $x_j^k - x_j^p$ where p represents the simulation with the simultaneous, i.e. all land-use changes (Fig. 2.2-6b). The difference between these two quantities is termed Δ given by (Mölders 2000)

$$(n-1)x_j^k - \sum_{i=1}^{n} x_j^i + x_j^p = \Delta \quad \begin{cases} > 0 & \textit{enhancement} \\ = 0 & \textit{superposition} \\ < 0 & \textit{diminuation} \end{cases} \qquad (2.2\text{-}22)$$

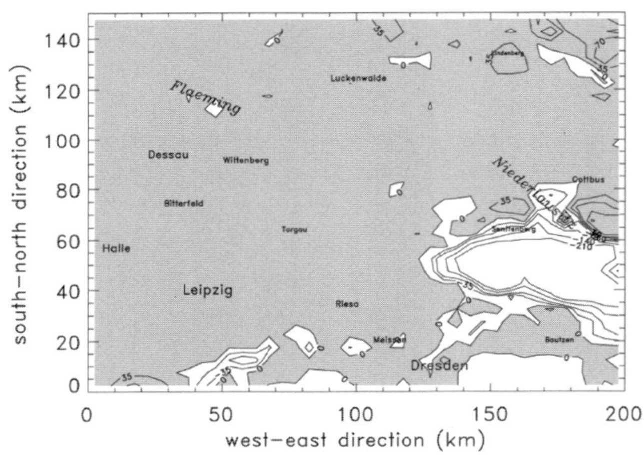

Fig. 2.2-7 Differences in noon latent heat-flux changes, Δ (in W m^{-2}) obtained by subtracting the change caused by simultaneous replacement of all land-use types from the sum of the changes caused by replacing one land use at a time. The gray and white areas represent positive and negative deviations from superposition, respectively. Areas of relevant deviations from superposition are indicated by the contour lines (from Mölders 2000)

Fig. 2.2-7 illustrates that latent heat fluxes are diminished (by more than about 210 W m^{-2}) at noon due to non-linear effects. Further investigations showed that the location of non-linear responses changes within the diurnal course. At night, for instance, latent heat fluxes deviate negatively (up to -69 W m^{-2}) in the conurbation of Dresden (see Mölders 2000), which may be partly attributed to the non-linear relationship between air temperature and saturation mixing ratio. These quantities are important for evaporation and cloud formation. Deviations from superposition can be induced by primary (land-use changes) and secondary (e.g. altered cloud distributions as a response to the land-use modifications) differences. They rely on (1) the size of the resulting new patch as well as the uniformity of its land use, (2) the kind of change in the hydrological and thermal characteristics, and (3) the type of land use prevailing in the environments of land-use conversion. According to the results of this case study, areas dominated by grassland and forests are more sensitive to concurrent land-use changes than areas used mostly for agriculture (Mölders 2000).

Box 2.2-2 ***Example of Non-linear Behaviour: Response of Vegetation on Water Stress***

Axel Bronstert

Transpiration through vegetation connects soil-physical, bio-physical and atmospheric water transport processes. Hence, transpiration T is driven and limited by different physical gradients $\partial\psi$ in the soil, biosphere or atmosphere:

$$T \approx \frac{\partial\psi}{\partial s}\text{, i.e.}\qquad T = g\frac{\partial\psi}{\partial s} \tag{2.2-23}$$

In principle, the reaction of the water flux (through soil, biomass or air) responding to a change in the corresponding gradient is linear, as long as the coefficient g is constant or at least constant for the domain of interest. For each compartment of the soil-vegetation-atmosphere system there is a specific coefficient characterising the conditions of the soil, vegetation and atmosphere respectively: for the soil it is the soil hydraulic conductivity, for the vegetation it is stomatal conductivity (or the stomata resistance, which is the inverse value of the conductivity), and for the near-canopy/atmosphere it is the aerodynamic conductivity (or the aerodynamic resistance, which is the inverse value of the conductivity). The non-linearity in transpiration is caused by (at least) two different aspects:

- Some conductivity values (for soil, vegetation, lower atmosphere, etc .) are not constant but a function of some other state variable. E.g. the soil hydraulic conductivity is strongly dependent on the soil moisture content.

- Each of the different conductivity values may reach a value where it diminishes (or where the resistance approaches infinity), e.g. the soil hydraulic conductivity reaches 0 if the soil equals the residual moisture content, and the aerodynamic conductivity reaches 0 if air moisture approaches saturation.

The non-linear response of transpiration as a function of matric head in the soil (which can be transformed into soil water content, given the soil-water retention function of the soil) has been summarised by Feddes et al. (1988) into the diagram given below. It shows the quotient α ($0 \leq \alpha \leq 1$) of potential / actual transpiration as a function of matric head *h*. For different ranges of matric head, α shows different values. *Below* a threshold h_{m4} (which usually is termed the *permanent wilting point PWP*), α is equal to zero, because of the inability of the vegetation to take water out of a soil dryer then the PWP. Soil with a moisture value *above* threshold h_{m1} (termed *anaerobis point*) will also not permit full transpiration for many plant types due to an oxygen deficit in the soil. If the soil moisture is between the values h_{m2} and h_{m3}, transpiration is not hindered by soil moisture deficit or plant physiological limitations, i.e. $\alpha=1$. The actual value of h_{m3} (*critical water content*) is controlled by the plant type *and* by the maximum transpiration possibly absorbed by the actual atmospheric conditions, expressed here as T_p.

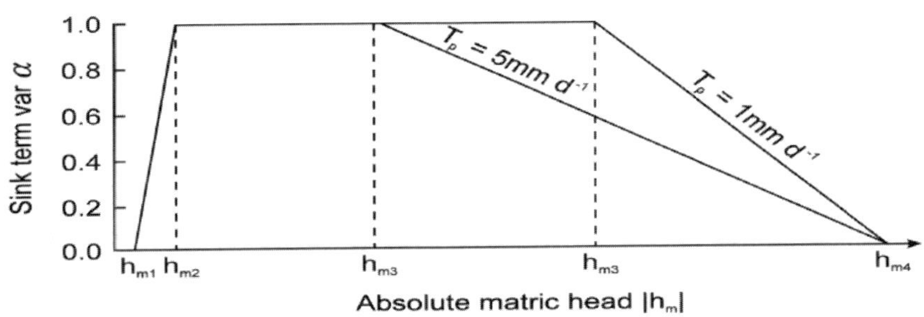

Fig. 2.2-8 Transpiration reduction α as a function of matric head *h* (according to Feddes et al. 1988)

To summarise, Fig. 2.2-8 gives an example for a non-linear behaviour due to non-constant or even diminishing nature of the governing parameters. Besides water or oxygen shortage in the soil, transpiration can further be influenced by additional factors, such as – among others – air temperature T_{air}, leaf area available for transpiration (parameterised as leaf area index *LAI*), air humidity H_{rel} and others. Following the approach of Jarvis (1976), the actual transpiration is controlled by the surface conductivity g_{surf} of the vegetation canopy, which is parameterised by a combination of reduction factors $F_{1\text{-}7}$. In this concept, the factor α mentioned in Fig. 2.2-8 can be regarded as the first reduction factor F_1. Each one of the factors $F_{1\text{-}7}$ may act as a threshold value if the specific physical conditions are constraining the water vapour transport within the soil-vegetation-atmosphere continuum:

$$g_{surf} = g_{max} * F_{1\text{-}7}; \qquad F_{1\text{-}7}\text{: LAI}, \theta, T_{air} \ldots \qquad (0 \leq F_{1\text{-}7} \leq 1)$$

$F_{1\text{-}7}$: reduction factors for transpiration (Jarvis 1976)

Box 2.2-3 ***Reduction in Rainfall Due to Deforestation in the Amazon Forest***

Ronald W.A. Hutjes

The Amazon Basin contains the largest extent of tropical forest on Earth, over 5 x 10^6 km^2. The annual discharge of the Amazon River into the Atlantic Ocean of more than 200,000 m^3 s^{-1} contributes about 18% of the global flow of fresh water into the oceans. In recent decades the Amazon has been deforested at a fast rate, with potentially large consequences for local and regional ecosystems, as well as for global water and carbon cycles and for atmospheric composition and functioning.

By the 1950s only about 1% of the forests of Amazonia had been cleared. Since then, the occupation and development of Amazonia has changed dramatically and is associated with a huge increase in extent and rate of deforestation. As a result of government plans to develop and integrate the economy of Amazonia, the population of Brazilian Amazonia grew from 3,5 million in 1970 to almost 20 million in 2000, leading to the deforestation of over 500,000 km^2 in Brazil alone. Current rates of annual deforestation range from 15,000 to 20,000 km^2 in Brazilian Amazonia (INPE 2001).

The concerns this raised stimulated a lot of research, both observational and in modelling. Observational studies played a dual role. On the one hand they documented the effects of deforestation on water and energy balances (especially at local scales), while on the other hand provided quantitative estimates for model parameters of both the original forest and of the pastures it is replaced by.

The ARME (Shuttleworth 1988) and ABLE (Garstang et al. 1990) campaigns of the early and mid-eighties documented surface-atmosphere exchange processes, and atmospheric transport and composition, respectively, of undisturbed forest. The succeeding ABRACOS (Gash et al. 1996) and RBLE (Nobre et al. 1996) campaigns addressed both forest and pastures and also extended the spatial and temporal scale of measurements. The most recent LBA-WETAMC (Silva Dias et al. 2002b) focussed on both the local effects of deforestation and on the regional response to the larger scale forcing. Combined, these field studies have given us a much more complete picture of the hydrometeorological effects of tropical deforestation, on the radiative balance, on evapotranspiration and energy partitioning, and on precipitation-producing processes.

Forest albedos range from 0.11 to 0.13 and correlate with soil moisture. In contrast, pasture albedo ranges from 0.16 to 0.20, depending primarily on Leave Area Index (LAI). Together with higher surface temperatures leading to more longwave emissions, this leads to a reduction of net radiation over pasture of about 11% compared to forests (Culf et al. 1995).

Also, for evaporation, a marked contrast exists between forest that shows little or no seasonal variation, and pastures. In the dry season pastures exhibit moisture stress because the shallower roots cannot reach water, whereas forest vegetation may absorb water from depths greater than 8 m (Nepstad et al. 1994). Forest trees do not appear to close stomata due to soil moisture stress, though they may do so in response to higher humidity deficits in the afternoon (Shuttleworth 1988). As a result, boundary layer development is similar in wet seasons but differs in the dry season, when they reach greater heights over the pasture (Fisch pers. comm.).

This rapid growth of the Planetary Boundary Layer (PBL) over pasture in the dry season has been shown to lead to earlier formation of cumulus and a consequent reduction in solar irradiation (Culf et al. 1996). Cutrim et al. (1995) used satellite imagery to show that shallow cumulus formed mostly over the cleared areas along the roads of Rondônia. In the wet season, a reverse situations may occur (Silva Dias et al. 2001) where warmer night time temperatures over forest lead to rainfall production from local moisture earlier in the day than over pasture. In such cases the PBL over forest may not grow as high as over the pasture due to the early onset of convective downdrafts. Finally, differences in aerosol dynamics between more pristine and deforested situations have also recently been revealed. In pristine situations (currently only found in the wet season with westerly regimes) aerosols and thus Cloud Condensation Nuclei (CCN) concentrations are very low leading to warm convective processes (few, large raindrops in low clouds) and relatively easily triggered rainfall. However, deforestation and associated biomass burning leads to very high aerosol and CCN concentrations. As a result, drops initially are many and small and aggregate to precipitable raindrops only when clouds become very deep. (Silva Dias et al. 2002a; Rickenbach et al. 2002; Andreae et al. 2002).

Over the past decade, most deforestation experiments have been performed with existing GCMs (Nobre et al. 1991; Dickinson and Kennedy 1992; Lean and Rowntree 1993, 1997; Polcher and Laval 1994a,b; Manzi and Planton 1996; Zang et al. 1996; Hahmann and Dickinson 1997). Most of these assessed the possible impact of complete removal of Amazonian rainforest and replacing it with pasture. This implied that surface albedo was changed from about 0.12 to 0.18, aerodynamic roughness from about 2 to 0.02 m and evaporation parameters (LAI, stomatal conductance) to lower values. All studies simulated a reduction in evaporation, although with different seasonality and a wide range of values, together with a reduction in net radiation. These GCM experiments agreed much less with respect to impacts on moisture convergence. In contrast to the other studies, Polcher and Laval (1994b), Lean and Rowntree (1997) and Manzi and Planton (1996) found an increase in moisture convergence. Polcher (1995) showed that the number of convective events was very sensitive to the precise values of pasture parameters (varying within observed uncertainties). The change in surface sensible heat flux changed with the number of intense convective events and thus large-scale moisture convergence. Surface evaporation affected the recycling rate and thus the precipitation brought by the weaker convective events, without affecting moisture convergence. All these GCM studies were unrealistic in the sense that they imposed complete deforestation of the Amazon. The higher (spatial) resolution of mesoscale models allows analysis of the effects of more realistic patterns of deforestation. Such studies suggest that with limited, relatively dispersed deforestation, rainfall reduction is much less severe and that rainfall might even increase, at least temporarily (see Fig. 2.2-9). This is probably due to enhanced triggering of convection by contrasts in surface energy partitioning (forest vs. pasture), that occur in this particular area at a length scale similar to that at which convection is generated (Avissar and Liu 1996; Silva Dias and Regnier 1996).

The coupling between the land surface and the climate, as mediated by energy and water cycles, can be strong but these interactions are highly non-linear and cover a wide range space and time scales, making analysis of process chains far from trivial. Much progress has been made in recent decades, both in observations and modelling (Kabat et al. 2004). Nevertheless, more quantitative understanding is needed of the possible impacts of human alterations of the landscape on climate. In particular, such work should be aimed at the identification of thresholds in the system. Which level of deforestation will affect climate to such an extend that rainforest can no longer be sustained? Such thresholds are defined by various aspects of vegetation-physiology and hydrology and meteorology.

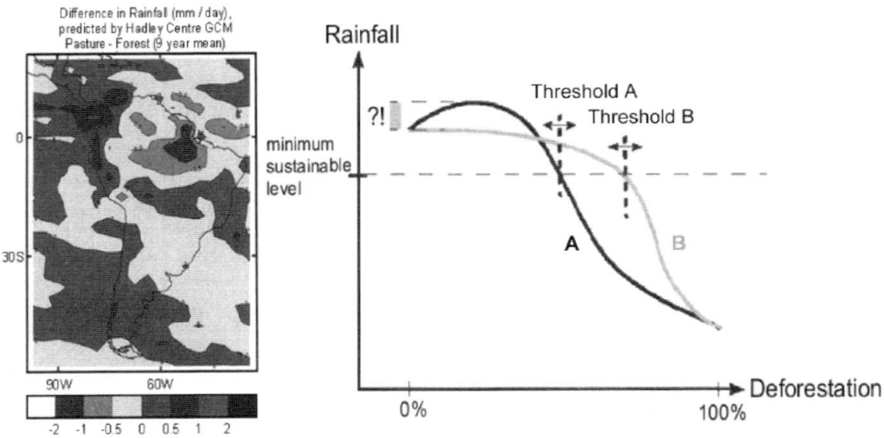

Fig. 2.2-9 Possible reduction in rainfall due to deforestation
Left: Example of a GCM study predicting a reduction in rainfall due to a complete Amazonian deforestation. Recent mesoscale modelling studies suggest that partial clearing of forests in more realistic scenarios might initially lead to an increase in rainfall, only decreasing with more extensive forest clearing
Right: If the latter is true, identifying threshold deforestation areas are crucial in the definition of sustainable management policies for the region (source: IGBP 1998)

2.3 Parameterisation of Complex Hydrological Systems

Jesus Carrera, Luis A. Bastidas

The very concept of coupled modelling carries implicitly a mechanistic flavour. Coupled modelling arises from the belief that understanding the parts and their interactions will lead to understanding the whole. Complexity, spatial and time variability and uncertainty cause this to be false. The objective of this section is to introduce some formalisation into the modelling process so as to properly formulate such issues.

The section follows two parallel lines of development: main text and boxes. The main text consists of three sections. The first one covers the process of model construction (conceptualisation, discretisation, calibration, error analysis and model selection). Conceptualisations and parameterisation of SVATS (Soil Vegetation Atmosphere Transfer Schemes) and groundwater models are the topic of two boxes. A third box is devoted to time integration of linear systems, which provides some justification for breaking complex systems into compartments. The second section is devoted to probabilistic models. Evaluation of prediction errors, the effect of spatial variability and stochastic models are reviewed. The third section is devoted to the issue of how to deal with complexity. Calibration and data assimilation are reviewed and a unified framework is presented.

2.3.1 Mechanistic Modelling

A model is an entity that represents some aspects of the behaviour of a real system in a simplified manner. The mechanistic (deterministic) approach to modelling is based on the assumption that one knows the physicochemical processes governing the behaviour of (each part of) the real system. This allows one to write equations purported to represent reality. These equations are then parameterised and discretised, so that they can be solved by computers. Since some of the parameters may not be easy to measure or may even lack a well-defined physical meaning (i.e. fudge factors) the model has to be calibrated. These steps are the core of what can be termed the modelling process, which is outlined in Fig. 2.3-1. This section is devoted to analising them in some detail. The structure follows that of Carrera et al. (1993), which is not essentially different from those of Hill (1998), Zheng and Bennet (2002) and many others.

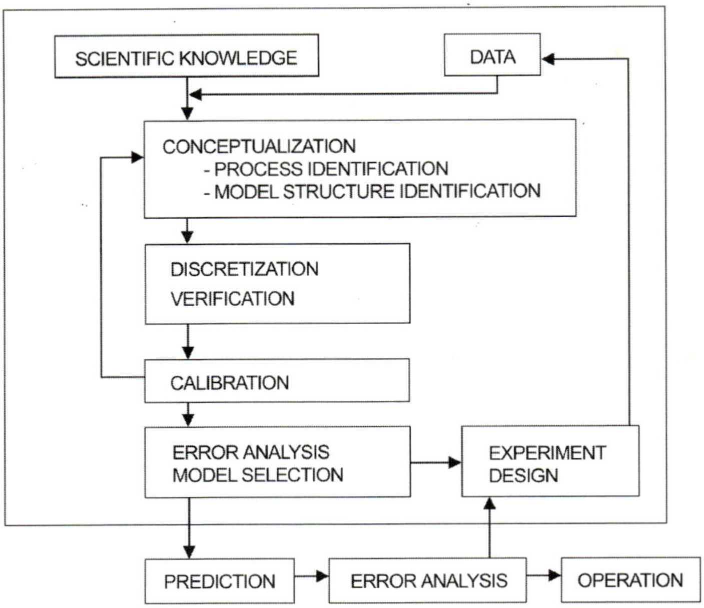

Fig. 2.3-1 The model development process. Notice the feedback for updating the conceptual model when new data become available or when calibration forces modellers to think back

Definition of Conceptual Model: Processes and Parameters

A conceptual model is a qualitative, often verbal, but more conveniently mathematical, description of the behaviour of a physical system. Since neither the physicochemical processes controlling water momentum, energy and solute fluxes nor the so-to-speak geometrical structure of those processes are always well known, conceptualisation is a difficult and hard to systematise task. For the purpose of further discussions, it is convenient to distinguish between process and model structure identification.

Definition of the relevant processes is ter med "process identification" (sometimes also called "perceptual model") and it is needed for several reasons. First, the number of processes that may affect fluxes is very large; for practical reasons, the modeller is forced to select those that affect most significantly the phenomenon under study. Second, not all the processes are well understood and they have to be treated in a simplified manner. In short, process identification involves numerous assumptions, both in the choice of the processes and in the way they are implemented in the model.

Writing equations for those processes usually entails some conservation principles (mass, momentum, energy) and some, more or less empirical, laws expressing fluxes in terms of potentials (see Boxes 2.3-1 and 2.3-2). Examples of these laws include Fourier's law expressing conductive heat flux as a function of thermal gradient, Manning's equation expressing water flow through a duct in terms of the hydraulic head gradient or Dalton's law expressing evaporation fluxes. These equations are left in terms of phenomenological coefficients (parameters). Sometimes, these coefficients refer to well-defined measurable quantities such as thermal conductivity. On the other hand, these parameters may reflect a large number of phenomena that are not treated in detail in the model. As a result, these parameters lack a well-defined physical meaning and are not easy to measure. They are often called "fudge factors". Examples include the exchange factor in evaporation processes, which may include all processes in the lower atmosphere (turbulence, plants, solar radiation, etc). While the factor has remained essentially unchanged since Dalton proposed it in 1802, a large number of factors have been added to enrich it (see Box 2.3-1). In fact, detailed evaluations of this factor sometimes yield order of magnitude variations over small distances in response to soil properties variability, variations in stomatal resistances or the like. Similarly, the leakage factor in aquifer–river interaction includes a large number of hard-to-represent processes (ranging from bacterial growth in the riverbed to eutrophication or sediment transport).

Some of the parameters may be spatially variable, some may not be easy to model and their meaning may be dependent on the level of detail and on the structure of the model. Therefore, their values can only be defined through model calibration. This topic will be addressed later. Before, however, we must revise the model structure identification stage.

Model structure identification refers to the definition of parameter variability, type of boundary conditions, model dimension, time regime, etc. In a somewhat narrower but more systematic sense, model structure identification implies expressing the model in terms of a finite number of unknowns called model parameters. The parameters controlling the above mentioned processes are variable in space. In some cases, they also vary in time or depend on the state variables. As discussed earlier, data are scarce so that such variability cannot be expressed accurately. Therefore, the modeller is also forced to make numerous assumptions to express the patterns of parameter variations, boundary conditions, etc. These assumptions are reflected on what is denoted model structure.

The above paragraphs suggest that the conceptualisation step of any modelling effort is somewhat subjective and dependent on the modeller's ingenuity, experience, scientific background and way of looking at the data, not to mention the availability of reliable data.

In fact, conceptualisation difficulties arise from those of systematisation. In many practical situations, the modeller is faced with an enormous volume of data (this is not always true!), most of which are qualitative in nature. The data may be worth one or several report volumes. Even if the modellers have participated in the collection of all the data, which is highly desirable, they will probably be confused and feel overwhelmed. Yet, after processing all the information, they are supposed to come out with a representation of the system, which is consistent with the data.

We feel that selection of the physicochemical processes to be included in the model is only rarely the most difficult issue. The most important processes affecting the movement of water and solutes and the transfer of energy and momentum are relatively well known. Therefore, ignoring a relevant process will only be caused by misjudgments and should be pointed out by reviewers. Difficulties arise when trying to characterise those processes and, more specifically, the spatial variability of controlling parameters.

In spite of the large amount of data usually available, their qualitative nature prevents a detailed definition of the conceptual model. Thus, more than one description of the system may result from the conceptualisation step. Selecting one conceptual model among several alternatives is sometimes performed during calibration, as discussed later. Actually, the main difference between models aimed at validation of specific processes and those oriented towards site characterisation arises at this stage. The former often allow prior specification of the alternative conceptual models, while the latter are normally subject to minor modifications during calibration.

Discretisation

Strictly speaking, discretisation consists of substituting a continuum by a discrete system. However, we are extending this term here to describe the whole process of going from mathematical equations, derived from the conceptual model, to numerical expressions that can be solved by a computer. Closely related is the issue of verification, which refers to ensuring that a code solves accurately the equations that its authors claim it solves (not to be mixed with validation, which refers to model validity, a topic we will revise later). As such, verification is a code-dependent concept. However, using a verified code is not sufficient for mathematical correctness. One should also make sure that time and space discretisation is adequate for the problem at hand. Moreover, numerical implementation of a conceptual model is not always straightforward. These types of difficulties became apparent in code intercomparison exercises such as INTRACOIN or INTRAVAL (Larsson 1992), in which a number of modelling teams solved several problems independently. Even well posed mathematical problems often led to widely different solutions because of slight variations in the solution methodology or misinterpretations in the formulation. The reasons behind these differences and ways to solve them only became apparent after discussions among the various teams involved.

The main concern during discretisation is accuracy. In this sense, it is not conceptually difficult, although it can be complex. Accuracy is not restricted to numerical errors (differences between numerical and exact solutions of the equations involved) but also refers to the precision with which discretisation reproduces the natural system.

Calibration

The choice of numerical values for model parameters is made during calibration, which consists of finding those parameter values that grant (1) a good fit between observations and model outputs and are (2) consistent with prior independent information.

Calibration is rarely straightforward. Data come from various sources, with varying degrees of accuracy and levels of representativeness. Some parameters can be measured directly in the field, but such measurements are usually scarce, prone to error, and dependent on the measurement method used. Furthermore, since measurements are most often performed on scales and under conditions different from those required for modelling purposes, they tend to be both numerically and conceptually different from model parameters.

Calibration can be tedious and time-consuming because many combinations of parameters have to be evaluated, which also makes it prone to be incomplete. This, coupled with difficulties in taking into account the reliability of different pieces of information, makes it very hard to evaluate the quality of results. Therefore, it is not surprising that significant efforts have been devoted to the development of automatic calibration methods. Reviews of groundwater hydrology inversion are presented by Yeh (1986) and Carrera (1987), and by Sorooshian and Gupta (1995) and Bastidas et al. (2002) for surface water.

A number of formulations are possible for model calibration. In this section, we will follow the maximum likelihood (ML) framework a presented by Carrera and Neuman (1986a). This formulation requires distributional assumptions about model output, which may lead to errors in statistical predictions, such as confidence intervals, if the assumption turns out to be incorrect. Yet, we feel that the ML formulation is advantageous for several reasons. First, ML is most often used with a normality assumption for the model output, which leads to a sum of squared residuals objective function. If this assumption is not correct, then statistical byproducts of the estimation should be used with caution, but the estimation itself is the same as the one obtained by least squares, which is known to be rather robust. Second, ML leads to criteria for comparing different conceptual models, which is one of the most important components of the modelling process.

Definition of the likelihood function requires specifying the data vector (measurements of state variables, and prior estimates of model parameters) and the error structure. We will assume that the residuals follow a zero-mean multivariate Gaussian distribution. Therefore, the error structure is fully defined by the covariance matrices. These matrices are never known accurately. In order to relax the need to specifying them, it is sometimes convenient to assume that they are of the form (Neuman and Yakowitz 1979):

$$\mathbf{C}_h = \sigma_h^2 \mathbf{V}_h \qquad \mathbf{C}_p = \sigma_p^2 \mathbf{V}_p \tag{2.3-1}$$

where $\mathbf{C}_h$ and $\mathbf{C}_p$ are the covariance matrices of measurements and parameters, $\mathbf{V}_h$ and $\mathbf{V}_p$ are known positive definite matrices (they represent the reliability that the modeller assigns to the data) and σ_h^2 and σ_p^2 are unknown scalars. With these definitions, the objective function $S = -2\ln(likelihood)$ becomes (Edwards 1972):

$$S = \frac{J_h}{\sigma_h^2} + \frac{J_p}{\sigma_p^2} + \ln|\mathbf{V}_h| + \ln|\mathbf{V}_p| + n_h \ \ln \sigma_h^2 + n_p \ \ln \sigma_p^2 + N \ln 2\pi \tag{2.3-2}$$

where

$$J_h = (\mathbf{h} - \mathbf{h}^*)^t \mathbf{V}_h^{-1} (\mathbf{h} - \mathbf{h}^*) \tag{2.3-3a}$$

$$J_p = (\mathbf{p} - \mathbf{p}^*)^t \mathbf{V}_p^{-1} (\mathbf{p} - \mathbf{p}^*) \tag{2.3-3b}$$

and $\mathbf{h}^*$ and $\mathbf{p}^*$ are the vectors of measurements and prior estimates of model parameter, whose dimensions are n_h and n_p, respectively; and $N=n_h+n_p$ is the total number of data entering the model.

Solution to the inverse problem is given by the minimum of S (Eq. 2.3-2) with respect to model parameters p and statistical parameters. However, if the statistical parameters are fixed, all but the first two terms in Eq. 2.3-2 become constant, so that minimising it is equivalent to minimising:

$$J = J_h + \lambda\ J_p \tag{2.3-4}$$

where $\lambda = \sigma_h^2 / \sigma_p^2$. This is the conventional sum of squared residuals objective function, whose minimum is the well-known Least Squares Estimate. Following Neuman (1973), this type of objective function has been widely used in hydrology. The first term measures model fit; it is small when residuals are small (recall Eq. 2.3-3b). The last term is sometimes called plausibility criterion and measures whether model parameters are reasonable.

Automatic calibration is not as simple as might be suggested by this section. It is complicated by two sets of problems. The first is related to conceptualisation and the second to instability and convergence difficulties. The initial conceptualisation derived by the modeller is rarely satisfactory. When calibrating the corresponding model, one is usually forced to revise the initial assumptions, often in the direction of increasing their complexity and the number of model parameters. A large number of model parameters leads to the second set of problems, namely instability and lack of convergence. These can be assessed through the error analysis, which is discussed next, while model selection is reviewed afterwards.

Estimation Errors

An important aspect of parameter estimation is the analysis of estimation errors. Asymptotically, ML estimates are normally distributed, which implies that the confidence region at a reasonable level is small enough to approximate state variables as linear functions of the parameters. Under the assumption of normality, the joint probability distribution of the parameter estimates is fully described by their mean and covariance matrix, where the mean is the estimate itself. On the other hand, it can be shown that the minimum variance bound, asymptotic limit of the covariance matrix, is the inverse of the Fisher information matrix, **F**, defined as Edwards (1972):

$$F_{ij} = \frac{1}{2} E\left[\frac{\partial^2 S}{\partial \beta_i \partial \beta_j} \right] \tag{2.3-5}$$

where S is the log-likelihood function in Eq. 2.3-2, and β_i is the i-th component of the parameter vector, β, including both model parameters, **p**, and statistical parameters, θ.

The covariance matrix (actually its minimum bound) is given by:

$$\Sigma_p = \mathbf{F}^{-1} \tag{2.3-6}$$

Note that Σ_p is positive-definite whenever prior information is available for all the parameters. As discussed later, this is important to ensure local uniqueness of the solution and, depending on the relative weighting of J_h and J_p, stability.

It is also worth mentioning that Σ_p depends on the derivatives of state variables with respect to p, which are a function of both the method of measurement (sampling locations and intervals) and the prevailing flow pattern. If the method of measurement and/or the flow pattern can be changed so as to make smaller and more diagonally-dominant, the estimates will become more reliable. Inverse modelling may thus be useful for the design of experiments and sampling schemes (Knopman and Voss 1989; Usunoff et al. 1992).

Model Selection

We stated earlier that the first step in any modelling effort involves constructing a conceptual model, describing it by appropriate governing equations, and translating the latter into a computer code. Indeed, one cannot start modelling without having first defined a conceptual model. However, qualitative data used for conceptualisation rarely suffice for unambiguous definition of a model. Many questions often remain unanswered. These refer to whether a process is relevant (Should osmotic effects be taken into account? Is surface retention important? Should matrix diffusion modelled explicitly?) or to the level of detail that is needed to represent each process (Can evapotranspiration be taken as a lumped term or does it need to be simulated in detail? Can one represent river-aquifer interaction by means of a leakage factor or should one simulate the hyporheic zone explicitly?). The answers to these questions depend both on specific site features and on model objectives. Hence, many answers are possible a priori. As a result, a number of models may have been proposed during the conceptualisation stage. In this section, we describe methods to choose between alternative models.

Model selection methods can be classified into three broad categories. The first category is based on a comparative analysis of residuals (differences between measured and computed system responses) using objective as well as subjective criteria. The second category is denoted parameter assessment and involves evaluating whether or not computed parameters can be considered as "reasonable". The third category relies on theoretical measures of model validity known as "identification criteria". In practice, all three categories will be needed: residual analysis and parameter assessment suggest ways to modify an existing model and the resulting improvement in model performance is evaluated on the basis of identification criteria. If the modified model is judged to be an improvement over the previous model, the former is accepted and the latter discarded.

The most widely used tool of model identification is residual analysis. The spatial and time distributions of residuals (differences between measurements and computations) are very useful in pointing towards aspects of the model that need to be modified. For example, in groundwater, a long tail in the breakthrough curve not properly simulated by a single porosity model suggests a need for incorporating matrix diffusion or a similar mechanism. These modifications should, whenever possible, be guided by independent information. Qualitative data such as lithology geological structure, geomorphology and hydrochemistry are often useful in aquifer modelling. A particular behavioural pattern of the residuals may be the result of varied causes that are often difficult to isolate. Spatial and/or temporal correlation among residuals may be a consequence of improper conceptualisation, but it may also be a consequence of measurement or numerical errors. Simplifications in simulating the stresses exerted over the system are always made and they also lead to correlation among residuals. For example, errors in flow rate measurements, variations in drop volume, neglecting flow rate variations, etc., will lead to correlated errors. To distinguish between correlations caused by improper conceptualisation and by measurement errors is not an easy matter. This makes analysis of residuals a limited tool for model selection.

An expedient way of evaluating a model concept is based on assessing whether or not calibrated parameters representing physico-chemical properties can be considered "reasonable"; that is, whether or not their values make sense and/or are consistent with those obtained elsewhere. Meaningless parameters can be a consequence of either poor conceptualisation or instability. If a relevant process is ignored during conceptualisation, the effect of such process may be reproduced by some other parameter. For example, the effect of sorption in solute transport problems is to keep part of the solute attached to the solid phase, hence retarding the movement of the solute mass; in linear instantaneous sorption, this effect cannot be distinguished from standard storage in the pores. As a result, ignoring retardation when modelling transport of a sorbing tracer will lead to increased, possibly above unity, estimates of porosity. This example suggests that parameter assessment may provide hints on how to modify the conceptual model. Indeed, such is the case in some instances: an unacceptably large porosity indicates a need for some retardation mechanism, a dispersivity comparable to or larger than the sample size points towards increasing the complexity of the flow field, etc. However, in our experience, parameter assessment tends to be more useful for ruling out some model concepts than for giving a hint on how to modify an inadequate model. Residual analysis is usually more helpful for this purpose.

Instability may also lead to unreasonable parameter estimates, despite the validity of the conceptual model. When the number of data or their information content are low, small perturbations in the measurements or deviations in the model may lead to drastically different parameter estimates. When this happens, the model may obtain equally good fits with widely different parameter sets. Thus, one may converge to a senseless parameter set while missing other perfectly meaningful sets. This type of behaviour can be easily identified by means of a thorough error analysis and corrected by fixing the values of one or several parameters.

The previous paragraph leads us to one of the crucial issues in model selection: the compromise between accuracy in representing the real system and robustness in model parameters.

The likelihood axiom provides a basis for comparing different hypotheses "within the framework of a specific model structure". As such, it is of little use in comparing models based on different structures. In fact, using the likelihood approach to choose among several models often results in selecting the model with the largest number of parameters. Alternative model selection criteria are thus required. Several such criteria have been used in the groundwater literature (Carrera and Neuman 1986b). Here, we describe the one proposed by Kashyap (1982), which has proven useful in several cases. It reads:

$$d_k = S + n_p \ln(N/2\pi) + \ln|\mathbf{F}| \qquad (2.3\text{-}7)$$

where $|\mathbf{F}|$ is the determinant of Fisher information matrix, as defined in Eq. 2.3-5, and n_p is the number of estimated parameters. In comparing two models, the one with the smallest value of d_k should be chosen.

It should be noted that this criterion supports the principle of parsimony in that, everything being equal, the simpler model (with lesser number of parameters) is chosen. Another interesting property of Eq. 2.3-7 is that, while it tends to select the simpler model, more complexity may be built into the model as the database increases. This is because as N (number of data) grows, so does the relative importance of the term S. Since minimisation of this term leads to increasingly larger number of parameters, the overall minimum of Eq. 2.3-7 is displaced towards models with growing complexity as the database increases.

A note of caution must be introduced here. Criteria such as Eq. 2.3-7 allow the modeller to evaluate the extent to which models are indeed supported by data. This is particularly important when using automatic and systematic methods for model construction and calibration, which may lead to perfectly meaningless overparameterised models. As such, model selection criteria are useful to test the quantitative prediction capabilities of the model under conditions similar to those of calibration. Yet, modellers bear ultimate responsibility to choose what they feel is relevant for prediction, especially when scales change.

2.3.2 Uncertainty and Probabilistic Models

Deterministic models, such as the ones described in Sect. 2.3.1, display two kinds of problems:

1. Uncertainties in the data. Model parameters (i.e. those describing physical properties) and initial and boundary state variables (i.e. external forcings) are not known. As a result, they contain errors that are propagated to model predictions, making them uncertain.

2. Model errors. Model equations may be inappropriate for a number of reasons. In fact, they are often based on processes defined at a given space/time scale and used on a different scale, where the dominant processes may be different.

In hydrology, these errors may be so large that they may render model prediction useless. This is why it is so important to formalise error analysis and to evaluate prediction errors. This section is devoted to analysing how error propagates and how it can be taken into account. The closely linked topic of spatial variability is also addressed in this section.

Sensitivity

Sensitivity analysis refers to the assessment of the dependence between model inputs and outputs and, specifically, how the latter respond to perturbations in the former. In mathematical terms, sensitivity of an effect (model output), f, to a factor (model input, parameter), p, is usually defined as the derivative of f with respect to p. If this derivative is very large, then small perturbations in p will lead to large perturbations in f. On the contrary, if this derivative is small, then f is said to be insensitive to p. Recognising that both model inputs and outputs are numerous, we will denote them by vectors, **f** and **p**.

Defining sensitivity as a derivative fails when the dependence between model outputs **f** and inputs **p** is non derivable (recall definitions of the types of non-linearities in Sect. 2.2). In coupled models, this will often be the case. Even in cases when the dependence is continuously differentiable, non-linearity may be so marked that the very concept of sensitivity as a derivative is very limited. In these cases, it is convenient to define sensitivity sensu latto as the variation in model output in response to changes in model parameters within their range of reasonable values.

The term sensitivity analysis is sometimes loosely equated with uncertainty analysis. It should be apparent, however, that sensitivity carries no strict uncertainty measure. While sensitivity may be purported to give some idea about the uncertainty in **p**, the term sensitivity is usually restricted to the dependence of **f** on variables one-at-a-time. When the effect of all model input uncertainties on model outputs are considered together, it is then more appropriate to speak of uncertainty analysis, which will be discussed in the next section (see also Box 2.3-4).

Computation of sensitivities is important for several reasons. On one hand, it is the first step towards a formal evaluation of uncertainty, which will be discussed in subsequent sections. On the other, knowledge of sensitivity allows modellers to assess qualitatively the uncertainty of both model predictions and calibrations. In general, high sensitivities, relative to the knowledge of parameters **p**, imply high uncertainties. In fact, when modellers say that a magnitude is unpredictable, they usually mean that the sensitivity of this magnitude to model input is extremely high.

A number of issues is associated with sensitivity analysis. These include (1) which are the effects and factors for which sensitivity needs to be evaluated, (2) how is it evaluated, (3) how does one deal with thresholds and other discontinuities.

Effects for which sensitivity needs to be evaluated. Leave room for surprises

Model outputs can be very extensive. This is especially true in coupled models in which one or several compartments may be simulated with distributed models and may represent time evolutions. It is clear that one cannot think of processing the derivative of every state variable at every cell and every time with respect to every input variable. One must constrain the desired outputs. A first shot at them is given by the model objectives (for example, total runoff and peak flow are natural outputs for a rainfall-runoff model). However, one should leave room for surprises because the process of analysing model output and sensitivities may itself suggest additional important outputs. For example, one may find out that, under certain conditions, runoff of a specific river becomes surprisingly small. Then, it may be appropriate to study the sensitivity of that particular model output.

In short, three points should be stressed. First, the importance of some variables may only become apparent during the sensitivity analysis. Second, a detailed examination of model output is needed. Finally, values and judgment are important when analysing these issues. As a result, expert participation and open attitudes are required at this stage.

Selecting the factors with respect to which compute sensitivity. Scenario analysis

As with model outputs, the number of input variables may be so large that it is not practical to compute the sensitivity of model outputs to all input variables separately. In fact, some of the variables should not be defined independently. For example, saturated hydraulic conductivity of a granular soil cannot be changed if entry pressure is not changed as well. Therefore, it is also important to group input variables when computing sensitivities.

Model inputs that deserve special attention are external forcings. As with internal inputs, their number can be huge. Even worse, their values can be highly uncertain. Think, for example, in the case of the human CO_2 emissions forcing in climate models. Distributed models would require specifying global CO_2 input, their spatial distribution and their time evolution. Analysing the sensitivity to the input in every surface cell at every time increment would be a Herculean task. Worse, it would be uninformative. A large short-term growth in CO_2 input is likely to be the effect of social trends that would carry many other associated changes. For example, it might be associated with a reduction in forest surface, a significant CO_2 sink in itself and also an important element in the hydrological cycle. Scenarios are motivated by the need to analyse these factors in a consistent and easy-to-present manner.

A scenario is a situation that can be foreseen, that is easy to understand by the layman and that allows specifying external forcings to the model in a consistent way. In the CO_2 emissions example, an integrated model would require specifying changes in population, migratory movements, use of energy and technology, social environmental concerns, etc. A number of scenarios can be built to define plausible but alternative trends for these variables (see IMAGE case study Sect. 4.13). Scenarios are widely used in the field of geological disposal of nuclear waste, where coupled models are used for evaluating the potential migration and distribution of pollutants. One possible scenario in this field would be a glaciation, where water recharge, temperature, mechanical stresses on the surface and other variables are specified together.

The above examples point out several issues:

1. Developing a scenario involves a variable degree of conjecturing. This is why they are often developed by expert judgment.

2. Scenario development is a multi- and interdisciplinary field in itself. A global change expert would argue that a glaciation is not a scenario, but a rather well known expectation. Still, this would be of little use for the design of a disposal facility.

3. Beside expert judgment, scenarios can be built from the outputs of other models. This is clear in the nuclear waste example. In the social trends example, however, it is clear that climate change will impact society. Therefore, treating them as a fixed scenario may not be appropriate, which explains the trend towards incorporating social models as compartments in coupled global change models (e.g., Pielke and Carbone 2002)

Error Analysis and Propagation

Formulating a model prediction is not just a matter of running a code. One must acknowledge errors. To evaluate them, one must start by analysing the prediction process. For one thing, formulation of predictions involves a conceptualisation of its own. Quite often, model development is motivated by analysing the response to some external action. Yet, this action leads to significant changes in the natural system, so that the structure used for calibration is no longer valid. For example, a basin model built under current conditions may not be appropriate to evaluate the downstream impact of a dam if the reservoir is so big that it changes significantly the moisture conditions in the basin. While numerical models can be used for network design or as investigation tools, most models are built to study the response of the medium to various scenario alternatives. Therefore, uncertainties on future natural and man-induced stresses also cause model predictions to be uncertain. In the example above, dam construction is sometimes accompanied by reforestation for sediment control. Reforestation, a man-induced stress (impact) on the system, leads to significant changes in the basin response. Finally, even if future conditions and conceptual model are known exactly, errors in model parameters will still cause errors in the predictions. In summary, three types of prediction uncertainties can be identified:

- Conceptual model uncertainties;
- Uncertainties in natural and man-induced stresses;
- Parameter and state uncertainties.

The first group includes two types of problems. One is related to model selection during calibration. That is, more than one conceptual model may have been properly calibrated and data may not suffice to distinguish which one is the closest to reality. It is clear that such indetermination should be carried into the prediction stage because all "valid" models may lead to widely different results under future conditions. The second type of problems arise from improper extension of calibration to prediction conditions, that is, from not taking into account changes in the natural system or in the scale of the problem. The only way we think for dealing with this problem consists of evaluating carefully whether or not the assumptions in which the calibration was based are still valid under future conditions. In our experience, model uncertainties can be very large.

The second type of uncertainties (i.e., those associated with future stresses and external forcings) are hard to evaluate. Future stresses may affect the validity of the model, as discussed earlier. This type of uncertainty is evaluated by carrying out simulations under a number of scenarios. Definition of these scenarios is an important subject in itself and was discussed earlier.

The last set of prediction uncertainties is the one associated with parameter and initial state uncertainties. The effect of these uncertainties can be quantified quite well. If f is a prediction to be made with the model (f is a function of parameters, $\mathbf{p}$), a lower bound of its variance is given by:

$$Var(f) = \left(\frac{\partial f}{\partial \mathbf{p}}\right)^t \Sigma_p \left(\frac{\partial f}{\partial \mathbf{p}}\right) + \sigma_{pO}^2 \qquad (2.3\text{-}8)$$

where σ_{p0}^2 represents model errors independent of parameter uncertainties. This equation represents a lower bound, so that actual uncertainty is larger than that. This is why more sophisticated approaches may be needed (see Box 2.3-4). Still, this equation allows one to spell out quantitatively how the different factors affect prediction uncertainty. In essence, prediction uncertainty grows with:

1. Sensitivity of predictions to model parameters and initial conditions. Obviously, a parameter is a source of concern only if predictions are sensitive to it, as measured by $\partial f / \partial \mathbf{p}$. This is why sensitivity analyses are sometimes performed instead of a formal error analysis.

2. Uncertainty in model parameters (and/or initial state). This is measured by the covariance matrix Σ_p .

Risk Assessment and Vulnerability

Risk is loosely defined as the chance of suffering damage from a hazard. Quantitatively, it is defined as the expected value of a loss. In this definition, loss is a quantitative evaluation of the damage (e.g. reduction in economic value, life expectancy or similar magnitudes). Let us assume that we have a model with *N* possible outcomes and that we can evaluate the probability of each, $\mathbf{p}_i$, and its loss, L_i. Then, risk is:

$$R = \sum_{i=1}^{N} P_i L_i \qquad (2.3\text{-}9)$$

Since both the magnitude of the harm (possibly evaluated as a magnitude easy to compare with others, such as economic loss, loss of life expectancy, etc.) and the probability of occurrence are taken into account, it is a very appealing concept. Politicians and decision managers like it very much because it purports objectivity.

Unfortunately, a single number carries no principles or moral values and does not convey any of the subtleties involved. For this reason, recognising that decision-making is always a multi-objective process, in general it will be a lot more appropriate to present the outputs of many simulations. This may be done in different ways: distribution of output, outcomes ensemble, envelope of all possible simulations, etc.

Risk is the sum, over all possible outcomes, of the product of loss function times the probability of occurrence of the outcome. Actually, evaluation of the loss is somewhat arbitrary and evaluation of the probability is uncertain. In fact, not all sources of uncertainty are known or taken into account. As a result, model output uncertainty and risk tend to be (badly) underestimated.

As discussed in the previous section, there are many sources of uncertainty. Some of them represent the current status of models. Others are caused by our limited knowledge of nature and its current state. Only a few represent truly unforeseeable magnitudes. One of the limitations of the risk concept is that, by mixing these concepts, it reduces its usefulness for proper decision making. A high (low) probability may be caused by insufficient knowledge of the system, which can be improved by additional measurements, by model uncertainties, which can be reduced by further research, or by controllable variables, which may be acted upon if there is a sufficient political will (e.g. emissions of CO_2 or other greenhouse effect gases).

Separating these sources of risk is an additional motivation for presenting ensembles of model simulations. These can be grouped so as to facilitate understanding and decision-making. Terms such as sensitivity studies, scenario analyses are often used in this context.

Risk assessment involves comparing the risks associated with different actions (control variables). While risk evaluations will always be uncertain and normally lead to underestimations, one would expect that the relative trends are less subject to error. For example, evaluating the full impact of changes in the emissions of CO_2 may be highly uncertain and so is the global warming associated with them. Yet, the trends are much less uncertain and that may be sufficient for action.

Another benefit of using the risk concept is that it leads modellers to concentrate on the impact of human actions on nature and society. This is a significant change in the way modellers are used to think that has been brought about by the emphasis on vulnerability

Vulnerability implies the need for protection. From an anthropogenic viewpoint,

Vulnerability is the characteristic of a person or group or component of a natural system in terms of its capacity to resist and/or recover from and/or anticipate and/or cope with the impacts of an adverse event (Blaikie et al. 1994; Downing et al. 1999).

Vulnerability therefore invariably embraces an

- *External dimension* (Vogel 1998), i.e. the threat of an event, that may increasingly predispose people to risk (e.g. climate change and its impacts on water resources), as well as an

- *Internal dimension*, i.e. the internal capacity to withstand or respond to an event, such as the defenselessness to cope with a hazard (e.g. poor people living on a flood plain) or the lack of means to cope with the aftermath of damaging loss.

People may thus face the same potential risk, but are not equally vulnerable because they may face different consequences to the same hazard. A more detailed summary of the vulnerability concept is presented in Guenni et al (2004).

Spatial Variability and Upscaling

Variability is an essential feature of some hydrological variables. Rainfall, porosity, soil texture, and vegetation cover among others are highly variable in space and/or in time. They are so variable that one cannot realistically aim at ever knowing them accurately. Increased sophistication and effectiveness of measurement devices causes an increasing level of resolution. Still, the apparently fractal nature of all those phenomena and parameters leads us to feel that there will always be some level of unresolved variability.

Variability is important not only because of the uncertainty it brings about, but also because large-scale behaviour of a spatially variable phenomenon may be significantly different from the small-scale behaviour. That is, changes in scale may (1) lead to changes in the effective parameters, (2) cause the governing equations to change, or (3) cause new processes to emerge. This section is motivated by the recognition that spatial and temporal variability is (1) very important, and (2) impossible to describe accurately. Spatial variability affects many hydrological processes (Blöschl and Sivapalan 1995; Famiglietti and Wood 1994; Bronstert and Bárdossy 1999; Woolhisher et al. 1996). Yet, the rest of this section is based on groundwater hydrology, a field where significant results have been achieved and which can be considered as a paradigm. Gelhar (1986) and Dagan (1989) provide thorough accounts of the topic.

The framework for dealing with spatial variability is the one of geostatistics (see Box 2.3-5) and random functions, which can be viewed as functions whose outcomes are random variables at every point or, inversely, as random entities whose realisations are space functions. A full characterisation of these functions would require defining the joint probability density function of the variable at all points in space. This task would be no easier than precise definition of the variable of interest and, what is worse, would probably be futile and meaningless. Therefore, simplifying assumptions have to be made. The most common one is to treat the variable (say $Y=\log T$, T being transmissivity, see Box 2.3-2) as a second order stationary random function (or simply stationary, for short), which means that its expected value and its variance are constant and that the spatial covariance between $Y(x_1)$ and $Y(x_2)$ is a function of $(x_1\text{-}x_2)$, but not of x_1 and x_2 separately.

Because of the above, the partial differential equations governing water flow and solute transport now become stochastic differential equations. That is, the coefficients in these equations are random functions. Therefore, the solutions (head and concentrations) are also random functions. Stochastic differential equations can be solved using Monte Carlo simulations (see Box 2.3-4) and small perturbation methods, which are described below.

Monte Carlo simulations consist of (1) generating many replicas (realisations) of the spatially variable field (say, transmissivities); (2) solving the flow and/or transport equation for each realisation; and (3) extracting relevant statistics. Some of the early applications of this approach include the works of Freeze (1975) or Smith and Schwarz (1980).

Small perturbations are based on writing the random function Y as the sum of a mean value $\overline{Y}$ and a perturbation Y'. Presumably, one can write the solution also as the sum of a mean and a perturbation. The point is how to relate the perturbation in the solution with the perturbation in the original random function. For this purpose, let us consider, for instance, the steady-state groundwater flow equation

$$\nabla(e^{Y}\nabla h) = 0 \qquad (2.3\text{-}10)$$

Assuming that Y' is small, we can expand $\exp(\overline{Y} + Y')$ as $T_g + f + f^2 +$ ...., where $T_g = \exp(\overline{Y})$. Similarly, we expand h as $\overline{h} + h^{(1} + h^{(2} +$ .... The motivation behind these expansions is to find $\overline{h}$ as a function of $\overline{Y}$ (actually T_g), which would represent a zero order approximation to h. From there, obtain $h^{(1}$ as a function of f (the first order approximation to h would be $\overline{h} + h^{(1)}$), and so on. To do so, we substitute these expansions into the flow equation (2.3-10), leading to:

$$\nabla(\overline{Y}\nabla\overline{h}) + \nabla(f\nabla\overline{h}) + \nabla(\overline{Y}\nabla h^{(1)}) + \ldots\ldots = 0 \qquad (2.3\text{-}11)$$

We can now choose $\overline{h}$ to satisfy the zero order terms:

$$\nabla(\overline{Y}\nabla\overline{h}) = 0 \qquad (2.3\text{-}12)$$

which yields $\bar{h}$ as a function of $\bar{y}$. To obtain $h^{1)}$, we choose it so as to eliminate the first order terms:

$$\nabla\left(\bar{Y}\,\nabla\, h^{(1}\right) = -\nabla\left(f\,\nabla\,\bar{h}\right) \tag{2.3-13}$$

Notice that the solution of $h^{(1}$ is a function of f and $\bar{h}$, which should have been computed before. Similarly, we can write $h^{(2}$ as a function of f, $h^{(1}$ and $\bar{h}$. Since f is a random function, an exact solution is not possible. However, $h^{(1}$ is linear in f, so that its properties as a random function can be derived from those of f, which is what we were seeking. Such derivation can be done in a number of ways ranging from spectral to Green's function methods.

In what follows, we present a few examples of results that have been obtained with this type of approach. Specifically, we illustrate the derivation of effective parameters and scale effects.

Effective parameters are those that represent the large-scale behaviour. Earliest work on stochastic hydrogeology was aimed at seeking effective conductivity. Gelhar (1976), Bakr et al. (1978) and Gutjahr et al. (1978) found that the relationship between effective and local permeability depends on the spatial dimension:

$$1-D:\; K_{eff} = K_g(1-\sigma_y^2/2) \tag{2.3-14}$$

$$2-D:\; K_{eff} = K_g \tag{2.3-15}$$

$$3-D:\; K_{eff} = K_g\left(1+\sigma_y^2/6\right) \tag{2.3-16}$$

where K_g is the geometric average of local values and σ_y^2 is the variance of $Y(ln\ K)$.

Still more interesting is the case in which local conductivity is isotropic but its covariance function is anisotropic. In this case, effective conductivity is given by (Gelhar and Axness 1983; Dagan 1989):

$$\mathbf{K}_{eff} = K_g\left[\mathbf{I} + \sigma_y^2\left(\frac{1}{2}\mathbf{I} - \mathbf{M}\right)\right] \tag{2.3-17}$$

where $\mathbf{M}$ depends on the assumed covariance function for $ln\ K$. The most interesting feature of Eq. 2.3-17 is that, as shown by Dagan (1989) for the case of axisymmetric covariance, the term in parenthesis in Eq. 2.3-17 is positive-definite. This implies that directional effective permeability is larger than its local counterpart. In summary, the effective conductivity can display anisotropy even if the local conductivity is isotropic.

The scale dependence of dispersion has been one of the most spectacular results of stochastic hydrogeology. An effective dispersion coefficient has been found for large distances, even when no dispersion is assumed at the small scale. The effective dispersion – macrodispersion – coefficient is proportional to the variance of *lnK* and to its integral distance (Gelhar et al. 1979; Gelhar and Axness 1983; Dagan 1982, 1984; Matheron and de Marsily 1980; Neuman et al. 1987; and many others).

Impressive as it is, this result (i.e. finding a macrodispersion coefficient) agrees poorly with the observed scale dependence of dispersivity and, specifically, with the fact that its growth may be unbounded. This prompted Sposito et al. (1986) to conclude: "the stochastic convection-dispersion model has not yet proved useful in shedding fundamental light on the scale effect". Actually, there are number of models that not only predict the growth of dispersivity but lead to expressions of this growth as a function of travel distance (Dagan 1988; Neuman et al. 1987; Neuman and Zhang 1990). Yet, with the only exception of Matheron and de Marsily (1980), all of them tend to an asymptotic macrodispersivity for sufficiently long travel distances. In this context, the conjecture of Neuman (1990) becomes particularly appealing. He conjectures that natural variability of hydraulic conductivity can be viewed as "a continuous hierarchy of log-hydraulic conductivity fields with mutually uncorrelated increments". This leads him to view hydraulic conductivity as a self-similar process with a fractal dimension of approximately n+0.75 (n being the topological dimension of interest; n=1 for 1-D problems, etc.).

Stochastic hydrogeology has been successful in explaining a number of other non-trivial observations and in finding large-scale effective equations (see e.g. Tartakovsky and Neuman 1999). However, rather than specific results, we want to stress that the approach is possible in most fields. While analytical perturbation-type methods can be mathematically cumbersome, Monte Carlo methods are not difficult conceptually.

2.3.3 Validation and Model Performance Evaluation

The main conclusion from the modelling approach presented in Sect. 2.3.1 is that, leaving aside steps that involve conceptualisation, most modelling steps allow a systematic approach to be followed. Methods are available for objective discretisation, verification, calibration, sensitivity and error analysis. These may be more or less difficult to do, but they can be tested and, although different authors may prefer different methods, results will not be very sensitive to the approach chosen, unless the problem is mathematically ill posed. In this context, building a model resembles the scientific method in the sense that it follows similar steps. First, some data are used for making conjectures about reality. These are then stated as hypotheses, which allows derivation of theories. Such theories can, in turn, be validated by comparing their predictions to actual data taken from appropriately designed experiments (appropriately designed means that they are not trivially similar to those on which the theory is based). These steps are not fundamentally different from the modelling steps discussed above. Therefore, it is natural to regard models as theories about the behaviour of the natural system.

In this context, then, validation refers to independent testing of a model. Validation of theories has been used by scientists since long ago. In fact, after emergence of a theory, it is standard practice to perform experiments aimed at disproving it (i.e. experiments that would lead to results inconsistent with theoretical predictions if the theory is wrong). Similarly, models are validated by testing them against conditions significantly different from those of calibration. In this context, validating a model (i.e. showing that it displays non-trivial prediction capabilities) should be taken as increased support for the model that allows the modeller to "feel good" about model predictions. The fact that a model has been validated does not imply that it will always lead to good predictions, in the same way that the fact that a theory has overcome many tests does not ensure that it will not fail in the next one: theories can be proven wrong, but they cannot be proven right.

Actually, we think that the main problem with validation arises as a result of the way it is perceived by different people. Many modellers feel that decision makers sometimes view validation as a fool-proof check of model validity. On the other hand, experience has taught scientists that long-standing theories can be proven wrong (or their range of applicability can be bounded) the moment an ingenuous person devises a mechanism that had been previously ignored. This is particularly true for coupled hydrological models, where experience is limited and where a large number of processes are being put together. Hence, one should not be surprised to find that processes that have normally been neglected as irrelevant become dominant now. As a result, scepticism towards long-term model simulations should not be taken as arrogance from people living in "ivory towers", but rather as a necessity for keeping active the search for new, previously ignored, processes and new understanding.

Box 2.3-1 ***Soil-Vegetation-Atmosphere Transfer Schemes (SVATS)***

Jim Shuttleworth

Soil-Vegetation-Atmosphere Transfer Schemes (SVATS) are a class of interface models whose purpose is to provide coupling between the near-surface atmosphere and the hydro-ecological processes that take place in the zone that extends typically from a few metres below the ground, through the vegetation into the lower atmospheric boundary layer. They normally represent processes as operating in one dimension. The upper boundary conditions are incoming solar and long-wave radiation, precipitation, atmospheric variables such as the temperature, humidity and wind speed of the air and, if relevant, the concentration of atmospheric constituents. In most SVAT models, the lower boundary conditions are weakly specified: often gravity drainage of soil water to a remote, unspecified groundwater table is assumed.

SVATS vary greatly with respect to the complexity with which they represent the processes that control the interface between the atmosphere and the ground. However, at a minimum, they must calculate (a) the portion of incoming solar radiation reflected back to the atmosphere; (b) the emission of long wave radiation by the ground; (c) the transfer of momentum between the atmosphere and the ground; (d) the energy fluxes outgoing to the atmosphere in the form of latent and sensible heat; (e) the sensible heat flux into the soil; and, if required, (e) the net flux of carbon dioxide and other atmospheric constituents between the atmosphere and the ground. The equations that are central to all SVATS are the laws that describe conservation of energy and water, and the conservation of other entities (such as carbon) if their exchange is represented in the model. All SVATS have at least one state variable, i.e. the water stored in a specified portion of soil just below the surface. However, most modern SVATS include representation of several soil layers and at least one layer of vegetation. The capacity of these layers to store water and (usually) energy is represented in the form of state variables in the model.

Diffusion processes are used to describe the movement of water, energy, and, if required, other entities into and out of the uppermost store, between the several stores within the model (if present), and into and out of the lowest store. In the case of movement through the atmospheric portions of the model, turbulent atmospheric diffusion and, in modern SVATS, molecular diffusion are represented. The latter is the primary transfer mechanism for latent and sensible heat between the inside and outside of plant leaves and through the atmospheric boundary layer that surrounds leaves. Soil water and heat diffusion is often also represented in SVATS, and occasionally also the diffusion of liquid water within plants. Diffusion is usually represented by the integrated forms of diffusion coefficients in the form of so-called "resistances". These coefficients control the total exchange between the stores represented in the model or between modelled stores and the overlying atmosphere. Recently, some SVATS have begun to include representation of the coupling between some of the exchange processes within the soil-vegetation-atmosphere system itself. Both carbon and water vapour enter or leave leaves by diffusing through the stomata and several models now seek to describe the strong coupling that consequently exists between carbon and water vapour exchange.

Box 2.3-2 ***Groundwater Models***

Jesus Carrera

Groundwater flow is governed by Darcy's law, which is a way of stating momentum conservation for slow flow. Aquifers are much more extensive than thick (length to depth ratios normally exceed 1000). Therefore, ever since Dupuit proposed it in the 19th century, most hydrologist have neglected vertical variations. This implies that the flow of water is expressed per unit width of aquifer. This flow is proportional to head gradient (head, h, being the energy per unit weight of water, which equals the elevation of water in wells). The constant of proportionality is called transmissivity, T.

Mass balance per unit surface of aquifer yields the flow equation:

$$S\frac{\partial h}{\partial t} = \nabla(T\nabla h) + r \qquad (2.3\text{-}18)$$

where S is the storage coefficient (volume of water storage change per unit surface of aquifer and per unit change in head) and r is recharge/discharge (per unit surface). Solving this equation requires knowing the initial head, h_0, as well as water fluxes along the boundaries. These may be expressed as:

$$Q = \alpha(h_e - h) + \beta \qquad (2.3\text{-}19)$$

where Q is flow per unit length of boundary, h is aquifer head, h_e is the head of an outer body of water (river, sea, etc.), α is a leakage factor and β is a constant flux. Depending on whether α is constant, time-dependent or a function of h, this equation may represent a broad range of hydrological fluxes both along the boundaries (Q) and per unit surface (writing r instead of Q). These include aquifer-river interaction, springs discharge, phreatophyte transpiration, etc.

Coupling to other hydrological elements is represented through the r and Q terms. The coupling can be both ways (i.e. the aquifer may impact the surface system and vice versa) and can be linear or non-linear. For example, growth of phreatophytes depends on evapotranspiration as a function (r), which, in turn, affects groundwater levels. Hence, it is usual to write evapotranspiration as a function of head. Yet, this term becomes zero when the head falls below the deepest roots. Another threshold occurs when the head reaches the land surface and water starts to flow on the soil surface. However, despite of these interactions, time scales relevant for groundwater systems are usually so much longer than those of surface water or atmospheric systems that they are normally treated separately.

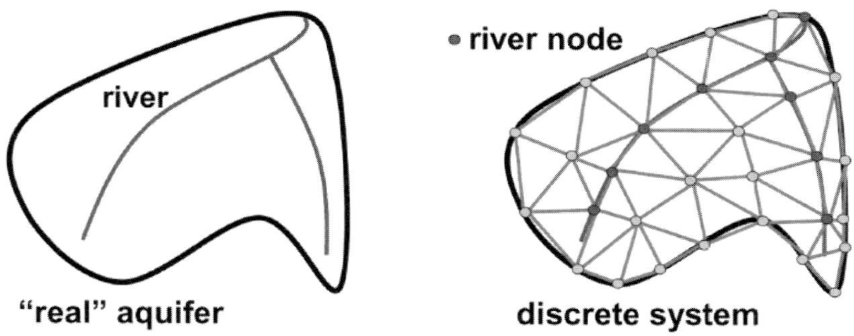

Fig. 2.3-2 A "real" groundwater system and its spatial discretisation

Solving the above equations requires discretising them, that is, substituting the aquifer continuum by a set of cells or elements covering the flow domain (Fig. 2.3-2). Many numerical methods (i.e. techniques for solving partial differential equations) do exist (e.g. Pinder and Gray 1977). The most widely known are the finite element and finite differences methods. All approaches lead to linear systems of equations, where each row expresses the mass balance of one of the cells in which the aquifer was discretised. These systems can be written in matrix form as:

$$\mathbf{D}\frac{\mathbf{dh}}{\mathrm{dt}} + \mathbf{Ah} = \mathbf{b} \qquad (2.3\text{-}20)$$

where the first term expresses the rate of change in storage in every cell (**D** is the storage matrix and **h** is the vector of heads at every cell); the second term represents exchange between adjacent cells (**A** being the conductance matrix) and the right hand side expresses external inflows and outflows).

Eq. 2.3-20 is a system of ordinary differential equations that can be integrated in time using different numerical approaches (Euler, Runge-Kutta, etc.; see e.g. Ralston and Rabinowitz 1978 or Lambert 1991). This will be further discussed in Box 2.3-3. It is important to bear in mind, however, that a groundwater model expresses nothing but the mass balance at each of the cells into which the aquifer has been divided.

Box 2.3-3 ***Time Integration and Systems Theory***

Jesus Carrera

After discretising the governing equations of linear systems, one obtains conservation equations such as (recall Box 2.3-2):

$$\mathbf{D}\frac{d\mathbf{h}}{dt} + \mathbf{A}\mathbf{h} = \mathbf{b} \qquad (2.3\text{-}21)$$

It is important to emphasise that this type of equation is valid for describing the time evolution of any linear system. The actual meaning of the terms in Eq. 2.3-21 depends on the phenomenon under study. But these equations usually represent some conservation principle (mass, momentum, energy etc.) and h is the vector of state variables (heads, concentrations, velocities, temperatures) at all cells in which the medium has been discretised. Thus, the first term represents the rate of change in storage (of mass, momentum, energy); the second term represents flow rates between cells and the right hand side represents inputs/outputs from/to the outside.

One important feature of Eq. 2.3-21 is that it makes it apparent that the future evolution of the system only depends on the past history through the current state. That is, if one knew the state and future actions, one could predict the evolution of **h**. Systems theory has shown than in many systems (certainly, those governing most natural processes), small perturbations in **b** or in the initial state will lead, in the long run, to large errors in **h**. Therefore, predictive capability is limited. Still, one may compute the evolution trend.

Time integration of Eq. 2.3-21 can be tricky in non-linear (**A** and **D** functions of the unknown **h**) and/or unstable cases. It is usually achieved by breaking the time axis in time steps Δt. In its simplest form, h^k (state variables at time *k*) can be written as a function of h^{k-1}:

$$\mathbf{h}^k = \mathbf{h}^{k-1} + \Delta t\left[\mathbf{D}^{-1}\left(\mathbf{b} - \mathbf{A}\mathbf{h}^{k-1}\right)\right] \tag{2.3-22}$$

It may be interesting to point out that, when **A**, **D** and **b** are constant (linear case) and symmetric, it is possible to write an analytical solution to Eq. 2.3-21 (e.g. Sahuquillo 1983):

$$L(t) = \sum_{i,j} u_{ij} e^{-\alpha_i t} h_{oj} + \sum_{i,j} v_{i,j} \frac{1 - e^{-\alpha_i t}}{\alpha_i} b_j \tag{2.3-23}$$

where *L(t)* is any linear combination of *h* values, α_i is the *i-th* eigenvalue of $\sqrt{\mathbf{D}^{-1}}\mathbf{A}\sqrt{\mathbf{D}^{-1}}$, u_{ij} and v_{ij} are combinations of its eigenvectors and h_{oj} are the initial values of *h*. Here, *i* is the eigenvalue counter and *j* is the nodes (or cells) counter.

The importance of Eq. 2.3-23 stems from its similarity with the solution of the simple linear deposit (Fig. 2.3-3). Assume that the initial stage of the deposit is x_0, that water enters with a constant recharge rate *r*, and that discharge is proportional to the stage *x*. Then, the evolution of the stage is governed by:

$$\frac{dx}{dt} = r - \alpha\, x \tag{2.3-24}$$

The solution of this equation is:

$$x = x_0 e^{-\alpha t} + r\left(\frac{1 - e^{-\alpha t}}{\alpha}\right) \tag{2.3-25}$$

Comparing Eq. 2.3-23 and Eq. 2.3-25 makes it apparent that any first order linear system can be written as linear combinations of linear deposits. Each α_i is sometimes referred to a mode of response (actually, $1/\alpha_i$ is a response time) and the matrix v_{ij} expresses the way in which an external action b_j activates a response in the *i*-th mode with respect to the control variable *L*.

This view lends some support to treating subsystems as boxes, especially when the subsystem is well approximated linearly. Most terms in Eq. 2.3-23 can be simplified (either because $\alpha_i t$ is very large, in which case $e^{-\alpha_i t} \approx 0$, or because $\alpha_i t$ is very small, in which case $(1 - e^{-\alpha t})/\ \alpha_i \approx t$). Therefore, one should not be surprised to find that indeed most of the behaviour of the subsystem is well captured by one or two modes of response. Still, proper definition of those modes may be difficult.

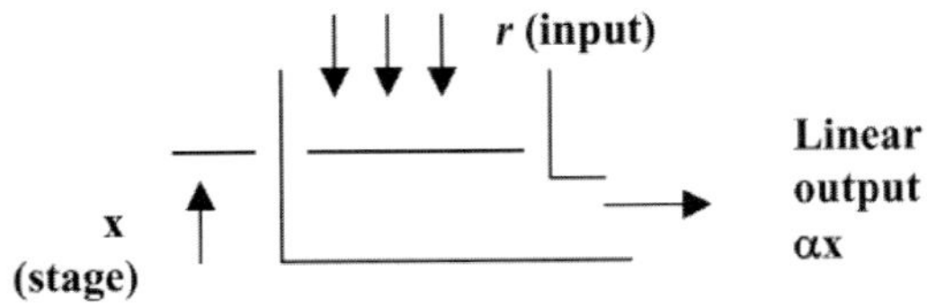

Fig. 2.3-3 The linear reservoir with input r and output $\alpha\, x$. Any linear system, as given in Eq. 2.3-21, can be represented by a combination of such linear reservoirs

Box 2.3-4 ***Monte Carlo Method***

Jesus Carrera

One of the most important issues in coupled models is to evaluate the uncertainty of predicted effects (models outputs, **f**). This uncertainty is caused by the fact that model inputs are uncertain. We will denote model inputs by **p**. These may refer to external forcing terms, parameters representing physical properties, or initial conditions. They may also represent the model itself (for example, one component of **p** may represent whether a given chemical reaction in the soil is treated as in equilibrium or as kinetic).

The methods presented in Sect. 2.3.2 to deal with this problem are appropriate for linear, or nearly so, systems. Those methods can be extended to continuous non-linearities (recall Sect. 2.2) if the uncertainty in model inputs is moderate. Otherwise, one must seek alternative methods to evaluate uncertainty. Of these, the most widely used and simplest to implement is the Monte Carlo method (MCM).

In essence, MCM consist of evaluating **f** repeatedly for a large number of combinations of model inputs **p**. In algorithmic form:

Step 1. Initialisation: Set i=0

Step 2. i= i+1

Step 3. Generate (simulate) a vector of model inputs, $\mathbf{p}_i$, with the appropriate statistics (one needs to specify the distribution of **p**, its mean, variance and the like)

Step 4. Use the model(s) to compute output $\mathbf{f}_i = \mathbf{f}\,(\mathbf{p}_i)$

Step 5. Compute and store the appropriate statistics

Step 6. Return to step 2, unless statistics have become stable, i has become huge or some other stopping criterion has been reached.

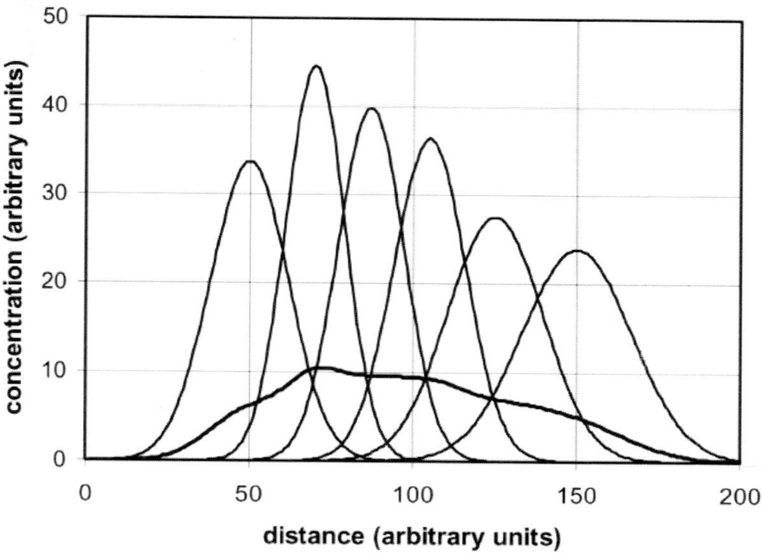

Fig. 2.3-4 Six Monte Carlo simulations of 100 days (actually, units are arbitrary) transport of a 1000 g mass pulse initially injected at x=10. Both velocity and diffusion coefficient are random. Notice that the average concentration field does not share any of the features of each realisation. It spans a much wider area and its maximum concentration is much smaller. While the simulations convey an idea of the uncertainty in travel time and maximum concentration, their ensemble average does not, and one must seek appropriate statistics

While the method is relatively simple, successful application requires that care be exercised in selecting the appropriate statistics and output and input variables.

Selecting appropriate statistics and output variables

Basically, one needs to ascertain what needs to be known and that not much information is lost when computing output statistics in step 5 above. When averaging (spatial distribution and/or time evolution) the output of many simulations, the salient features may be lost. For example, if river hydrographs or concentrations breakthrough curves are averaged, the result will bear none of the salient features of any of the realisations. In the example of Fig. 2.3-4, it is clear that one must keep track of maximum values and, possibly, peak times. But this is not always clear a priori. Therefore, the basic recommendation is to think ahead of time as to what can be expected. This can be aided by thoroughly examining the model output of some simulation. Specifically, extreme simulations (e.g. those with highest peak or earliest arrival in Fig. 2.3-4) should be thoroughly studied. As an additional benefit, this will help the modeller to gain insight on how the model behaves. This discussion might be reduced if model objectives were well defined beforehand. The fact is that they rarely are. In fact, MCM aims at "what if". Therefore, one should leave "room" to be surprised and to change representative output statistics, as one learns about the system.

Selection of input variables and their statistics

As with the output, selection of input variables requires some thinking beforehand. One must avoid specifying variables subject to feedbacks (e.g. rainfall in a coupled atmosphere-river basin model). One should not simulate independently variables that are correlated (e.g. saturated conductivity and entry pressure in a soil). One should consider usually the possibility of including spatial variability (see Box 2.3-5). One must check for absurd combination of model inputs (e.g. if albedo is treated as a function of soil moisture, $a = p_1+p_2w$, some combinations of p_1 and p_2 may lead to negative or larger than 1 albedo).

In order to learn, incremental runs may be appropriate. That is, one starts by simulating the effect of what are thought to be the main variables and keeps adding new variables to the simulation. In this way one learns about the system behaviour, about the interactions of input variables and also about the sensitivity of model output to input data. In the long run, it is this learning that counts.

An issue hard to resolve is the choice of statistical properties for input variables. It is difficult and important. At the very least, output variance grows linearly with that of the input. Normally, it grows more than linearly. Traditionally, these statistics have been derived from historical records (e.g. rainfall). Unfortunately, climate change studies make this approach somewhat inappropriate. Expert elicitation is sometimes purported to be the best approach when data are lacking. However, this only works when expert experience can be of use, again a rare event when studying the response to climate change. In cases where spatial variability is the issue, then spatial sampling coupled to geostatistical tools (see Box 2.3-5) may be appropriate. However, there is no general rule. The problem has to be approached separately for each variable. In fact, one could argue that the motivations for coupled models is precisely the need to substitute the uncertainty of some variable by a model (or rather a compartment).

Box 2.3-5 ***Geostatistics, Kriging and Spatial Variability***

Jesus Carrera

Matheron (1971), who is considered to be the father of geostatistics, defines it as the study of random functions (RFs). A spatially varying variable $Z(\boldsymbol{x})$ is called a random function if its value at every point can be viewed as a random variable. Conversely, each realisation of a random function is a function of space. This type of description is appropriate for reflecting the knowledge associated with many natural variables (rainfall, porosity, sand content in a soil may all be viewed as random functions). One can view geostatistics as the extension of time series analysis to 2 or more dimensions. This extension is tricky because it causes measurements to be no longer evenly distributed and domains to be bounded.

The implicit assumption in geostatistics is that knowing $Z(\boldsymbol{x})$ at a point x conveys information about its values nearby. That is, $Z(\boldsymbol{x})$ displays a certain structure. Structural analysis is the part of geostatistics aimed at characterising the structure of random functions. This entails stating the properties of $Z(\boldsymbol{x})$, specifically in which way do its statistics (mean, variance, etc) vary over space. The most singular tool to do that is the variogram, which yields the variability of $Z(\boldsymbol{x})$ at a distance $\boldsymbol{h}$:

$$\gamma\,(h) = \frac{1}{2} E \quad \left([Z(x) - Z(x+h)]^2 \right) \tag{2.3-26}$$

E stands for expectation. A typical variogram is shown in Fig. 2.3-5. It is clear that the variogram plays the same role as the covariance function, only it is a bit more general. In fact, for stationary RFs, (i.e. those with constant mean, σ_z^2) : $c(h) = \sigma_z^2 - \gamma(h)$

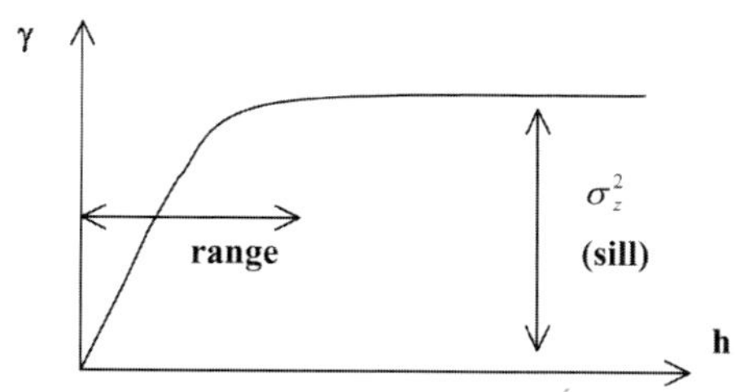

Fig. 2.3-5: Variogram

The variogram describes the spatial dependence of variability. A point measurement of a variable conditions the likely values within the range. At longer distances, the uncertainty on the variable is independent of this

One of the early problems of geostatistics was to obtain the minimum variance estimate of $Z(x)$ as a linear function of measurements $Z_i=Z(x_i)$. The resulting method is called Kriging, after Krige, the South African mining engineer who first used this framework. Kriging yields a linear estimate of the form:

$$Z_K(x) = \sum_{\mathrm{i}} \lambda_{\mathrm{i}} Z_{\mathrm{i}} \qquad (2.3\text{-}27)$$

where $Z_K(x)$ is the Kriging estimate and λ_i are the Kriging weights. Under rather general conditions, this type of linear relationship can be derived from the conditional expectation formula:

$$Z_K(x) = E\left[Z(x)/\mathbf{Z}_m\right] = E(Z(x)) + \mathbf{c}_{zm}\mathbf{C}_{mm}^{-1}(\mathbf{Z}_m - E(\mathbf{Z}_m)) \qquad (2.3\text{-}28)$$

where $\mathbf{Z}_m=\{Z_i\}$ is the measurement vector, $\mathbf{C}_{zm}$ is the covariance between the estimated variable and the measurements and $\boldsymbol{C}_{mm}$ is the covariance among measurements.

Kriging has a limited usefulness when one is interested in spatial variability and data are scarce. The result of Kriging is the conditional expectation of Z. As such, it is much smoother than the original function $Z(x)$. The process of generating RFs that reproduce the variability of Z, yet honour measurements Z_i, is called conditional simulation (i.e. generate simulations of $Z(x)$ that are conditioned upon Z_i). As shown in Fig. 2.3-6 emphasis is now placed on reproducing the variability pattern of $Z(x)$.

The above methods can be extended to the case where there are measurements of not only Z, but also of other variable(s) Y that is (are) correlated with Z. The corresponding methods are called coKriging and cosimulation. They are especially interesting in cases in which one has extensive coverage of variable(s) Y (e.g topography, satellite images, etc), even if they are only weakly correlated to Z.

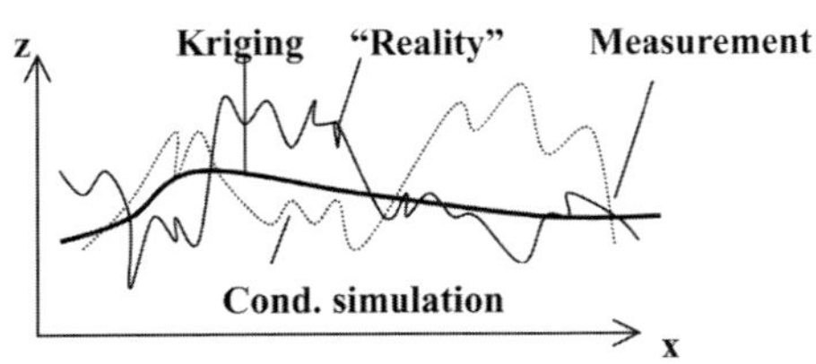

Fig. 2.3-6 Kriging is much smoother than reality. Conditional simulation aims at reproducing the variability of reality and coinciding with it at measurement points

Box 2.3-6 ***A Common Framework for Calibration and Data Assimilation***

Jesus Carrera, Luis A. Bastidas

Calibration and data assimilation can be formulated in a unified manner, provided that one is willing to formulate the problem statistically and to make a few simplifying assumptions.

The calibration problem can be formulated as follows. Let **h** be the vector of measurements of state variables (i.e., **h** may comprise water, flow rates, concentration measurements). Let **h**(**p**) be the vector of corresponding computations (i.e. values of **h** computed with model parameters, **p**). Let $\mathbf{V}_{hh}$ be the covariance matrix of prior errors (i.e. errors that can be attributed to measurement, but not to errors in parameters). Similarly, let $\overline{\mathbf{p}}$ be the vector of prior estimates of model parameters and $\mathbf{V}_{pp}$ be its covariance matrix. Assuming that **h** and **p** can be considered approximately jointly gaussian, then the maximum likelihood and maximum a posteriori estimates of **p** are obtained by minimising:

$$f(\mathbf{p}) = (\mathbf{h} - \mathbf{h}(\mathbf{p}))^t \mathbf{V}_{hh}^{-1} (\mathbf{h} - \mathbf{h}(\mathbf{p})) + (\mathbf{p} - \overline{\mathbf{p}}) \mathbf{V}_{pp}^{-1} (\mathbf{p} - \overline{\mathbf{p}}) \qquad (2.3\text{-}29)$$

It can be shown (simply taking derivatives) that the minimum of this function satisfies the conditional expectation formula:

$$\mathbf{p} = \overline{\mathbf{p}} + \mathbf{V}_{ph} \mathbf{V}_{hh}^{-1} (\mathbf{h} - \overline{\mathbf{h}}) \qquad (2.3\text{-}30)$$

where $\mathbf{V}_{ph}$ is the cross covariance between **h** and **p**. That is,

$$\mathbf{V}_{ph} = \mathbf{V}_{pp} \mathbf{J}_{ph} \qquad (2.3\text{-}31)$$

where $\mathbf{J}_{ph}$ is the Jacobian matrix, containing the derivatives of h with respect to p.

The data assimilation problem can equated to that of Eq. 2.3-30 by stating it as that of finding the best estimate of state variables at time k resulting from incorporating (assimilating) the measurements at such time. In the conventional Kalman filter notation, the system is written as

$$\mathbf{p}_{k+1} = \mathbf{F}_k \mathbf{p}_k + \mathbf{G}_k \mathbf{w}_k \qquad (2.3\text{-}32)$$

$$\mathbf{h}_k = \mathbf{H}_k^t \mathbf{p}_k + \boldsymbol{\upsilon}_k \qquad (2.3\text{-}33)$$

Eq. 2.3-32 represents the system equation, $\mathbf{p}_k$ is now the vector of state variables at time k, $\mathbf{F}_k$ is the system matrix, $\mathbf{w}_k$ is a model error noise with covariance matrix $\mathbf{Q}_k$, $\mathbf{h}_k$ is the vector of measurements, linear combination of the state variables with weight matrix $\mathbf{H}_k$, and $\boldsymbol{\upsilon}_k$ is a vector of measurement errors, with covariance matrix $\mathbf{R}_k$.

The expected value of $\mathbf{p}_k$ prior to the measurement $\mathbf{h}_k$ would be $\mathbf{p}_{k/k-1}$ (p_k conditioned upon what was known up to time k-1). Similarly, one would expect the measurement to be $\mathbf{H}_k^t\,\mathbf{p}_{k/k-1}$. The covariance matrix of h_k would then be the sum of the uncertainties coming from $\boldsymbol{\upsilon}_k$ and $\mathbf{h}_k/_{k-1}$, that is:

$$\mathbf{V}_{hh} = \mathbf{H}_k^t \mathbf{V}_{k/k-1} \mathbf{H}_k + \mathbf{R}_k \tag{2.3-34}$$

similarly

$$\mathbf{V}_{ph} = \mathbf{V}_{k/k-1} \mathbf{H}_k \tag{2.3-35}$$

Substituting these last two equations into Eq. 2.3-30 yields the Kalman filter equation:

$$\mathbf{p}_{k/k} = \mathbf{p}_{k/k-1} + \mathbf{V}_{k/k-1}\mathbf{H}_k(\mathbf{H}_k^t\mathbf{V}_{k/k-1}\mathbf{H}_k + \mathbf{R}_k^{-1})(\mathbf{h}_k - \mathbf{H}_k^t\mathbf{p}_{k/k-1}) \tag{2.3-36}$$

while it looks imposing, Eq. 2.3-36 is simply a restatement of the conditional expectation equation Eq. 2.3-30.

Notice that we have used this conditional expectation equation not only for calibration and data assimilation, but also for Kriging (recall Box 2.3-5). This points to the generality of this equation and its usefulness for linear systems. Generalisations come when the system is non-linear or when the dimensions of the problem are so big that these equations cannot be applied in their standard form. In this context, think of coupled models with millions of nodes, where millions of measurements (e.g. satellite data) have to be integrated in each time step). Obviously, special equations have to be used in those cases. Anderson and Moore (1979) provide an excellent account of these topics.

References Chapter 2

Anderson BDO, Moore JB (1979) Optimal filtering. Prentice Hall, Englewood Cliffs, New Jersey, p357

Anderson MP, Woessner WW (1991) Applied groundwater modelling. Academic Press, San Diego, p381

Andreae MO, Artaxo P, Brandao C, Carswell FE, Ciccioli P, da Costa AL, Culf AD, Esteves JL, Gash JHC, Grace J, Kabat P, Lelieveld J, Malhi Y, Manzi AO, Meixner FX, Nobre AD, Nobre C, Ruivo MDLP, Silva-Dias MA, Stefani P, Valentini R, von Jouanne J, Waterloo MJ (2002) Biogeochemical cycling of carbon, water, energy, trace gases, and aerosols in Amazonia: The LBA-EUSTACH experiments. J Geophys Res Special Issue LBA, 107(D20): Art. No. 8066

Argyros IK, Szidarovszky F (1993) The theory and applications of iteration methods. CRC Press, Boca Raton, London

Avissar R, Liu Y (1996) Three-dimensional numerical study of shallow convective clouds and precipitation induced by land surface forcing. J Geophys Res 101: 7499-7518

Bakr AAM, Gelhar LW, Gutjahr AL, McMillan JR (1978) Stochastic anaylysis of spatial variability in subsurface flow: Comparison of one-and three-dimensional flows. Water Resour Res 14:262-271

Basson MS, Allen RB, Pegram GGS, van Rooyen JA (1994) Probabilistic management of water resources and hydropower systems. Water Resources Publications, Highlands Ranch, Colorado, p434

Bastidas LA, Sorooshian S, Gupta HV (2002) Emerging paradigms in the calibration of hydrologic models. In: Singh VP, Frevert D (eds) Mathematical models of large watershed hydrology. Water Resources Publications, LLC, Highlands Ranch, Colorado, Vol I, pp25-56

Beven K (1993) Prophecy, reality and uncertainty in distributed hydrological modelling. Adv Water Resour 16(1):41-51

Blackie JR, Eeles CWO (1985) Lumped catchment models. In: Anderson MG, Burt TP (eds) Hydrological forecasting. Wiley, Chichester, pp311-345

Blaikie P, Cannon T, Davis I, Wisner B (1994) At risk: Natural hazards, people´s vulnerability and disasters. Routledge, London

Blöschl G, Sivapalan M (1995) Scale issues in hydrological modelling- a review. Hydrol Process 9:251-290

Brandt A, Fernando HJS (eds) (1995) Double-diffusive convection. Am Geophys Union, Geophys. Monograph 94, p334

Bronstert A, Bárdossy A (1999) The role of spatial variability of soil moisture for modelling surface runoff generation at the small catchment scale. Hydrol Earth Syst Sci 3(4):505-516

Carrera J (1987) State of the art of the inverse problem applied to the flow and solute transport equations. In: Custodio E, Gurgui A, Lobo Ferreira JP (eds) Analytical and numerical groundwater flow and quality modelling. NATO-ARW Ser. C: Mathematical and Physical Sciences, D. Reidel, Norwell, Mass, Vol 224, pp549-583

Carrera J, Neuman SP (1986a) Estimation of aquifer parameters under transient and steady state conditions. Part 1: Maximum likelihood method incorporating prior information. Water Resour Res 22(2):199-210

Carrera J, Neuman SP (1986b) Estimation of aquifer parameters under transient and steady stated conditions. Part 3: Application to synthetic and field data. Water Resour Res 22(2):228-242

Carrera J, Mousavi SF, Usunoff E, Sanchez-Vila X, Galarza G (1993) A discussion on validation of hydrogeological models. Reliability Engineering and System Safety 42:201-216

Chartrand G (1977) Introductory graph theory. Dover Publ, NY, p294

Culf AD, Fisch G, Hodnett MG (1995) The albedo of Amazonian forest and range-land. J Climate 8:1544- 1554

Culf AD, Esteves JL, Marques Filho A de, da Rocha OHR (1996) Radiation, temperature and humidity over forest and pasture in Amazonia. In: Gash JHC, Nobre CA, Roberts JM, Victoria RL (eds) Amazonian deforestation and climate. John Wiley, Chichester, pp175-191

Cutrim E, Martin DW, Rabin R (1995) Enhancement of cumulus clouds over deforested lands in Amazonia. B Am Meteorol Soc 76:1801-1805

Dagan G (1982) Stochastic modeling of groundwater flow by unconditional and conditional probabilities. Part 2. The solute transport. Water Resour Res 18(4): 835-848

Dagan G (1984) Solute transport in heterogeneous porous formations. J Fluid Mech 145:151-177

Dagan G (1988) Time-dependent macrodispersion for solute transport in anisotropic heterogeneous aquifers. Water Resour Res 24:1491-1500

Dagan G (1989) Flow and transport in porous formations. Springer, New York, NY, p465

Dickinson RE, Kennedy P (1992) Impact on regional climate of Amazon deforestation. Geophys Res Lett 19:1947-1950

Dooge JCI (1973) Linear theory of hydrologic systems. U.S. Department of Agriculture, Technical Bulletin No. 1468, Washington, D.C., p327

Dooge JCI (2003) Linear theory of hydrologic systems. EGU Reprint Series, 1, Katlenburg-Lindau

Downing TE, Olsthoorn AJ, Tol RSJ (1999) Climate change and risk. Routledge, London

Eagleson PS (1970) Dynamic hydrology. McGraw-Hill, New York

Eagleson PS (2003) Dynamic hydrology. EGU Reprint Series, 2, Katlenburg-Lindau

Edwards A (1972) Likelihood. Univ Press Cambridge, Mass

Famiglietti JS, Wood EF (1994) Multiscale modeling of spatially variable water and energy balance processes. Water Resour Res 30:3061-3078

Fath BD, Patten BC (1999) Review of the foundations of network environ analysis. Ecosystems 2:167-179

Feddes RA, Kabat P, VanBakel PJT, Bronswijk JJB, Halbertsma J (1988) Modeling soil-water dynamics in the unsaturated zone - state of the art. J Hydrol 100(1-3):69-111

Freeze RA (1975) A stochastic-conceptual analysis of one-dimensional groundwater flow in nonuniform homogeneous media. Water Resour Res 11:725-741

Freeze RA, Cherry JA (1979) Groundwater. Prentice Hall, Englewood Cliffs, NJ, p604

Garstang M, Ulanski S, Greco S, Scala J, Swap R, Fitzjarrald D, Browell E, Shipman M, Connors V, Harriss R, Talbot R (1990) The Amazon Boundary-Layer Experiment (ABLE 2B): A meteorological perspective. B Am Meteorol Soc 71:19-31

Gash JHC, Nobre CA, Roberts JM, Victoria RL (eds) (1996) Amazonian deforestation and climate. John Wiley, Chichester

Gelhar LW (1976) Effects of hydraulic conductivity variations on groundwater flows. In: Hjorth P, Jönsson L, Larsen P (eds) Hydraulic problems solved by stochastic methods. Proc Second International Symposium on Stochastic Hydraulics, Lund, Sweden, Water Resources Publications, Fort Collins, CO, pp409-431

Gelhar LW (1986) Stochastic subsurface hydrology from theory to applications. Water Resour Res 22(9):135S-145S

Gelhar LW, Axness CL (1983) Three dimensional stochastic analysis of macrodispersion in aquifers. Water Resour Res 19:161-180

Gelhar LW, Gutjahr AL, Naff RL (1979) Stochastic analysis of macro-dispersion in a stratified aquifer. Water Resour Res 15:1387-1397

Guenni LB de, Schulze RE, Pielke RA Sr, Hutchinson MF (2004) The vulnerability approach. In: Kabat P, Claussen M, Dirmeyer PA, Gash JHC, Guenni LB de, Meybeck,M, Pielke RA Sr, Vörösmarty CJ, Hutjes RWA, Lütkemeier S (eds) (2004) Vegetation, water, humans and the climate: A new perspective on an interactive system. Springer-Verlag, Berlin, pp499-514

Gupta RS (1989) Hydrology and hydraulic systems. Prentice-Hall, Englewood Cliffs, New Jersey

Gupta RS (2001) Hydrology and hydraulic systems. 2nd Edition, Waveland Press, Long Grove, IL, p867

Gupta HV, Bastidas LA, Sorooshian S, Shuttleworth WJ, Yang ZL (1999) Parameter estimation of a land surface scheme using multicriteria methods. J Geophys Res Atmospheres 104(D16):19491-19503

Gutjahr AL, Gelhar LW, Bakr AA, McMillan JR (1978) Stochastic analysis of spatial variability in subsurface flow. Part 2: Evaluation and application. Water Resour Res 14(5):953-960

Hahmann AN, Dickinson RE (1997) RCCM2-BATS model over tropical South America: Application to tropical deforestation. J Climate 10(8):1944-1964

Hill MC (1998) Methods and guidelines for effective model calibration. USGS Water Resources Investigations Report 98-4005, p97

Holzbecher E (1998) Modeling density-driven flow in porous media. Springer Publ, New York, p286

Holzbecher E (2003) What is a coupled model. In: Gnauk A, Heinrich R (eds) The information society and enlargement of the European Union. Proc. 17th Int. Conf. Informatics for Environm. Protection, Metropolis Verlag, Marburg, pp208-215

Huyakorn P, Pinder G (1983) Computational methods in subsurface flow. Academic Press, San Diego, p473

IGBP (1998) LBA is moving forward. Global Change Newsletter 33:1-9

Imbe M, Nakano Y, Nakashima N, Nakamura S, Ogawada D (2000) An engineering approach towards appropriate hydrological water cycle in urban areas: First report. In: Sato K, Iwasa Y (eds) Groundwater updates. Springer Publ, Tokyo, pp313-318

INPE (2001) http://sputnik.dpi.inpe.br:1910/col/dpi.inpe.br/lise/2001/05.16.09.55/doc/html/pag_8.htm

Jarvis PG. (1976) Interpretation of variations in leaf water potential and stomatal conductance found in canopies in field. Phil Trans R Soc Lond B 273(927): 593-610

Kabat P, Claussen M, Dirmeyer PA, Gash JHC, Guenni LB de, Meybeck M, Pielke RA Sr, Vörösmarty CJ, Hutjes RWA, Lütkemeier S (eds) (2004) Vegetation, water, humans and the climate: A new perspective on an interactive system. Springer-Verlag, Berlin, p566

Kashyap RLD (1982) Optimal choice of AR and MA parts in autoregressive moving average models. IEEE Trans Pattern Anal Mach Intel (PAMI) 4(2):99-104

Knopman DS, Voss CI (1989) Multiobjective sampling design for parameter-estimation and model discrimination in groundwater solute transport. Water Resourc Res 25(10):2245-2258

Lambert JD (1991) Numerical methods for ordinary differential systems. Wiley, Chichester, p293

Lapwood ER (1948) Convection of a fluid in a porous medium. Proc Camb Phil Soc A 225:508-521

Larsson A (1992) The international projects INTRACOIN, HYDROCOIN and INTRAVAL. Adv Water Resourc 15:85 – 87

Lean J, Rowntree P (1993) A GCM simulation of the impact of Amazonian deforestation on climate using an improved canopy representation. Q J Roy Meteor Soc 119:509-530

Lean J, Rowntree P (1997) Understanding the sensitivity of a GCM simulation of Amazonian deforestation to the specification of vegetation and soil characteristics. J Climate 10:1216-1235

Loucks DP, Stedinger JR, Haith DA (1981) Water resource systems planning and analysis. Prentice-Hall, Englewood Cliffs, New Jersey

Manzi AO, Planton S (1996) Calibration of a GCM using ABRACOS and ARME data and simulation of Amazonian deforestation. In: Gash JHC, Nobre CA, Roberts JM, and Victoria RL (eds) Amazonian deforestation and climate. John Wiley, Chichester, pp505-529

Matheron G (1971) The theory of regionalized variables and its applications. Les Cahiers du CMM, Fasc. N° 5, ENSMP, Paris, p211

Matheron G, de Marsily G (1980) Is transport in porous media always diffusive? - A counterexample. Water Resour Res 16:901-917

Mölders N (2000) Similarity of microclimate as simulated over a landscape of the 1930s and the 1980s. J Hydrometeorol 1:330-352

Mölders N, Raabe A, Tetzlaff G (1996) A comparison of two strategies on land surface heterogeneity used in a mesoscale-β meteorological model. Tellus A 48:733-749

Nepstad DC, Carvalho CR, Davidson EA, Jipp PH, Lefebvre PA, Negrelros GH, da Silva ED, Stone TA, Trumbore SE, Vieira S (1994) The role of deep roots in the hydrological and carbon cycles of Amazon forests and pastures. Nature 372:666-669

Neuman SP (1973) Calibration of distributed parameter groundwater flow models viewed as multiple objective decision process under uncertainty. Water Resour Res 9(4):1006-1021

Neuman SP (1990) Universal scaling of hydraulic conductivities and dispersivities in geologic media. Water Resour Res 26(8):1749-1758

Neuman SP, Yakowitz S (1979) A statistical approach to the inverse problem of aquifer hydrology. Part 1: Theory. Water Resour Res 15(4):845-860

Neuman SP, Zhang YK (1990) A quasi-linear theory of non-fickian and fickian subsurface dispersion. Part 1: Theoretical-analysis with application to isotropic media. Water Resour Res 26(5):887-902

Neuman SP, Winter CL, Newman CM (1987) Stochastic theory of field-scale fickian dispersion in anisotropic porous-media. Water Resour Res 23(3):453-466

Nield DA, Bejan A (1992) Convection in porous media. Springer Publ, New York, p408

Nobre C, Sellers P, Shukla J (1991) Amazoman deforestation and the regional climate change. J Climate 4:957-988

Nobre CA, Fisch G, da Rocha HR, Lyra RF, da Rocha EP and Ubarana V N (1996) Observations of the atmospheric boundary layer in Rondônia. In: Gash JHC, Nobre CA, Roberts JM, Victoria RL (eds) Amazonian deforestation and climate. John Wiley, Chichester, pp413-424

Patten BC, Higashi M, Burns TP (1990) Trophic dynamics in ecosystem networks: Significance of cycles and storage. Ecol Model 51:1-28

Petoukhov V, Ganopolski A, Brovkin V, Claussen M, Eliseev A, Kubatzki S, Rahmstorf S (2000) CLIMBER-2: A climate system model of intermediate complexity. Part I. Clim Dyn 16:1-17

Pielke R, Carbone RE (2002) Weather impacts, forecasts, and policy - An integrated perspective. B Am Meteorol Soc 83(3):393-403

Pinder, GF, Gray WG (1977) Finite element simulation in surface an subsurface Hydrology. Academic Press, New York, p295

Polcher J (1995) Sensitivity of tropical convection to land surface processes. J Atmos Sci 52(17):3143- 3161

Polcher J, Laval K (1994a) The impact of African and Amazonian deforestation on tropical climate. J Hydrol 155:389-405

Polcher J, Laval K (1994b) A statistical study of regional impact of deforestation on climate of the LMD-GCM. Clim Dyn 10:205-219

Ralston A, Rabinowitz P (1978) A first course in numerical analysis. 2nd Edition, McGraw-Hill, p556

Rayleigh Lord (1916) On convection currents in a horizontal layer of fluid when the higher temperature is on the under side. Phil Mag XXXII:529-546

Reilly TE, Goodman AS (1985) Quantitative analysis of saltwater-freshwater relationships in groundwater systems - a historical perspective. J Hydrol 80: 125-160

Rickenbach TM, Ferreira RN, Halverson JB, Herdies DL, Dias MAFS (2002) Modulation of convection in the southwestern Amazon basin by extratropical stationary fronts. J Geophys Res Special Issue LBA, 107(D20): Art. No. 8040

Saaltink MW, Carrera J, Ayora C (2001) On the behavior of approaches to simulate reactive transport. J Contam Hydrol 48:213-235

Sahuquillo A (1983) An eigenvalue numerical technique for solving unsteady linear groundwater models continuously in time. Water Resour Res 19(1):87-93

Seigneur C (1993) Multimedia modeling. In: Zannetti P (ed) Environmental modeling I. Comp Mech Publ, Southampton and Elsevier Science, London, pp225-272

Sellers PJ, Mintz Y, Sud YC, Dalcher A (1986) A simple biosphere atmosphere model (SiB) for use within general circulation models. J Atmos Sci 43:505-531

Seth A, Giorgi F, Dickinson RE (1994) Simulating fluxes from heterogeneous land surfaces: Explicit subgrid method employing the biosphere-atmosphere transfer scheme (BATS). J Geophys Res 99D:18651-18667

Shuttleworth W J (1988) Evaporation from Amazonian rainforests. Phil Trans R Soc Lond B 233:321-346

Silva Dias MAF, Regnier P (1996) Simulation of mesoscale circulations in a deforested area of Rondonia in the dry season. In: Gash JHC, Nobre CA, Roberts J.M, Victoria R. (eds) Amazonian deforestation and climate. John Wiley, Chichester, pp531-547

Silva Dias MAF, Nobre C, Marengo J (2001) The interaction of clouds and rain with the biosphere. Global Change Newsletter 45:8-11

Silva Dias MAFS, Rutledge S, Kabat P, Dias PLS, Nobre C, Fisch G, Dolman AJ, Zipser E, Garstang M, Manzi AO, Fuentes JD, Rocha HR, Marengo J, Plana-Fattori A, Sa LDA, Alvala RCS, Andreae MO, Artaxo P, Gielow R, Gatti L (2002a) Cloud and rain processes in a biosphere-atmosphere interaction context in the Amazon region. J Geophys Res Special Issue LBA, 107(D20):Art. No. 8072

Silva Dias MAFS, Petersen W, Dias PLS, Cifelli R, Betts AK, Longo M, Gomes AM, Fisch GF, Lima MA, Antonio MA, Albrecht RI (2002b) A case study of convective organization into precipitating lines in the Southwest Amazon during the WETAMC and TRMM-LBA. J Geophys Res Special Issue LBA 107(D20):Art. No. 8078

Smith L, Schwartz FW (1980) Mass transport 1. A stochastic analysis of macroscopic dispersion. Water Resour Res 16:303-313

Sorooshian S, Gupta VK (1995) Model calibration. In: Singh VP (ed) Computer models of watershed hydrology. Water Resources Publications, LLC, Highlands Ranch, Colorado, USA, pp23-63

Sposito G, Jury WA, Gupta VK (1986) Fundamental problems in the stochastic convection-dispersion model of solute transport in aquifers and field soils. Water Resourc Res 22(1):77-88

Subramanian S, Balakotaiah V (1997) Analysis and classification of reaction-driven stationary convective patterns in a porous medium. Phys Fluids 9(6)1674-1695

Tartakovsky DM, Neuman SP (1999) Extension of "Transient flow in bounded randomly heterogeneous domains, 1, Exact conditional moment equations and recursive approximations". Water Resour Res 35(6):1921-1925

Usunoff E, Carrera J, Mousavi SF (1992) An approach to the design of experiments for discriminating among alternative conceptual models. Adv Water Resour 15(3):199-214

Vogel C (1998) Vulnerability and global environmental change. LUCC Newsletter 3:15-19

Wang HF, Anderson MP (1982) Introduction to groundwater modelling. Freeman and Co., San Francisco, p237

Ward RC, Robinson M (1990) Principles of hydrology. McGraw-Hill, London.

Woolhiser DA, Smith RE, Giraldez. JV (1996) Effects of spatial variability of saturated hydraulic conductivity on Hortonian overland flow. Water Resour Res 32:671-678

Yeh WW-G (1986) Review of parameter identification procedures in groundwater hydrology: The inverse problem. Water Resour Res 22(2):95-108

Zang H, Henderson-Sellers A, McGuffie K. (1996) Impacts of tropical deforestation. Part I: Process analysis of local climate change. J Climate 9:1497-1517

Zheng C, Bennet GD (2002) Applied contaminant transport modeling. 2nd Edition, John Wiley & Sons, New York

CHAPTER 3

Systematisation of the Interactions between Hydrological and related Cycles

3.0 Introduction

Ekkehard Holzbecher

In the preceding chapter the concepts of compartments, fluxes, interactions, links, coupling, feedback and interfaces are tackled at a very abstract, general, theoretical level. Here these terms are applied to the hydrological cycle and to related processes. The concept can be applied to the entire global hydrological cycle, which is a closed system as far as water is concerned, with atmosphere, ocean, land surface and subsurface as subsystems; or to a regional hydrological cycle, which is an open system with processes such as precipitation, evaporation, evapotranspiration, surface runoff, groundwater recharge etc., as far as water is concerned. It can also be applied to processes related to the water cycle, i.e. mass and heat transfer in general. In the following chapter the subject is presented even more specifically in case studies.

The subsystems (or compartments) of the hydrological cycle are connected in general, but the connection may be quite different for regional systems and when it is not concerned with water. A systematisation enables the examination of the interactive and complex behaviour within the system of compartments within the water cycle. Phenomena can be studied in an objective way. What is most relevant here is that the systematic approach allows us to characterise a system with respect to the "coupling property": is it coupled or not?

In the first section several approaches of systematisation are shown. The interaction can be visualised in a flow diagram that is set up in a similar structure as that already presented in Sect. 2.1. The fundamental influences between processes in the different compartments are presented in a matrix of interactions and feedback. The major and minor links between the subsystems, shown in the matrix, provide an insight into the complexity of the entire environment.

It should be stressed that the matrix would become even more extended and complicated, if economic and social aspects were also included. Surely economic and social processes have an important impact on the processes and on the interactions within the hydrological cycle. At the regional scale, examples can easily be found where the anthropogenic impact on hydrological processes can outweigh by far the impact of climatic change (changing river beds, damming, irrigation, sealing, extraction of surface water or groundwater). It is clear that is does not make much sense to include all possible processes and interactions. Depending on the goal of the study, one has to restrict the system under study. Thus in the following sections selected topics on water, energy and mass fluxes are highlighted. First, the focus is on transport processes for mass and heat, which itself turns out to be a subject of high complexity. Finally, energy and mass transfer at the hydrometeorological interface is described as a specific example for a subsystem of transfer within the hydrological cycle.

The development of a model can be subdivided into several steps. Roughly seven steps can be distinguished.

1. The goal has to be defined clearly (application, scale).

2. One has to derive a conceptual model.

3. One has to identify the processes that are important to achieve the aim for which the model structure is built.

4. These processes have to be described mathematically and have to be parameterised depending on the computer capacity.

5. The processes have to be linked by transfer equations that have an impact on the parameterisation as well (so step 4 may have to be repeated).

6. The input data necessary to run the model have to be assimilated (also the data availability may have an impact on step 4!).

7. A strategy has to be developed and applied as to how the model can be validated by independent datasets.

3.1 Coupled Processes and Interaction Matrix

Axel Bronstert, Ekkehard Holzbecher, Pavel Kabat, Oleg F. Vasiliev

The purpose of this section is to systematise the coupling requirements of hydrological models with atmospheric, soil and biosphere models. The first step in such a systematic analysis of these interacting systems is to identify the governing internal processes (i.e. influencing each other by feedbacks) and to identify the external forces of the system (i.e. boundary conditions which are relevant for the system but which are not significantly influenced by the system itself).

After the identification of the processes and external forces, one may transform those to a flow diagram depicting model structure and exchange of information between the various components of the model. In this flow diagram, one can distinguish between pools of mass or energy and interactions between those pools, which are the mass or energy fluxes.

This conceptual model of the interacting natural systems (let us call it *box-and-arrow-diagram model*) might then be transferred into a matrix showing all possible interactions in-between the system(s) under study (let us call it *interaction matrix*). Note that the *box-and-arrow-diagram model* and the *interaction matrix* contain the same information. Nevertheless, the *interaction matrix* forms the basis for a further analysis, ranking the importance of the various process interactions.

This procedure – derivation of the *box-and-arrow-diagram model* and the *interaction matrix* – are the principal steps for a *systematic* development of a coupled model. In the following section we describe this principle for the hydrological and biogeochemical cycles where the nucleus or main area of interest is the hydrological cycle.

Summarising, it has to be stressed that for the construction of a coupled model the following points have to be clear from the beginning:

- What are the questions to be covered by the coupled model?
- What are the internal processes and which are external forces influencing the system modelled?
- What are the main coupling processes/feedbacks?
- What is (are) the relevant space and time scale(s), which have to be handled by the model (see Sect. 1.2)?

3.1.1 Box-and-arrow Diagram of the Hydrological and related Cycles

As mentioned above, the first step in a systematic development of a coupled model is to identify the governing internal processes (i.e. influencing each other by feedbacks) and to identify the external forces of the system (i.e. boundary conditions which are not significantly influenced by the system itself). In Fig. 3.1-1, a flow diagram is shown depicting model structure and exchange of information for the hydrological cycle and related biogeochemical cycles. This diagram gives just a sketch of the related cycles and does not aim to be complete. Indeed, in a strict sense, it can never be complete.

Here boxes represent pools (or compartments) of mass or energy, ovals represent processes which define the fluxes between pools (and arrows represent the direction of flow), state variables or threshold triggers are shown by dashed line boxes (state variables) and ovals (triggers to different processes). Note that in this flow diagram external forces are not given.

However, one has to keep in mind that this system is strongly driven by external forces, such as, e.g. large-scale atmospheric processes, human activities altering the landscape, or socio-economic processes with an even more cross-cutting character such as greenhouse gas emissions, pricing of energy or population development.

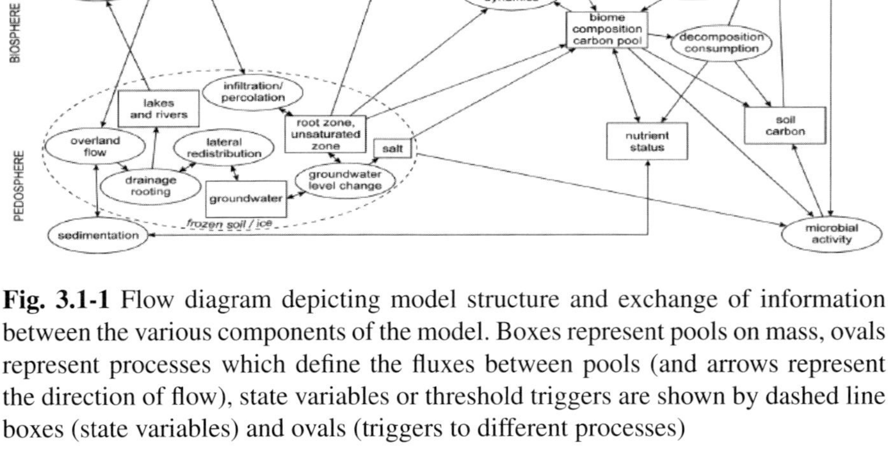

Fig. 3.1-1 Flow diagram depicting model structure and exchange of information between the various components of the model. Boxes represent pools on mass, ovals represent processes which define the fluxes between pools (and arrows represent the direction of flow), state variables or threshold triggers are shown by dashed line boxes (state variables) and ovals (triggers to different processes)

3.1.2 Interaction Matrix of the Hydrological and related Cycles

This conceptual model of the interacting natural systems might then be evaluated by various experts with different scientific background to determine the primary or direct ("first order") processes, those that act through intermediate ("second order") steps (such as evapotranspiration contributing to the atmospheric pool of water vapour and then returning as precipitation) and those that operate through tertiary or higher order levels of controls. This expert analysis results in a parsimonious set of processes that define the coupled model without introducing undue complexity. The outcome of this expert analysis ranks the importance of various processes in respect to feedbacks of energy, hydrological and biogeochemical cycles, and places emphasis on which interlinkages between different pools of matter and energy inputs are important.

For a further systematisation, the processes are divided into the four categories:

- Hydrological processes;
- Atmospheric processes;
- Land-surface processes;
- Biogeochemical transport.

(1) Hydrological processes	**Label**
- runoff generation	H 1
- drainage/routing by streams and rivers	H 2
- storage in lakes and river impoundments	H 3
- water storage in soil and land cover	H 4
- infiltration and physical processes in soil	H 5
- lateral transfer of soil moisture	H 6
- flow in root zone/unsaturated zone	H 7
- groundwater flow	H 8
- snow and ice storage	H 9
- soil/ground freezing	H10

(2) Atmospheric processes	**Label**
- precipitation	A 1
- evaporation and energy fluxes	A 2
- ABL transport processes	A 3
- free troposphere dynamics	A 4

(3) Land-surface processes	**Label**
- diurnal changes	L1
- seasonal dynamics (aggregation)	L2
- inter-annual changes	L3
- composition and succession of vegetation	L4

(4) Biogeochemical transport	**Label**
- sediments and suspended matter	T 1
- nutrients (N, P, K)	T 2
- salinisation	T 3
- carbon fluxes (CO_2 and methane)	T 4

How to read the Interaction Matrix:

A line influences a column, (i.e. a column is influenced by a line). For example, the "x" in the matrix element (line A1, column H4) means that the precipitation influences directly the storage in the soil, the "-" in the matrix element (line H8, column T2) means that the ground water influences the nutrient cycle by a secondary link. Direct two-way couplings can be identified by comparison of entries in corresponding matrix fields on both sides of the diagonal. Additionally these are indicated by a capital "X".

3.1.3 Discussion of the Interaction Matrix

The intention of the matrix is twofold. One is to provide an appropriate tool to present an overview and to structure ideas and thoughts. The other is to provide some insight by utilising it in the form of an adjacency matrix, in order to identify cycles and feedback more easily (for further details see also Sect. 2.1 and 3.2). It is important to realise, that the composition of the matrix presented here is the result of a discussion among scientists with different expertise within hydrology and related scientific branches. It is impossible to comment on all entries within the matrix. Some of the entries are clear, some others are less clear. This discussion was focused on influences between processes in general; it was not directed at a particular site or a special scale. Thus site-specific conceptual models may lead to a somewhat different matrix.

The matrix gathers the main processes which form the hydrological cycle and which contribute from related cycles. Elements in the matrix indicate links. An entry for one process in line i and another process in column j indicates a link from i to j, i.e. process i has an influence on process j.

Matrix of interactions and feedbacks in the hydrological cycle with atmospheric, ecological and biogeochemical transport processes

		Hydrological Processes H										Atmospheric Processes A				Land-surface Processes L				Biogeochem. Transport T				Sum		
		1	2	3	4	5	6	7	8	9	10	1	2	3	4	1	2	3	4	1	2	3	4	Pr	Se	To
Runoff generation	H1		x		x		X									-				x	x			5	1	6
Drainage/routing by streams and rivers	H2			X	X		x													x	x			5	0	5
Storage in lakes and river impoundments	H3		X		x			X	X			-	X	X						x	x			8	1	9
Water storage in soil and land cover	H4		X			X		X				-	X	-		X	-	-		-	-	-	-	5	6	11
Infiltration and physical processes in soil	H5	x			X		x	x					-			-				-	-	-		4	5	9
Lateral transfer of soil moisture	H6	X						X	X								x	x	-					5	1	6
Flow in root zone / unsaturated zone	H7	X		X	X	x	X		X			-	X	-		X	-	-			x	x	-	10	5	15
Groundwater flow	H8	x		X				X								-					-			3	2	5
Snow and ice storage	H9	x	x	x	x	x		x	x				X	-		x	x	x	-					11	2	13
Soil/ground freezing	H10	x	-		x	x		x	x				X											6	0	6
Precipitation	A1	x	-	x	x	x	-	-	-	x	x		-		X	-	-	-			-	-		5	10	15
Evaporation and energy fluxes	A2			X	X			X		X	X	x		X		X	X							8	0	8
ABL transport processes	A3			X	-			-		x	x	-	X		X	x	x				-			5	4	9
Free troposphere dynamics	A4				-							X	x	X		x	x							5	1	6
Diurnal changes	L1			x	X			X				-	X	-	-		x	x					x	7	2	9
Seasonal dynamics (aggregation)	L2	-	-		-			-		-	x	-	X	-	-			x			x	x	-	4	8	12
Inter-annual changes	L3	-	-		-			-		-	-	-	x	-	-				X		-	-	-	2	10	12
Composition and succession of vegetation	L4															x	x	X			x			4	0	4
Sediments and suspended matter	T1					x															x			2	0	2
Nutrients (N, P, K)	T2																-	-						0	2	2
Salinisation	T3					x											-	-		x				2	2	4
Carbon fluxes (CO_2 and methane)	T4												-		-	-	-	-			-			0	5	5
Number of links:	Primary links (Pr)	6	4	8	9	6	4	8	4	3	4	1	8	2	2	6	6	5	1	4	7	2	1			
	Secondary links (Se)	2	3	0	3	0	1	4	1	2	1	7	3	6	4	5	6	6	2	2	7	4	4			
	Total number (To)	8	7	8	12	6	5	12	5	5	5	8	11	8	6	11	12	11	3	6	14	6	5			

x → primary link, - → secondary link, X → direct two-way coupling

There are three types of matrix entries. “x” denotes a usual one-way link. Two-way links, which are entries in corresponding elements on both sides of the main diagonal of the matrix, are additionally highlighted by using the capital “X” symbol. A “-” entry indicates a secondary link. Note that a secondary link is understood here to be a link of minor importance in relation to major links.

The number of primary links, secondary links and total number of links are shown in the last lines and columns of the table. Large numbers at the end of lines indicate highly influential processes, i.e. those which have an influence on many others. Large numbers at the end of columns indicate manifold influenced list members.

The interpretation of this matrix results in some principle statements concerning the ordering of influences between the disciplines considered:

- Hydrology has an important impact on the atmosphere (mainly on evaporation) and the atmosphere has an even stronger effect on hydrology (mainly by evaporation and precipitation). Of greatest influence are: Precipitation and fluxes in the unsaturated zone, followed by storage processes, either as snow and ice or in soil and land cover. It turns out that processes at and near the ground surface are of greatest influence for the entire hydrological cycle.

- Hydrology has also an important impact on land-surface processes and land cover, the impact of land cover on hydrology is also strong, but mostly these are secondary links (only the diurnal fluxes effect directly some hydrological processes).

- The impact of hydrology on the biogeochemical fluxes is very important. On the other hand, there is almost no (direct or secondary) influence of biogeochemical fluxes on hydrology (only sedimentation and salinisation have direct impact on the soil infiltration properties).

- There are strong linkages and feedbacks between atmospheric processes and land cover in each direction.

- A number of atmospheric processes influence the biogeochemical-transport processes by secondary links (mainly via the hydrological fluxes). On the other hand there are no relevant primary and secondary feedbacks (influence of biogeochemical processes on the atmospheric processes).

- Land-surface processes and land cover has a considerable impact on the biogeochemical transport processes, and there are also some feedbacks.

Summarising, one can identify the processes with the most linkages to others (both inter- and intra-disciplinary):

Hydrological processes:

- unsaturated zone / root zone
- snow and ice processes
- water storage in soil and vegetation

Atmospheric processes:

- precipitation,
- energy balance / evaporation

Land-surface processes:

- water fluxes in the vegetation (partly also CO_2-fluxes), separately for all mentioned time scales (diurnal, seasonal, inter-annual)

Biogeochemical cycles:

- nutrient fluxes

This leads to the hypothesis that a coupled model has to be composed of at least the linkages of these processes to cover most of the internal dynamics of the system. Note, that we speak about the linkages between the processes, not of the processes themselves. This leaves room for a process description in the appropriate detail in case the specific process is important for the problem under investigation.

A further challenge would be to introduce some kind of weights to characterise intensities of various interactions, based on results of field experiments. For example: the impact of precipitation (A1) on runoff generation (H1) is much higher than on soil/ground freezing (H10) or on the free troposphere dynamics. A "weighted" sum of influences could better explain the role of interactions among various investigated processes.

The discussion sketched above is a rather general one. In the case of an explicitly defined interacting system to be modelled, some filters can be applied on this general methodology, such as

- Size of area (catchment) (e.g. rainfall feedback is irrelevant on plot scale);
- Time horizon (e.g. vegetation succession is not relevant in a time horizon of one year);

- Hydro-climatological conditions (e.g. snow and ice is irrelevant in hot climates).

If a site-specific conceptual model leads to a rather different matrix, then this matrix cannot be understood as a generic framework to which filters can be applied but only as a guide or as an example of a general methodology. From this discussion, and also the examples given in Chapter 4, one understands that hydrological models have many different applications and structures. They have to be tailored to the specific application accordingly. Consequently, it is probably irrational to develop a generalised model from which all applications can be deduced.

3.2 Coupling Aspects of Heat and Mass Transfer

Ekkehard Holzbecher, Oleg F. Vasiliev

While in the preceding chapter a general overview on the linkage between processes connected to the hydrological cycle was given, the focus here is on heat and mass transfer. Although the subject is more specific, it is still impossible to try a complete description in a few pages. The examples given provide an idea of the variety of approaches and concepts. The structure of the section makes a reference back to Sect. 2.1., using the concept introduced there of compartments, of inter-compartmental and intra-compartmental coupling.

What is understood as a compartment is noted in the section mentioned above. Most examples given here concern the ground surface as interface between atmosphere and pedosphere or the bottom of surface water bodies as interface between free water and sediments, i.e. saturated porous media. Depending on the very specific conceptual model, the porous medium may be part of the bottom sediment compartment or of the groundwater compartment.

Another distinction is made between flow and transport. "Flow" concerns the medium itself, which in rivers, lakes and aquifers is water and in the atmosphere is air. "Transport" refers to components, which are contained in the medium in some way, such as chemical components, suspended matter or biomass. Transport is always induced by flow in a compartment, and therefore heat and mass transfer have to be studied in combination with flow, when there is flow.

3.2.1 Intra-compartmental

Flow

Flow within a compartment is usually described by differential equations. The differential equations are obtained starting from basic principles and empirical laws. Using numerical methods, models compute a certain set of variables from the set of differential equations under consideration of additional boundary and initial conditions.

The differential equations for flow depend strongly on the type of compartment. Models concerning surface water or the atmosphere are based on formulations derived from the Navier-Stokes equations. For incompressible fluid flow in the gravity field these read:

$$\nabla \cdot \mathbf{v} = 0 \tag{3.2-1}$$

$$\frac{\partial}{\partial t}\mathbf{v} + \mathbf{v} \cdot \nabla \mathbf{v} = -\frac{1}{\rho}\nabla p - \mathbf{g} + \frac{\mu}{\rho}\nabla^2 \mathbf{v} \tag{3.2-2}$$

where $\mathbf{v}$ denotes velocity, p pressure, ρ density, $\mathbf{g}$ the vector of acceleration of gravity and μ dynamic viscosity. The equations are valid in 1D, 2D and 3D. The unknown variables are p and the components of $\mathbf{v}$, i.e. 2 unknowns in 1D, 3 unknowns in 2D and 4 unknowns in 3D. Obviously the system is coupled: none of the variables can be calculated without taking the other variables into account. C oupling of variables within a compartment can be analysed by an examination of the describing differential equations[1].

In the subsurface porous media flow is based on empirical Darcy's Law:

$$\mathbf{v} = -\frac{\mathbf{k}}{\mu}(\nabla p - \rho \mathbf{g}) \tag{3.2-3}$$

or a generalised form, which is also valid in 1D, 2D and 3D. $\mathbf{k}$ denotes the tensor of permeabilities. In this case the set of equations can be de-coupled. For example, the velocity components in the steady-state continuity equation $\nabla \cdot \mathbf{v} = 0$, can be replaced with help of Darcy's Law and a differential equation for p results, which is independent of the velocity:

$$\nabla \frac{\mathbf{k}}{\mu}\nabla p = -\nabla \frac{\mathbf{k}\rho}{\mu}\mathbf{g} \tag{3.2-4}$$

After the differential equation is solved for p, the velocity components can be computed explicitly using Darcy's Law (Holzbecher 1998).

Within both groups of fluid flow, various different formulations can be found and different dependent variables are used. Formulations in terms of pressure can, under certain conditions, be replaced by formulations using hydraulic head or potential. Velocity terms can be used differently. In hydraulics, variables like stream function and vorticity are in use. Pioneering work was done by Prandtl (1952, 1965). Peyret and Taylor (1985) provide a mathematical introduction and an overview.

[1] The reader should not be confused by the fact that the subject in these paragraphs are "coupled variables", whereas in preceding chapters the term "coupled" was used for compartments and/or processes. Of course it is not the same. A clear distinction, in which situation-coupled processes lead to coupled variables, is not intended here. A system description, including coupled processes, leads to a more complex mathematical description, which in most cases will contain coupled variables.

Hydraulic models always take turbulence into account, although quite often in a relatively simple manner using Reynolds stresses (Peyret and Tayler 1985; Rajar and Cetina 1995). More complex approaches often utilise the so-called k-ε-method, where the Navier-Stokes equations for mean values of hydraulic variables (pressure, velocity) are closed by the introduction of k- and ε-variables (Rodi 1980).

In hydrological balances, turbulence plays no role. In balance models storage is related to all types of inflow and outflow (continuity) and turbulence in the interior can be neglected. For example, lakes have been modelled as a combination of connected parts, such as bays. In the sense of Sect. 2.1. these parts can be understood as compartments. There are also models in which lakes and rivers are treated as a network of compartments, in which each lake is represented by one compartment. In deep lakes limnologists distinguish the part above the density interface, called pycnocline[2], (epilimnion) and below the pycnocline (hypolimnion) (Imboden and Lerman 1978). More detailed is the division of lakes into a set of horizontal layers. A state-of-the-art report on this type of approach in hydrology is given by Orlob (1983). A more recent overview is given by Jørgensen (1995).

In river modelling, the 1D Saint-Venant equations are a common tool (Cunge 1989):

$$\frac{\partial}{\partial t}A + \frac{\partial}{\partial x}Q = q \tag{3.2-5}$$

$$\frac{\partial}{\partial t}Q + \frac{\partial}{\partial x}\left(\frac{\beta Q^2}{A}\right) + gA\frac{\partial}{\partial x}h - gA(S_0 - S_f) \tag{3.2-6}$$

with variables A (wetted area) and Q (discharge) and parameters: lateral inflow q, momentum coefficient β, water depth h, acceleration due to gravity g, bed slope S_0 and friction slope S_f. The latter depends on Q. Altogether it is a coupled system of two first order partial differential equations. They are derived from the mass and momentum conservation laws. In the continuity equation turbulence is not explicitly taken into account, as terms vanish by integrating over the river cross-section. Nevertheless there are phenomena in streams which cannot be understood without turbulence.

In groundwater, as mentioned above, turbulence is irrelevant. Flow is described by Darcy's Law, which in its most simple form states that flux is proportional to the gradient of hydraulic head. The factor of proportionality,

---

2 Usually the density jump is caused by the temperature change, so that the interface may be called thermocline. In general, salinity change can also be the reason for the stratification of a lake.

hydraulic conductivity, depends on the porous medium. A complete 3D description is obtained by combining Darcy's Law with the fluid mass balance equation (see Box 2.3-2).

Darcy's Law can be extended for flow in the unsaturated zone. The 1D Richard equation takes into account that saturation changes but requires a characteristic retention curve which describes the relation between saturation and suction. Bear and Verruijt (1987) introduce major basic concepts of subsurface water flow and transport.

Mass and Heat Transport

Component transport within a compartment in many cases can be described by the same type of differential equation independent of the medium. The equation is derived using the mass balance principle for component mass. Using the gradient operator (∇) the transport equation can be written as:

$$S\frac{\partial c}{\partial t} = \nabla \mathbf{D} \nabla c - \mathbf{v} \cdot \nabla c - v_{sett}\frac{\partial c}{\partial z} + q \tag{3.2-7}$$

where the vector field $\mathbf{v}$ represents the flow field and v_{sett} denotes the settling velocity. This is the so-called transport equation, also called advection-diffusion equation, which needs to be solved for the component concentration c. With different parameter values it is used for transport in the atmosphere, in surface water and in subsurface air and water. The parameters appearing in the Eq. 3.2-7 for different compartments may differ by orders of magnitude. They may even be derived from different processes.

The term on the left side represents the storage or removal from the pool. The coefficient S here is a general storage parameter which is a measure for the temporal response of storage processes. Depending on the flow type within a compartment, different phenomena are included in S. In porous media transport the porosity has to be considered ($S=\varphi$ with porosity φ). If there is an interaction between the fluid phase and an adjacent fluid or solid phase, the simplifying concept of retardation can often be used (Kinzelbach 1986). In cases of fluid-solid interaction the retardation factor R, defined by the sorption characteristic, has to be taken into account ($S=\varphi R$).

On the right side of the equation different processes are represented, which may all be active simultaneously. The first term describes diffusion and dispersion processes. These are driven by concentration gradients. The characteristic properties are gathered in the dispersion tensor $\mathbf{D}$. In the simplest case the dispersion tensor can be replaced by a scalar diffusivity parameter. But in many cases the situation is more complex as there is a dependence on velocity. In 1D the dispersivity is proportional to velocity, but in the more dimensional situations the tensor has to be used in a form which accounts for different dispersion characteristics longitudinal and transversal to the flow direction. For porous media the general form of the dispersion tensor was derived by Scheidegger (1961).

The second term on the right side of the equation represents advection. Advection is the genuine transport process, i.e. migration with the velocity field **v**. In the atmosphere **v** represents the wind velocity. While in pure fluids discharge per area and velocity coincide, both are different in multi-phase situations. If advection is the dominant transport process, as is often the situation in natural systems, models may run into numerical problems. Then the numerical algorithm delivers oscillations (in some parts concentrations above the possible maximum) or results show an additional dispersion (numerical dispersion) (Ehlig 1977; van Genuchten 1977).

Sediments, suspended matter and solid particles may have their own settling velocity v_{sett}, which introduces another velocity term in the vertical direction in the differential equation. The settling finally results from gravity and is directly responsible, for example, for sediment settling in surface water bodies and for precipitation in the atmosphere. It is indirectly relevant for the transport of components which are attached to settling particles, for example for transport by wet and dry deposition in the atmosphere towards the ground surface.

The last term q of the transport equation represents all other processes which act generally as sources or as sinks, i.e. processes, which produce or remove mass of the component in question. There are a lot of different processes, which can be described by this term and only few examples can be given. Biological, chemical, biochemical degradation as well as radioactive decay are often described as first-order losses: $q = -\lambda c$, with degradation (decay) parameter λ. For radionuclides, λ is a constant, but for biochemical degradation it is a variable itself, depending on the (bio-, geo-) chemical environment. Then the situation comes close to that of general biogeochemical reactions, for which the term q can also be utilised.

It often turns out that the behaviour of a component cannot be separated from the (bio-) chemical environment, i.e. the concentration and availability of species determine the fate of the entire system. In that case the production or loss of component mass due to several reactions has to be included in q. The computational requirements are extremely high, if the transport and interaction of a great number of interacting chemical components are to be modelled in a multi-dimensional flow field, as is done in atmospheric models.

When several species, linked with each other by reactions, are modelled, the transport equation has to be solved for each of these species. Two types of reactions have to be distinguished: equilibrium or kinetics. If for an application the time scale of interest exceeds the typical reaction time, the reaction is fast and it can be assumed that the equilibrium is reached – if the reaction is reversible. Equilibria are usually described by the *Law of Mass Action*. For the reaction $A + B \leftrightarrow C$ between species A, B and C the equilibrium is given by:

$$\frac{a_A a_B}{a_C} = K \tag{3.2-8}$$

where a denotes the activity of the species, and K is the reaction-specific equilibrium constant. The activity of a species is the product of the concentration and an activity factor γ, which depends on the charge of the species and the ionic strength of the solution (Parkhurst 1995).

In general, for a set of species the coupled problem for reactive transport is given by:

$$S\frac{\partial}{\partial t}\mathbf{c} = \nabla\mathbf{D}\nabla\mathbf{c} - \mathbf{v}\cdot\nabla\mathbf{c} - v_{sett}\frac{\partial}{\partial z}\mathbf{c} + \mathbf{S}_{eq}{}^{T}\mathbf{r}_{eq} + \mathbf{S}_{kin}{}^{T}\mathbf{r}_{kin} \tag{3.2-9}$$

where the vector $\mathbf{c}$ contains all species concentrations. The vectors $\mathbf{r}_{eq}$ and $\mathbf{r}_{kin}$ denote the reaction rates of equilibrium and kinetic reactions, and the matrices $\mathbf{S}_{eq}$ and $\mathbf{S}_{kin}$ relate reactions (in rows) and species (in columns) for equilibrium and kinetic reactions.

The problem with the reactive transport equation is that the rates of the equilibrium reactions are not known beforehand. Thus the entire set of equations is manipulated by linear transformations in order to make the term, corresponding with equilibrium reactions, vanish. Such a transformation is always possible, but not unique. It can be described by the multiplication with another matrix $\mathbf{U}$ from the left (Saaltink et al. 1998), which is equivalent to the transition from species concentrations to total concentrations, also called components. The system for the total concentrations $\mathbf{u} = \mathbf{U} \cdot \mathbf{c}$ is given by

$$S\frac{\partial}{\partial t}\mathbf{u} = \nabla\mathbf{D}\nabla\mathbf{u} - \mathbf{v}\cdot\nabla\mathbf{u} - v_{sett}\frac{\partial}{\partial z}\mathbf{u} + \mathbf{U}\mathbf{S}_{kin}{}^{T}\mathbf{r}_{kin} \tag{3.2-10}$$

In addition, reaction equilibria have to be taken into account, which can also be noted in matrix form:

$$\mathbf{S}_{eq}\cdot\mathbf{log(a)} = \mathbf{log(K)} \tag{3.2-11}$$

where the vector **a** denotes all activities, and the vector **K** all equilibrium constants. Altogether, a mathematical system results in which transport differential equations are combined with a set of algebraic equations, so-called algebraic differential equations. This example of reactive transport illustrates some problems which emerge when concentration state variables within a compartment are coupled. The non-linearity of such a system is mostly severe, so that advanced numerical methods are necessary for the solution of the algebraic differential equations. Further complications appear when additional processes are to be taken into account, such as adsorption and desorption or precipitation and dissolution, or for situations in which the described simplification leading to total concentrations is not possible.

Descriptions of heat transport can also be put in the formulation of the advection-diffusion equation given above. The differential equation is a formulation of the energy conservation principle and the unknown variable to be determined in numerical models is the temperature T (instead of c). Various formulations are given in the literature (Nield and Bejan 1992; Holzbecher 1998), which include "material" (media) properties such as specific heat capacity and thermal conductivity. If the equation is formulated with S=1, on the right side of the equation the thermal diffusivity appears at the right-hand side in place of **D**.

In the heat transport equation, the term q represents energy sinks or sources. Models of heat budget in the atmosphere include the following processes in that term: absorbed solar radiation, latent heat connected with precipitation in liquid or snow forms and sensible heat flux (Petoukhov et al. 2000). A more detailed diagram of influences in climate systems, particularly in the atmosphere, is given by Robock (1985).

Concerning reactive transport, the parameters describing the biogeochemistry often show a strong dependence on temperature, as for example the equilibrium "constants" introduced above. When temperature is also a state variable, this dependency has to be taken into account. Thus there is a link from the temperature to concentrations. In aqueous systems the opposite link can mostly be neglected, although all reactions are basically accompanied by an energy transfer. Special applications can be envisaged, in which there is an effect of concentrations on temperature, yielding a coupled system of concentration and temperature.

3.2.2 Inter-compartmental

Flow

Water fluxes between compartments usually depend on a set of variables in both compartments. For example, the flux through the sediment layer depends on water levels in the surface water and in the surrounding groundwater. The water level of the river or lake has to be related to hydraulic head in the connected aquifer. In the example there is outflow of surface water when the water level in the river is higher than in the aquifer. When the groundwater table is above the surface water level, the river obtains inflow of groundwater. When the water levels h_A and h_B in both compartments are given, the flux from one compartment to the other can be calculated using a transfer coefficient α_{AB}:

$$Q_{AB} = \alpha_{AB}(h_B - h_A) \tag{3.2-12}$$

In the example of exchange between the groundwater compartment and a surface water body, the transfer coefficient is a modified form expressing the ratio between hydraulic conductivity and thickness of the sediment layer. Note that the formula connecting the values h_B and h_A generally defines a feedback between the compartments A and B. The boundary condition depends on both h_B and h_A. The connection between compartments can be more subtle, as the following example shows.

A groundwater model is set up for an aquifer (A), which is hydraulically connected to a lake. The water level in the lake is usually used as a Dirichlet-type boundary condition for the computation of groundwater heads with a numerical model. The model also computes velocities, i.e. water fluxes across the lake bottom. Thus, based on water levels in one compartment, fluxes are obtained according to the formula given above, where the coefficient α_{AB} represents the hydraulic gradient divided by length. Vasiliev (1987) uses another more general formulation which is relevant for the connection between an unconfined aquifer and a river or lake, in which the coefficient itself is head-dependent.

If the water level of the lake is in fact controlled by surface outflow, the described procedure is sufficient, as there is an influence from groundwater to surface water but not vice versa. There is no feedback between the compartments. But if the lake's water level is changing due to fluctuating groundwater inflow, the system has feedback. The fluxes Q_{AB}, computed on the basis of a groundwater model, may then be used to make up a water balance for the lake – assuming that values for precipitation and evaporation are available. The lake water balance in that case provides a new value for the lake water level, which in the next iterative step has to be considered as boundary condition in the groundwater model. In this situation there is feedback.

Generally, a flux equation, as given above, describes the connection of flux on one side and head or pressure gradients on the other side. It is probably impossible to give a general formulation for all types of expressions that can be found in models. The Earth surface boundary of atmospheric models is especially rich in different approaches, concerning evaporation, evapotranspiration, surface runoff and groundwater recharge. The discussion of the example presented can only provide an idea of the complexity that can be found in the various disciplines involved.

Mass and Heat Transport

In the most simple formulation the interface component mass flux is determined by the interface velocity v_{AB} and the concentration c_A in the compartment *A* with outflow:

$$q_{AB} = -v_{AB} c_A \tag{3.2-13}$$

The formula, under certain conditions, quantifies advective transport and, if other processes can be neglected, constitutes a one-way flux from compartment *A* to compartment *B*. The component mass per area leaving a well-mixed surface water body is often calculated that way in hydrology.

Similarly, dry deposition from the atmosphere (compartment *A*) to ground or water surface (compartment *B*) can be treated according to the given formula, if c_A is the particle concentration in the atmosphere and v_{AB} is the velocity representing dry deposition per area. Wet deposition from the atmosphere can be calculated using the precipitation rate per unit square as characteristic velocity and the chemical concentration in the droplets c_A. Another example: the formula describes losses of particulate matter in surface waters to the sediment layers, where v_{AB} is the sedimentation rate per area and c_A the particle concentration in the water. To quantify the losses of a chemical substance attached to particles, the formula needs to be adjusted: the given product is multiplied by the chemical concentration in the solid (particulate) phase.

The simplest approach leading to a two-way exchange, and thus to a feedback, as set out in Sect. 2.1 (a two-way exchange between two compartments *A* and *B*), is described by the formula

$$q_{AB} = k_{AB}(kc_B - c_A) \tag{3.2-14}$$

where k_{AB} is a parameter quantifying the temporal dynamics of the exchange process between the compartments A and B. k denotes the equilibrium state, i.e. there is no exchange if the concentrations fulfill the equation $c_A / c_B = k$. When the concentration in compartment A is high, the exchange term becomes negative, describing a loss from A to B, and B gains from A. If, on the other hand, the concentration c_B is high, the exchange is directed in the opposite direction. The exchange increases and decreases with the deviation from the equilibrium. An example for this type of exchange can be found at phase interfaces. For most components which are contained in liquid and gaseous phase a chemical equilibrium determines the concentrations of components in the compartments. At the surface of water bodies for certain chemical components there is a permanent net exchange to compensate concentration changes due to other processes in the hydrosphere and in the atmosphere. Within the soil, gaseous and liquid phases can be regarded as two compartments, for which exchange (for example of oxygen) is governed by such a relationship.

Concerning heat flux only, one example should be mentioned. Energy exchange of the soil with the atmosphere typically consists of the resultant net thermal and solar radiation fluxes, turbulent heat and latent heat fluxes. The resulting heat exchange between soil and atmosphere is used in atmosphere GCMs. In the soil model such a flux formula is usually used as a boundary condition for temperature (Savijärvi 1992).

3.2.3 Analytical Solutions

Generally, the analytical description of hydrological systems leads to systems of partial differential equations for the different compartments, which are coupled by boundary conditions at the interfaces. Usually such multi-compartment systems do not possess an analytical solution and thus need to be solved by a numerical model on a computer. In the following a special situation is described, for which an analytical solution exists.

In the situation of no feedback, the system of compartments can be solved sequentially in an appropriate order. If, for example, a part of a stream is partitioned into a series of compartments, the simulation of water flux from one compartment to the other should follow the downstream course of the water. If there is no upstream effect in the entire system, the solution is found in that manner.

Multi-compartment systems with feedback can be solved analytically in very special cases only. One such situation is given if compartments are connected by linear flux terms. There is an analytical solution, if the entire system can be described by a linear system of ordinary differential equations:

$$\frac{\partial}{\partial t}\mathbf{c} = \mathbf{A} \cdot \mathbf{c} + \mathbf{b} \tag{3.2-15}$$

where vector **c** represents the unknown concentrations in the compartments. With the matrix **A,** fluxes between the compartments are described and vector **b** gathers all additional flux terms. Such systems may appear in the modelling of the component dynamics, if it is assumed that the interior of all compartments is simple. Chemists use the term "stirred tank reactor" (Aris 1961) – not only for such a simple design. In fact natural systems are seldom that simple that the linear approach is justified. According to that approach, neglecting all transport processes like advection, diffusion and dispersion, no gradients may occur within compartments; but there are gradients in real systems in the field. Usually, the state of compartments can be understood only by complex models.

Solutions of linear systems of the type noted can be found in textbooks on ordinary differential equations. They are given in terms of eigenvalues λ_i and eigenvectors $\mathbf{u}_i$ of the matrix **A**. If all eigenvectors are of first order and the system is homogeneous (**b**=0), the solution can be written as:

$$\mathbf{c}(t) = \sum_{i=1}^{N} c_i \exp(\lambda_i t)\mathbf{u}_i \tag{3.2-16}$$

with appropriate parameters c_i. In the case of multiple eigenvectors the formulation is slightly more complex, as coefficients of polynomial form have to be used (Walter and Contreras 1999). Analytical solutions can be given for non-homogeneous systems (**b**≠0), too. There are solutions even if functional relationships **b**(t) have to be taken into account. Some more detail for a similar set of equations is given in the Box 2.3-3.

It is seldom that the interaction of various compartments can be described by a linear system. Processes at compartment boundaries are often complex. Particular processes may have to be considered at almost every interface: evaporation at the free surface of water bodies, capillary forces at the groundwater table. The non-linearity may be introduced by an explicit formula, but can also be the result of a numerical simulation or an analytical solution for one of the compartments.

As an example for a non-linear response, Fig. 3.2-1 shows the non-linear dependence of one-dimensional heat (or mass) transfer through a system governed by advection and diffusion. It is derived under the assumption of a given temperature T_0 (or concentration) at one boundary and a prescribed gradient ∇T on the other boundary. The influence of the boundary conditions can be represented by a dimensionless number $T_0/(\nabla T \cdot L)$, where L is the length of the system (Holzbecher 2000). The relation between advection and diffusion is also represented as a dimensionless quantity: the Péclet-number. The response of the system, i.e. the heat (mass) flux j through the system, is related finally to the flux j_{diff} through the system with only diffusion (and the same boundary conditions). In that way heat or mass transfer is non-dimensionalised; in convection studies these quantities are termed Nusselt-number *Nu* (for the thermal case) and Sherwood-number *Sh* (for the solute case) (Holzbecher 1998).

Fig. 3.2-1 depicts the relation of advective flux in comparison to diffusive flux and can be used to describe heat (mass) transfer through an interface dominated by one-dimensional advection and diffusion. For example, a bottom sediment layer between the surface water and the groundwater or bedrock compartment could be characterised by such non-linear functional dependencies.

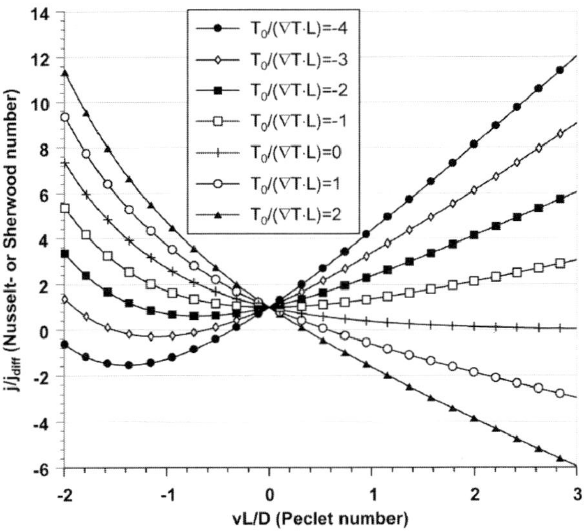

Fig. 3.2-1: Total mass transfer through an advection-diffusion system with Dirichlet condition on one and Neumann condition on the other boundary; Nusselt-number (for thermal transport) or Sherwood-number (for solute transport) in dependence of Péclet-number *Pe* and the dimensionless parameter $T_0/(\nabla T \cdot L)$ (T_0 = value of transport variable at one boundary, ∇T= gradient at the other boundary, L = length of the system) (Holzbecher 2000)

3.2.4 Numerical Aspects

Due to the non-linearities of different origin, the coupling of compartments representing natural systems can be solved analytically only in degenerate cases. For that reason there is seldom an alternative to the application of computer models.

The coupling aspect comes into play again in the design of the modelling software or the coupling of software for different compartments. Three approaches can be distinguished concerning the structure of the algorithms. One is the *direct substitution approach* (DSA), in which the entire system is treated simultaneously (Yeh and Cheng 1996). All unknown variables are gathered in one big, usually non-linear, system of equations. For that reason the approach is also referred to as one-step method (Steefel and MacQuarrie 1996).

In the two other approaches the entire system is split into different parts, which are treated separately one after the other. This can be done in an iterative manner (SIA – sequential iterative approach) or the steps can be performed only once (SNIA – sequential non-iterative approach). In reactive transport modelling, the two modules geochemistry and transport are called in separate steps; for that reason both procedures can be termed two-step methods (Steefel and MacQuarrie 1996).

The three approaches are visualised in a simple manner in the following Fig. 3.2-2. Only two compartments are considered for simplicity, *A* and *B*. In the sequential design the calculation of the state variables (for example water levels or concentrations) at a certain time level starts with one of the compartments (here *A*). The results of the model for this compartment are used in the next step for the calculation of the state variables in the other compartment (here *B*). Then the software proceeds with the next time level.

The disadvantage of the sequential approach is that it fails if there are direct short-term feedbacks between the compartments, i.e. if there is a strong coupling between the compartments. In that case an iterative design becomes necessary in which the results from compartment *B* are used to update compartment *A* at the same time level. This procedure is repeated (iteration) until a certain accuracy criterion is fulfilled.

Using a two-step method, the coupling can be captured with the iterative approach only. So in principle the iterative design should be used generally. The disadvantage is that the computer requirements concerning execution time are much higher, as a modelling loop through the compartments has to be made within each iteration, while it is made only once in the sequential approach. The sequential approach is a procedure, which partially deactivates the coupling between the compartments in order to enhance the computational performance of the algorithm.

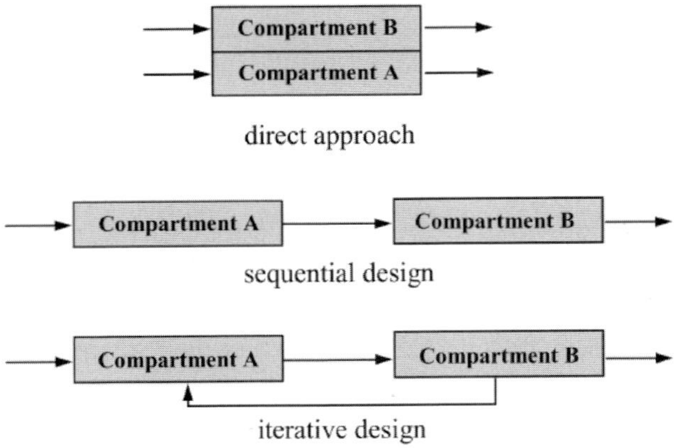

Fig. 3.2-2 Simple scheme for distinction between direct approach, sequential non-iterative and iterative software design (see text)

It is often difficult to judge if the errors made in the sequential approach are severe or not. As there is a dependence on the time-step, a double model run with differing timesteps may indicate if a main part of the coupling is captured in the model.

On the other hand, using an iterative software design, there are several means to reduce the number of iterations performed to a minimum. The accuracy criteria should not be too strict. A common requirement is that the iteration stops if the first three or four digits of all state variables are maintained within one iteration step.

Concerning the software design, it is better to use a model with iterative design than a code in which the sequential approach is implemented. It is possible to work sequentially with the iterative software design if the number of maximum iterations is set to 1. In modern user-friendly software design there should be an input-control-box available which allows altering this option.

The distinction between the three software designs makes sense only for coupled models. If there is no feedback in a system and the compartments are treated in the order given in the corresponding triangular connectivity matrix (see Sect. 2.1), an iterative procedure is unnecessary.

Numerical algorithms have to consider feedbacks in order to be efficient and not to produce artefacts. For general systems this surely is a subject which should be examined further – not only in computer sciences.

Box 3.2-1 ***Eutrophication Modules for a Surface Water Flow Code***

Ekkehard Holzbecher

DUFLOW (1998) is a user-friendly, advanced code for the simulation of flow and transport in a network of surface water bodies, such as rivers, canals and lakes. There are two eutrophication modules included in DUFLOW, which are called Eutrof1 and Eutrof2.

Both modules include the cycling of nitrogen, phosphorus and oxygen. The growth of phytoplankton can also be simulated. State variables, which both modules have in common, are: algal biomass; organic and inorganic phosphorus; organic, ammonia and nitrate nitrogen in the water column; oxygen, biochemical oxygen demand (BOD) in the water column and suspended solids concentration.

The interaction of state variables within the surface water can be represented by a box-and-arrow diagram (DUFLOW 1998). There are clearly cycles within the box-and-arrow diagram, indicating a coupling within the water column model. For example: organic nitrogen is produced during respiration and die-off of algae. Part of the organic nitrogen is associated with particulate matter and will settle to the ground. Due to mineralisation, part of the organic nitrogen will be released as ammonia, which is used for algae growth again. There is a similar cycle concerning phosphorus: organic and inorganic phosphorus are produced during various biomass transformation processes; inorganic phosphorus is also consumed during biomass growth.

The difference between the two modules lies in the connection between the free water column and the sediments. In Eutrof1 the user may specify exchange fluxes for oxygen, ammonia and phosphorus and these are taken as valid for the entire simulation. Eutrof2, in contrast, calculates these fluxes during the run. In order to do so, a module is extended by a model for two sediment layers: the top layer and the lower layer.

Three fluxes are considered at the water-sediment interface: diffusive exchange flux, sedimentation flux and re-suspension flux. Advective flux within the pore space of the sediments is neglected. The sedimentation flux for all chemical species depends on the concentration in the free water column; the re-suspension flux depends on the concentration in the pore water of the sediments; diffusive flux depends on both concentrations. Thus. there is a coupling between the sediment and the water column compartments.

The flux between the lower sediment layer and the top layer depends on the net velocity between sedimentation and re-suspension. If re-suspension exceeds sedimentation, there is an upward flux only, depending on the concentration in the lower layer. If sedimentation exceeds re-suspension, there is a downward flux, depending on the top sediment layer concentration, only. The connection between the two sediment layers is one way only. It may be a problem for the programmer that the direction of the dependency may change with time, because the order of computation has to be exchanged, when the dependency changes.

In summary, the Eutrof1 module can be recommended for studies on the short-term behaviour of eutrophy, for example to examine the impacts of discharge on oxygen dynamics, or to explore the effects of flushing on chlorophyll-a. Longer time scales have to be simulated with the Eutrof2 module. In particular for problems of remobilisation, in which nutrients are stored in sediment due to varying load in the water column, the changing flux at the sediment-water interface has to be considered and cannot be approximated by a constant.

Box 3.2-2 ***The Global Biogeochemical Nitrogen Cycle***

Ekkehard Holzbecher

Lin et al. (2000) present a model for the global biogeochemical cycle in terrestrial ecosystems. Although the terms used in the paper are not always the same as introduced in this volume, the simulation approach can be treated perfectly using the terms given in Sect. 2.1. Compartments are: atmosphere (A), biosphere (B), soil organic phase (O), soil inorganic phase (I). Sub-compartments are: leaf and trunk/root of the biosphere, detritus and humus of the soil organic phase and ammonium and nitrate of the soil inorganic phase.

The following fluxes across interfaces are considered:

- from A to O: nitrogen fixation
- from A to I: dry and wet deposition
- from B to O: litter-fall
- from O to I: mineralisation
- from I to B: plant uptake
- from I to A: ammonia volatilisation, N_xO gaseous losses, denitrification

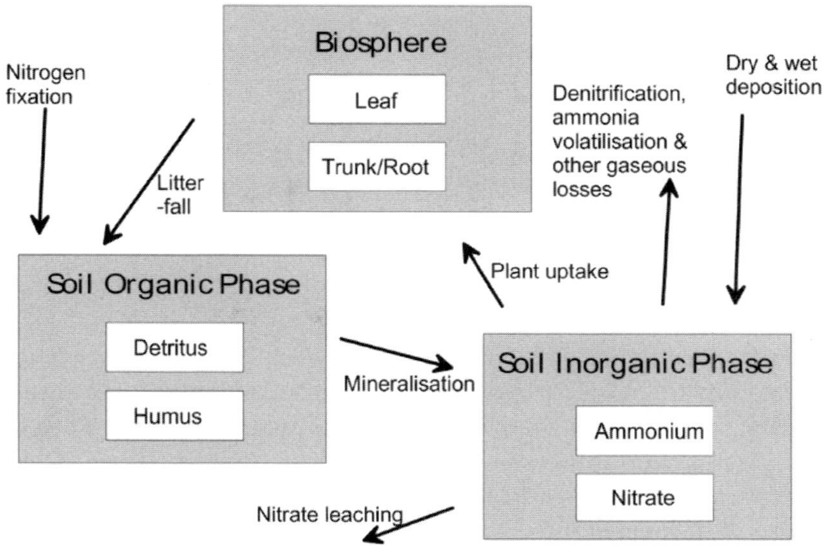

Fig. 3.2-3 Schematic view of the compartmental model for nitrogen

Explicit formulae are given for all fluxes, for example ammonia volatilisation is assumed to be $5.8 \cdot 10^{pH-10}$ as function of pH. Nutrient uptake, nitrification, denitrification and nitrate leaching all depend on soil parameters. The former three also depend on temperature.

Explicitly modelled in the simulation of Lin et al. (2000) are the compartments B, O and I with sub-compartments. The atmosphere needs only to be introduced as the origin or direction of relevant fluxes. For a complete reproduction of the model, another compartment would have to be added representing groundwater and surface water bodies (the hydrosphere) as receiver of mass due to nitrate leaching. This completion does not make a difference to the model set-up, as it is a compartment which is not treated explicitly, like the atmosphere.

From the aspect of coupling, the model is quite simple: the only two-way coupling is between ammonium and nitrate within the inorganic soil phase.

3.3 Feedbacks at the Hydrometeorological Interface

Nicole Mölders

Hydrologists and atmospheric scientists have a common interest in specific parts of the water cycle, namely the "land-surface/atmosphere interface"; this is the upper boundary for hydrological models and the lower boundary in meteorological models. Precipitation, for instance, is a source of water for hydrological models, yet is a sink for meteorological models. In the case of evapotranspiration[1], the opposite is true.

Due to the different aspects of the water cycle that are of major or minor interest to hydrologists or atmospheric scientists, only those processes are considered in detail that are relevant for the specific application while other aspects are neglected or crudely simplified. Such simplifications, however, can cause inconsistencies in the description of the water cycle, even in the case of uncoupled modelling. In hydrological modelling, for instance, neglecting the feedback of increased soil moisture on evapotranspiration, cloud- and precipitation-formation can lead to failures in predictions of flood intensity, because runoff from an individual river catchment may depend partly on precipitation and evapotranspiration originated in the basin (e.g. Liston et al. 1994). Water storage within the river basins, among others, depends on soil type, soil depth, surface heterogeneity and vegetation phenology (Miller et al. 1994). Vegetation phenology itself depends on microclimatological conditions and again affects them. Such interaction between vegetation phenology and the atmosphere is usually not included in the simulations despite models exist that describe plant growth and could provide the phenological parameters required. However, first attempts of linking the reaction of vegetation dynamics to climate changes are encouraging (e.g. Claussen 1997).

In General Circulation Models (GCM), river runoff provides the crucial link for returning water from continents to the ocean (Miller et al. 1994). River runoff is an important input for ocean models because freshwater flow affects the thermohaline circulation of the ocean. Thus, neglecting processes that are usually attributed to another scientific discipline may lead to false predictions in that part of the water cycle that one is interested in.

This example shows that in some cases, but definitely not all, Earth system modelling, climate modelling, water resource research, climate impact studies etc. require the coupling of hydrological and atmospheric

[1] Evapotranspiration stands for the sum of transpiration and evaporation. It should be determined concurrently.

models. The aim of coupling these models is to more appropriately consider the interaction of the atmospheric and land-surface component of the water cycle. This section evaluates recent concepts on such a coupling, points out the difficulties to be solved, and addresses the advantages and disadvantages. Finally, a concept of a hydrometeorological module to specifically represent the interface of hydrological and atmospheric models is introduced.

3.3.1 Introduction

Types of models

In coupling hydrological and atmospheric models, regional climate, mesoscale[2] meteorological or NWP (Numerical Weather Prediction) models should be applied as atmospheric models. The atmospheric model may be nested into a global model to allow large-scale integration. To derive a consistent parameterisation, the hydrological model should be based on the fundamental laws of hydro- and thermodynamics, which in hydrology has been labeled as "physically based hydrological model". Concerning the spatial discretisation, the so-called "distributed hydrological" models are of special interest for the coupling with atmospheric models, in order to account for the spatial heterogeneity of hydrology and atmospheric processes.

Different Relevance of Processes

Despite the common interest of atmospheric scientists and hydrologists in land-surface modelling, both by tradition and purpose, the land-surface parameterisations applied in hydrological and atmospheric models differ strongly (e.g. Mölders and Raabe 1997; Graham and Bergström 2000). Hydrological models require a precise partitioning of precipitation into evapotranspiration, infiltration, interception, retention and runoff to determine the water fluxes within the soil and groundwater recharge, i.e. the water budget is of main interest. Atmospheric models need a precise partitioning of precipitation between the aforementioned processes to determine the partitioning of incoming radiation between soil heat flux and the turbulent fluxes of sensible and latent heat, i.e. they additionally need the energy budget.

---

[2] The term mesoscale is used in the meteorological sense and addresses meteorological phenomena of 2 to 2000 km extension with time scales from several minutes to severals days (see also Sect. 1.2).

Coupling requires that the process description is compatible to both models. Thus, prior to coupling of hydrological and atmospheric models (1) it has to be defined which processes are of relevance for the hydrological or atmospheric model under consideration, and (2) the parameterisations have to be chosen that allow representation of the relevant processes in the coupled system. This means that simplifications, which are suitable for one of the models but not for the other, have to be replaced by the respective more complex description. For example, in atmospheric models, the water budget is often parameterised by using a bucket model (e.g. Manabe 1969) or a force restore method (e.g. Deardorff 1978). These parameterisations, however, are rarely suitable for the demands of hydrological models.

Spatial and Temporal Scales

Other differences between hydrological and atmospheric models are related to the typical scales of processes to be simulated (see Figs. 1.2-3 to 1.2-6). Therefore, the temporal and spatial resolutions of these models differ strongly. Atmospheric models usually apply grid cells of several square kilometres as their horizontal unit of area. Depending on the projection of the grid, they may be of rectangular, square or triangle type. Despite increasing computational power, decreasing the size of these grid cells is limited by the range of validity of the assumptions made in the parameterisations (e.g. the fetch-conditions from which the parameterisations of the surface fluxes are derived require a relationship of 1:100 for the ratio of the vertical to horizontal grid resolution), and the limited availability of initial data. Hydrological models have to represent adequately the river basin variability and its sub-basins which consequently define the model dimensions. Especially in complex terrain, the grid resolution may be very fine. Decreasing the horizontal resolution of hydrological models to match the resolution of atmospheric models makes it difficult to know how to represent the (then subgrid-scale) heterogeneity of land-surface properties, such as topography, geology and vegetation cover. These characteristics, however, play an important role in determining the water storage and runoff.

The solution of the energy balance demands time steps of several minutes in atmospheric models, while a lot of hydrological models are often satisfied with daily time steps (e.g. Graham and Bergström 2000). NWP models usually neglect lateral flows of soil water and groundwater, because these processes are slow as compared to the typical simulation period (1-3 days) of these models. Furthermore, they do not account for river flow dynamics as this does not match their spatial discretisation concept.

One main task in coupling hydrological and atmospheric models is to overcome these temporal and spatial gaps. Procedures for aggregation (upscaling) and disaggregation (downscaling) serve to bridge these spatial gaps (see e.g. Mölders and Raabe 1997). These procedures have been discussed in more detail in Sect. 1.2. For further information see Mölders et al. (1997) or Giorgi and Avissar (1997).

3.3.2 Different Levels of Coupling

The task to be addressed by the coupled model system determines the degree of coupling and its realisation. We may distinguish between parameterisations, one-way coupling, two-way coupling, and integrated (sometimes denoted as integrative) modelling. If the identity of the models vanishes (i.e. there are no two models that can easily run alone) or when the coupling time shrinks to zero (because the response times in the coupled parts become equal) one speaks of integrated models. In integrated models all processes are calculated simultaneously and as such, variables of state and fluxes are updated at the same place. In the following, the advantages and disadvantages of the three ways of coupling are discussed and examples are presented. It is shown that these different types of coupling are all justifiable depending on the intent of the application, as the task to be addressed determines the degree of coupling required. This means that integrated modelling is not necessarily the better choice than one-way coupling, and one-way coupling is not necessarily less advanced or smart than two-way coupling or integrated modelling.

Simple Parameterisations of Large-scale Runoff

Being aware of the difficulties in coupling hydrological and atmospheric models caused by the different spatial and temporal scales, several authors suggested a simple scheme in which all runoff within a river basin reaches the river mouth instantaneously: Miller et al. (1994) introduced a river model that allows the excess land-surface water (over saturated soils) calculated by a GCM to run off into a river if a river is available within a grid cell. The direction and speed of flow is either constant or depends on topographical gradient. Sausen et al. (1994) proposed a one-parameter model that represented each grid cell by a linear reservoir with different retention coefficients for the flows in the east, west, north and south direction. These coefficients depend on orography and grid size. This approach allows for discharge in the horizontal direction (2D-plane) in all directly adjacent grid cells, for which the approach is addressed to as being two-dimensional. As this approach does not distinguish the different types of flow processes, Hagemann and Dümenil (1998) introduced a parameterisation that realised the cascade of overland and river flow by equal linear reservoirs, and considered base flow by a one parameter model. The corresponding retention coefficients depend on topographical gradient between two neighbouring grid cells and on the grid cell size.

This type of approach gives only a very rough estimate of the river runoff component of the hydrological cycle and is much too uncertain to be used for hydrological purposes. Additional difficulties arise due to the coarse resolution of GCMs resulting in the disappearance of some large lakes in the land-surface scheme of the atmospheric model. If the river retention process in a lake (or large wetland system) is not represented by the GCM, the calculated runoff into the oceans may be overestimated significantly. Another difficulty is that the coarseness of the GCM grid cells also may introduce errors in the river network, consequently leading again to errors in the fresh-water supply to the oceans.

Despite the temporal and spatial gaps between the resolutions of hydrological and atmospheric models, several different methods have been developed and tested recently to couple these models directly, as introduced in the following paragraphs.

One-way Coupling

One-way coupling is the simplest way to couple hydrological and atmospheric models. This way of coupling was intended to drive hydrological models by the output of atmospheric models for flood prediction or water management. The models are coupled in a cascading manner, first running the atmospheric model and then running the hydrological model. Usually, the time-steps of the precipitation (computed by the atmospheric model) prescribes the time steps to be applied for the hydrological model.

Due to the cascading nature of one-way coupling, no feedback effects of the hydrological model can influence the simulation of the atmospheric model. Obviously, if simulating the hydrological cycle of large areas, applying a one-way coupling may lead to wrong estimations of the water fluxes from the land surface to the atmosphere which may be propagated in wrong cloud and precipitation distributions (e.g. Mölders 1999). However, this aspect plays a minor role in short-term predictions (several hours).

There are several short-comings of one-way coupling. Different parameterisations might be used in the coupled models. The soil-physical and plant physiological parameters or even topography, soil- and vegetation-type may differ (see Mölders and Raabe 1997) resulting in different soil wetness and infiltration within the one-way coupled model system, i.e. the soil and vegetation processes look different in the respective model. Such inconsistencies may lead to model artifacts that can yield wrong results or conclusions. To avoid discrepancies in the surface characteristics of the coupled modelling system, the parameterisations of the soil and vegetation processes should be similar or, even better, the same. The latter means a complete re-coding of parts in at least one of the coupled models.

If atmospheric and hydrological models are coupled to improve forecasting for flash floods (= flood caused by short, local and high intensive rain storms), a one-way coupling may be disadvantageous, because the errors in the predicted precipitation pattern and intensity will propagate in the hydrological model (e.g. Mölders et al. 1999). A false prediction of the position of the precipitation field of only one atmospheric grid cell may already lead to significant simulation errors in the magnitude and location of the river flood.

Two-way Coupling

As pointed out above, neglecting feedback processes of one part of the water cycle may cause errors in predicting the other part of the water cycle. On the typical time scales of NWP models, for instance, lateral soil moisture fluxes are negligible. Since no three-dimensional distributions of soil moisture exist, soil moisture in NWPs is initialised by the soil moisture distribution of the prediction of the previous day. Doing so, however, makes the assumption of negligible lateral soil water fluxes invalid. Thus, this kind of initialisation in combination with neglecting the lateral soil water movements and surface runoff may yield an underestimation of soil moisture in river valleys and an overestimation of soil moisture in the nearby mountainous regions, which usually receive more precipitation (e.g. Müller et al. 1995). Consequently, the local recycling of previous precipitation may be predicted incorrectly which may lead to wrong forecasts of convective clouds, showers and thunderstorms (for details see Mölders et al. 1999).

In NWP models and GCMs, for instance, a constant groundwater level may cause errors in the simulated water fluxes to the atmosphere (Fig. 3.3-1, left picture), because a fixed groundwater level neglects the water table increase during precipitation periods and the decrease during drought periods. On large-scales, during droughts, the lowered groundwater table may amplify the drought because evapotranspiration is reduced (Fig. 3.3-1, middle picture) and thus reducing soil moisture available for clouds and rainfall. On the contrary, long-lasting rainfall may trigger its persistence

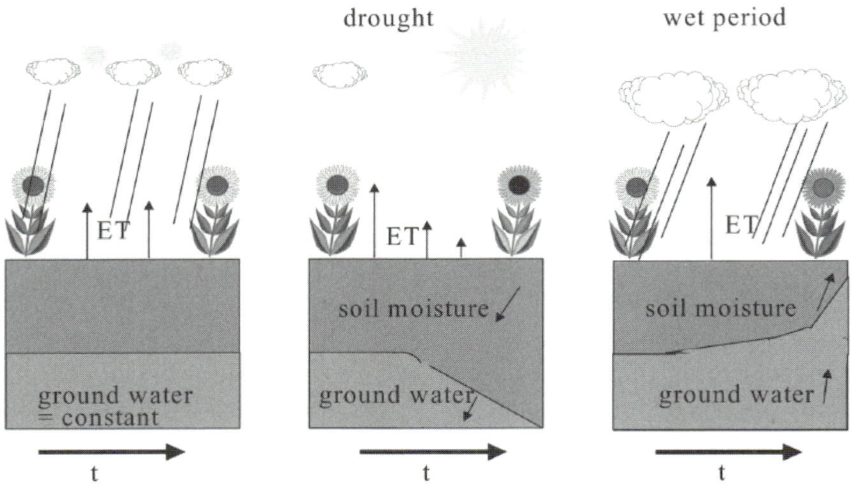

Fig. 3.3-1 Schematic illustration of groundwater level changes on evapotranspiration modelling, ET, with time, t

by recycling previous precipitation through enhanced evapotranspiration possible because of the high water table (Fig. 3.3-1, right picture). On local scales, the soil moisture decreases during drought, but may have little effect on the groundwater table in most cases. The groundwater is usually renewed by transverse movements in aquifers. Replenishing of the groundwater (e.g. from rivers) is not necessarily related to rain in the same area. In the case of wet periods, evaporation may be marginal or even zero on the local scale because of the increase in atmospheric humidity. Local groundwater rise may be caused by transverse movements in aquifers by rivers being connected with such aquifers. In general, on the local scale, floods can cause a temporary rise in groundwater tables downstream of areas that received huge amounts of precipitation.

Of course it is not necessary to include all processes that are, in principle, coupled. For example, in Mediterranean or maritime areas, precipitation mainly results from advection of cloud systems that formed over the oceans. The more continental an area is located, the smaller the total amount of precipitation. The fraction of the total precipitation that results from local recycling of precipitation increases with increasing distance from the oceans on the upwind side. In continental areas, soil moisture and the depth of the groundwater table play a dominant role in precipitation.

To include such prominent feedback processes, the idea of a two-way coupling was developed. The coupling is achieved through the exchange of simulated fluxes (e.g. evapotranspiration, precipitation, soil moisture fluxes, lateral soil water fluxes, groundwater recharge/discharge, etc.). The design of the models to be coupled determines the degree to which a coupling is possible.

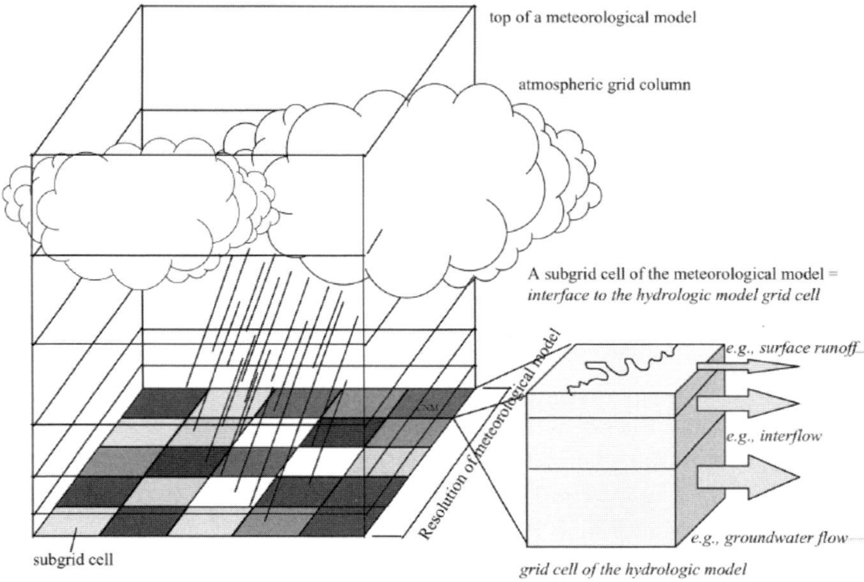

Fig. 3.3-2 Schematic view of the downscaling applied in the atmospheric model. The different grey levels represent different types of land use (after Mölders et al. 1999)

Mölders et al. (1999) coupled a conceptual hydrological model (NASMO, precipitation runoff model, Maniak 1996) and an atmospheric model (GESIMA, Geesthacht's simulation model of the atmosphere, Eppel et al. 1995) in a two-way mode. In doing so, the hydrological processes of the river catchment (translation, retention, and surface and subsurface lateral fluxes) are introduced into the atmospheric model. The atmospheric model drives the hydrological model by providing evapotranspiration and precipitation rates. The results provided by the hydrological model are used to modify soil moisture in the atmospheric model. The different spatial scales of the models were bridged by aggregation of lateral flow and runoff determined by the hydrological model and disaggregation of evapotranspiration, precipitation as well as soil wetness simulated by the atmospheric model. The module designed to couple the models was based on the explicit subgrid scheme (see discussion of disaggregation in Sect. 1.2), adapted for the mesoscale to downscale the hydrologically relevant quantities (e.g. Mölders et al. 1996). At the land-atmosphere interface, a high resolution grid is defined, i.e. several subgrid cells within one cell of the atmospheric model, to enable the exchange of the simulated fluxes of the two models (see Fig. 3.3-2). These subgrid cells may be covered by at least one vegetation- and/or soil-type. The energy and water budgets are calculated for each subgrid cell using the subgrid cell surface characteristics and the micro-climate. This means that, in each subgrid cell, the fluxes are individually calculated with their specific subgrid soil characteristics and near-surface meteorological forcing. The aggregation of the subgrid-cell results to be used in the atmospheric grid cell was performed by simple arithmetic averaging. The precipitation simulated for an atmospheric grid cell by the meteorological model was disaggregated by assuming that subgrid-cell precipitation is related to the ratio of subgrid cell and grid cell mean elevation (see Mölders et al. 1996).

The comparison of simulation results with and without the two-way coupling showed that even on a short time-scale surface runoff and lateral water flows have an impact on the spatial distribution of soil wetness, soil temperature, and therefore a feedback effect on cloudiness and the thermal regime of the atmospheric boundary layer within the catchment. A trend towards moister valleys and drier hills is predicted by the two-way coupled mode (Mölders and Raabe 1997, Mölders et al. 1999).

Compared to one-way coupling two-way coupling has the advantage of the consideration of the feedback between the land and atmospheric part of the water cycle. As for the one-way coupling, shortcomings of the two-way coupling are that (1) the parameterisations and surface characteristics can be inconsistent, and that (2) false predictions may propagate.

A technical disadvantage of the two-way coupling is that the data exchange requires that both models have to be run simultaneously and the data exchange requires a lot of storage capacity. In many cases, hydrological and atmospheric models apply different simulation time steps, so the exchange of simulation results has to be carefully adapted to the corresponding time step, i.e. an adequate time sequence for the data exchange has to be defined. If the time step for data exchange is large, information may be lost due to the aggregation procedure performed.

An Integrated Hydrometeorological Module

To avoid the before-mentioned shortcomings of a two-way coupling by exchanging simulation results of two rather independent hydrological and atmospheric models, the development of a two-way coupling by a fully integrated model (one single code) is presented. Such a realisation saves computer time and manpower, but needs interdisciplinary cooperation. It guarantees that the hydrological, soil and the atmospheric processes are represented in a consistent manner.

Recently, some developments in the direction of a fully integrated hydrometeorological module were carried out, for example – starting from the hydrological modelling perspective – by Famiglietti and Wood (1991), Band et al. (1993), Stieglitz et al. (1997). On the other hand, atmospheric scientists have included the most relevant hydrological processes into their existing surface modules of atmospheric models, such as OSULSM (Oregon State University land-surface model, Chen et al. 1996), HTSVS (hydro-thermodynamic soil vegetation scheme, Kramm et al. 1994, 1996; Mölders and Rühaak 2002), SEWAB (Mengelkamp et al. 1999), RAMS (Walko et al. 2000).

In general, if coupling hydrological and atmospheric models, the hydrologists should provide the modelling expertise for groundwater dynamics, lateral soil water fluxes and surface runoff. The atmospheric modellers should determine the fields of wind, pressure, air temperature, humidity, radiation, cloudiness and precipitation. The soil-vegetation module should be a common part (hydrometeorological module) that serves as the upper boundary condition for the hydrological and the lower boundary condition of the atmospheric processes. This soil-vegetation module has to fulfill the requests of both scientific disciplines to the highest degree of accuracy. In the next section, the special needs of atmospheric and hydrological models to simulate the boundary conditions are discussed to elucidate which physical processes have to be considered in the hydrometeorological module to which accuracy (for more details see Mölders 2001).

3.3.3 Specific Requirements for Modelling Water and Energy Fluxes at the Hydrometeorological Interface

Representation of Small-scale Variations in Surface Characteristics and Conditions

The heterogeneity of surface characteristics is of great importance in both hydrological and atmospheric modelling. Friedrich et al. (2000) examined the influence of surface heterogeneity on spatial distribution, temporal development and the domain-average of the ratio between sensible and latent heat flux (Bowen-ratio) for synthetic landscapes of differing degrees of surface heterogeneity by applying a mesoscale meteorological model. Their results substantiate that land-surface distributions will non-linearly influence the Bowen-ratio if patches of equal type exceed a certain size and that the surface type dominating a landscape does not necessarily determine the mean Bowen-ratio representative for this area. Thus, when applying the dominant surface type as the representative one for a grid cell, the margin of error in the Bowen-ratio depends on the horizontal resolution of the model (or of available data).

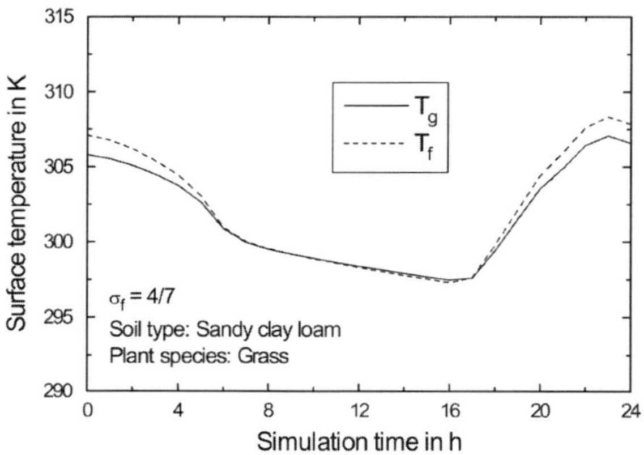

Fig. 3.3-3 Temporal development of leaf surface and ground temperature (T_g: Temperature of the ground; T_f: Temperature of the foliage)

Therefore, the heterogeneity on the microscale, which is of relevance for the near-surface stratification of the atmosphere (stability) and the atmospheric fluxes of sensible and latent heat, as well as the heterogeneity on the mesoscale, which affects the height of the atmospheric boundary layer, vertical mixing, and possibly cloud location, should be taken into account. A special feature of the hydrometeorological intersection module should be its ability to represent fine scale variations in surface characteristics, such as terrain slope, vegetation type, soil type and moisture or water bodies, which often vary considerably over short distances (e.g. Avissar and Pielke 1989).

The difficulty to bridge the spatial scales by aggregation/disaggregation can be addressed within the framework of considering subgrid-scale heterogeneity, i.e. the heterogeneity at the mesoscale (see discussion of disaggregation and aggregation issues in Sect. 1.2). Disaggregation of hydrologically relevant quantities provided by the atmospheric calculations are to be used as evapotranspiration and precipitation rates in the hydrological model part. The calculated hydrological processes have to be upscaled for use in the calculation of the atmospheric processes.

The heterogeneity at the microscale should be included by a mixture approach to consider simultaneously at least bare soil and/or vegetation within one grid cell (e.g. Deardorff 1978; Kramm et al. 1994, 1996). The surface temperatures of the leaf surface (foliage) and ground may differ strongly during the diurnal course (Fig. 3.3-3).

In the atmospheric model part used in the integrated model, a bulk-parameterisation (e.g. Cotton et al. 1982, Mölders et al. 1997) considering at least five water classes (water vapour, cloud water, rainwater, ice, graupel) should be applied to predict the mean precipitation within an atmospheric grid cell, because such a parameterisation is more physical and closer to the real processes than using only a cumulus-parameterisation (e.g. Mölders et al. 1994). The precipitation provided by a state-of-the art cloud module by the atmospheric model part has to be downscaled to the resolution of the hydrological model part. Recently, various downscaling procedures of different degrees of complexity and concept were developed to heterogenise precipitation (e.g., von Storch et al. 1993, Leung and Ghan 1995, Leung et al. 1996, Mölders et al. 1996).

Short vegetation should be considered by at least one layer. In the case of tall vegetation, multi-layer canopy models (e.g. Ziemann 1998) should be applied within the subgrid cells partly or totally covered by tall vegetation. Up til now, there exist no NWP models or GCMs that consider tall vegetation by some kind of multiple layer canopy model. Recently, Pyles (2000) coupled a multi-layer canopy model with a mesoscale meteorological model.

Interception loss should be included, at least for tall vegetation. It should be allowed that only parts of the vegetation are wet. The energy and hydrological budget equations have to be solved within a subgrid cell simultaneously.

Soil Physics

In the hydrometeorological module, all soil processes are calculated on the subgrid-scale, i.e. the same resolution as the hydrological model part. The treatment of the soil physics should be based on the fundamental principles of thermodynamics of irreversible processes, energy and mass conservation. An appropriate approach for the vertical water dynamics in the unsaturated zone should be applied, accounting for gravity and soil water tension, i.e. the unsaturated Darcy equation or a similar simplified approach should be applied. The hydrological model part should also account for the lateral soil water fluxes. The heat and moisture transport within the soil are to be solved by balance equations for soil temperature and volumetric water content. Solving the fully coupled equations means that the Ludwig-Soret effect (i.e. a temperature gradient can change soil volumetric water content) and Dufor effect (i.e. a moisture gradient may alter soil temperature) are taken into account, which may be especially important if high differences in temperature of rainfall and soil water exist. Within one cell, vertical varying soil types should be taken into account. The soil-module should include groundwater table dynamics to account for the impacts of changes in groundwater recharge and varying groundwater level on soil moisture, evapotranspiration and runoff. The impact of surface soil moisture on the atmosphere was discussed earlier in Sect. 3.2.1 as an example of feedback mechanisms that can require coupled models

Moreover, the soil-water uptake by roots has to be considered to ensure that evapotranspiration is related to the water content of different soil layers (not only the uppermost soil layer as often realised in atmospheric models). It should be ensured that the root density may vary with depth and time.

The parameterisation of infiltration has to be consistent with the calculation of the soil water dynamics. It may account for variable infiltration capacity, including macropore effects, and for differences in subgrid infiltration characteristics.

For coupling of a hydrological model part with an atmospheric model part, soil frost processes have to be considered because large parts of the continents are regularly frozen during winter in high and mid-latitudes or mountainous regions. From the hydrological point of view, soil frost leads to the freezing of soil-water, virtually totally restricting its mobility and capillarity, such that infiltration as well as percolation are almost ineffective. Since soil frost hinders infiltration of water into the soil, rain falling onto frozen soil or the melting of a snow layer over frozen soil will contribute strongly to runoff. Aspects of soil frost that affect the atmosphere are more indirect than those in hydrology. The thermal stability and low air temperatures, and the consequent low saturation pressure of water vapour, lead to low evaporation. The moisture will be stored in frozen ground and may increase spring peak flood events. In addition, transpiration plays a minor role because deciduous forests have lost their leaves and the stomatal conductivity of coniferous forests is also lower. If the freezing processes of soils are neglected, excessive water vapour fluxes into the atmosphere will be predicted as there is seemingly still "liquid" water available that, moreover, requires less energy for evaporation than ice (e.g., Mölders and Walsh 2004).

The depth of the interface between an unfrozen upper soil and a frozen deeper soil, for instance, may vary within the diurnal course. The determination of the surface water and energy fluxes is extremely uncertain when the exact freezing depth is unknown.

For the reasons discussed above, the inclusion of soil frost processes when coupling hydrological and atmospheric models is important for adequate calculation of runoff and water supply to the atmosphere in winter. The terms for soil freezing and thawing should be included in the coupled equations of heat and moisture transport within the soil and should be based on thermodynamics.

Snow

Another important process to be considered in the proposed hydrometeorological module is the treatment of snowmelt and previous snow accumulation. Snow is treated differently in hydrological and atmospheric models because of the different relevance of the various aspects of snow for atmospheric and hydrological processes. In hydrological modelling, namely, the retarded entry of precipitation into the land phase of the water cycle is the most prominent aspect of snow. Thus, a simple day-degree method is sufficient in most hydrological applications. In atmospheric models, besides the delayed water entry into the land phase of the water cycle, the insulating effect of snow that prevents the underlying soil from cooling and the high albedo of snow that affects the energy budget (e.g., Plüss and Ohmura 1997; Abdalati and Steffen 1997; Cline 1997; Baker et al. 1999; Robinson et al. 1992) are the most important aspects of snow to be considered. Albedo, for instance, dramatically changes when snow falls and rests on the ground, especially where the underlying ground has an albedo below 0.15 when wet. Since the albedo associated with snow cover typically ranges between 0.35 and 0.9, the energy exchange between the soil surface and atmosphere is therefore in winter generally weaker than in summer.

Disappearance of snow leads to runoff and removes a critical constraint on both water vapour pressure and surface temperature. As long as the snowpack exists, these quantities cannot rise above 610 Pa and 273.15 K. Therefore, the surface-atmosphere coupling becomes more relevant after the melting of snow. Exposed soil surfaces within a partly broken snow coverage lead to substantial sensible heat fluxes, convection and increased vertical mixing in the surface layer. If sufficient moisture is available, clouds may form. The cloud shadows may feed back to a reduced melt process. The strong spatial contrast in the energy budgets of snow-covered and snow-free areas may lead to significant advective flow similar to a sea breeze (Baker et al. 1999).

As a consequence of the aspects discussed above, when coupling hydrological and atmospheric model parts, the snow accumulation and melt processes should be considered by a multiple layer snow model (e.g. Anderson 1976; Foster et al. 1996; Cayan 1996; Fröhlich and Mölders 2002) if snow events were frequent and snow accumulation is high. In the case when the coupled model is applied mainly in regions of little snowfall and usually ephemeral snow coverage, it has to be examined whether a single layer snow model could be sufficient.

3.3.4 Conclusions

During the last decade, several attempts were made to consider the interaction between the land and atmospheric part of the water cycle in both long-term climate modelling and short-term numerical weather predictions. There are three different concepts of how to treat the complexity of the physical system "water cycle":

- Parameterisation of subsurface and surface hydrological processes in the atmospheric model;

- One-way or two-way coupling of hydrological and atmospheric models by data exchange between independent hydrological and atmospheric models; and

- Direct coupling of hydrological and atmospheric model parts by use of a common intersection module, here denoted as the hydrometeorological module within a fully integrated model.

In the latter, the hydrological and atmospheric model loose their stand-alone identity and become one hydrometeorological model. Such an integrated concept seems to be the most sophisticated and detailed way to realise the coupling between hydrological and atmospheric models, because it provides optimised technical and physical consistency within one model system.

Irrespective of the large amount of scientific effort in recent years, the current state of complex modelling is far from being satisfactory from the point of view of operational applications, e.g. for flood prediction, water resource management, or climate change assessment (see also IPCC 2001). The choice of the type and degree of coupling (parameterisation, one-way, two-way, or fully integrated) should depend on the task to be addressed. The requirements for the design of integrated coupling between an atmospheric and hydrological model was introduced earlier, Sect. 3.3.3. This hydrometeorological module may serve as the lower boundary condition for an atmospheric model. A schematic view for the data to be provided and exchanged by the hydrometeorological module is given in Figure 3.3-4. The hydrometeorological module is to be called up for each surface grid cell of the atmospheric model and should include subgrid-cell representation of prognostic snow-cover, prognostic equations for soil volumetric water content and soil temperatures (in z-direction only) considering the Ludwig-Soret- and Dufor-effect, treatment of soil freezing and thawing, water uptake by roots, local runoff of heavy precipitation and snowmelt, as well as energy and moisture budgets for soil, vegetation, canopy air, temporary surface water, respectively (e.g. intercepted water, flood, snow-cover). The subgrid cells should match the resolution of the hydrological model. Soil and snow-cover have to be divided into multiple vertical levels. Vegetation and canopy have to be represented by at least a single layer.

The hydrometeorological module provides the water and energy fluxes, surface temperature, and moisture, surface albedo and emissivity to the atmospheric model, while it provides the vertical soil water fluxes, soil volumetric water content, groundwater recharge, infiltration, melt water, and ponded water to the hydrological part. The ponded water can produce runoff. The atmospheric model delivers to the hydrometeorological module surface pressure, specific humidity, air temperature, wind and short- and long-wave downward radiation. The hydrological model provides to the hydrometeorological module the lateral soil water fluxes, surface and channel runoff (Fig. 3.3-4).

The practical aspects of the difficulties in coupling hydrological and atmospheric models are not only in identifying the feedback and bridging the spatial gaps of the models, but also setting up the coupled model packages in such a way that computational efforts, data exchange and required data are kept manageable in a time reasonable for forecasts and applications. A general challenge is to initialise the coupled model and provide the soil physical and plant physiological data needed. While for the atmospheric part the initial data can be obtained from radiosonde observations, there are no routine networks measuring soil moisture, soil ice, and soil temperature. Measurements of the level of groundwater tables exist only for selected catchments and not for large areas. Data sets of vertical profiles of soil types are not available in a spatial resolution and completeness as required, for instance, for NWP or GCMs.

One big challenge is not to couple what can be coupled all into one integrated model, but rather to couple only what makes sense to answer a question that could not be addressed without coupling.

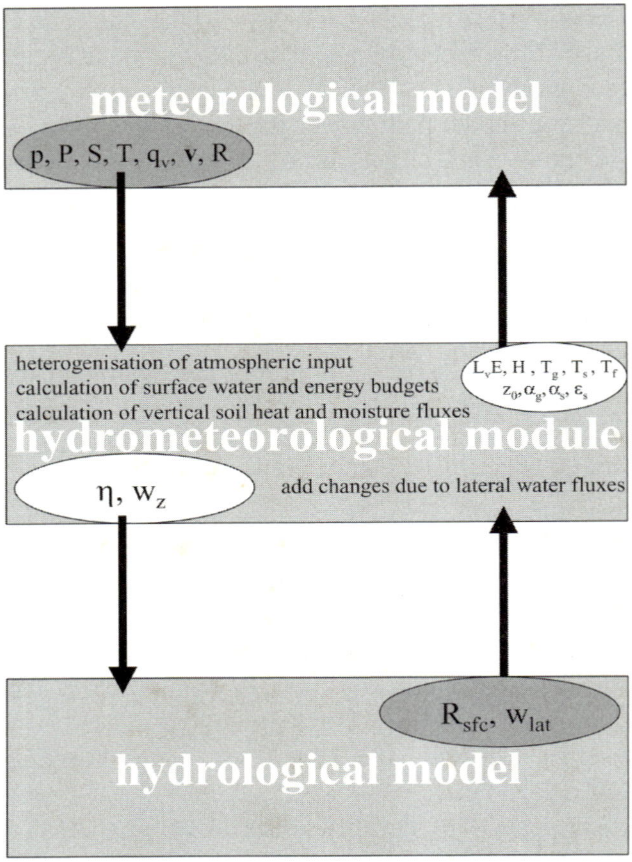

Fig. 3.3-4 Flow chart of calculated data exchange in a hydrometeorological module used as intersection in a fully coupled hydrological-atmospheric model (after Mölders and Rühaak 2002). Here, p is pressure, P stands for precipitation, S for snow, T, T_g, T_S, T_f denote to the temperatures of air, ground, snow and foliage, q_V and **v** are the specific humidity of air and the wind vector, respectively. The letter R represents the short-wave and long-wave downward radiation. Furthermore, η, w_{lat}, w_Z, and R_{sfc} are the volumetric water content, lateral and vertical water fluxes in the soil as well as (channel and/or surface) runoff. The latent and sensible heat flux densities are denoted as L_VE, and H, respectively. Finally, z_0, α_g, α_S, and ε_S represent the roughness length, albedo of the ground and snow as well as the emissivity of snow

References Chapter 3

Abdalati W, Steffen K (1997) Snowmelt on the Greenland ice sheet as derived from passive microwave satellite data. J Climate 10:165-175

Anderson EA (1976) A point energy and mass balance model of snow cover. NOAA Technical Memorandum NWS Hydro-17. US Department of Commerce, Silver Springs, MD, p217

Aris R (1961) The optimal design of chemical reactors. Acad. Press, New York, p191

Avissar R, Pielke RA (1989) A parameterization of heterogeneous land surface for atmospheric numerical models and its impact on regional meteorology. Mon Weather Rev 117:2113-2136

Baker JM, Davis KJ, Liknes GC (1999) Surface energy balance and boundary layer development during snowmelt. J Geophys Res 104D:19611-19621

Band LE, Patterson P, Nemani R, Running SW (1993) Forest ecosystem processes at the watershed scale: Incorporating hillslope hydrology. Agr Forest Meteorol 63:93-126

Bear J, Verruijt A (1987) Modeling groundwater flow and pollution. D. Reidel Publ, Dordrecht, p414

Cayan D (1996) Interannual climate variability and snowpack in the western United States. J Climate 9:928-948

Chen F, Mitchell K, Schaake J, Xue Y, Pan H-L, Koren V, Duan QY, Ek M, Betts A (1996) Modeling of land surface evaporation by four schemes and comparison with FIFE observations. J Geophys Res 101D:7251-7268

Claussen M (1997) Modeling bio-geophysical feedback in the African and Indian monsoon region. Clim Dyn 13:247-257

Cline DW (1997) Effect of seasonality of snow accumulation and melt on snow surface energy exchanges at a continental alpine site. J Appl Meteorol 36:32-51

Cotton WR, Stephens MA, Nehrkorn T, Tripoli GJ (1982) The Colorado State University three-dimensional cloud/ mesoscale model. Part II: An ice phase parameterization. J Recherche Atmos 16:295-320

Cunge J (1989) Review of recent developments in river modelling. In: Fallower RA, Goodwin P, Matthew RGS (eds) Hydraulics and environmental modelling of coastal, estuarine and river waters. Proc. Int. Conf. Bradford, Gower Technical, Aldershot, pp393-410

Deardorf JW (1978) Efficient prediction of ground surface temperature and moisture, with inclusion of a layer of vegetation. J Geophys Res 84C:1889-1903

DUFLOW (1998) Reference manual version 1.0. Stichting Toegepast Onderzoek Waterbeheer (stowa), Zoetermeer (NL)

Ehlig C (1977) Comparison of numerical methods for solution of the diffusion-convection equation in one and two dimensions. In: Gray WG, Pinder GF, Brebbia CA (eds) Finite elements. Water Res, 1. Conf., Pentech. London, pp71-102

Eppel DP, Kapitza H, Claussen M, Jacob D, Koch W, Levkov L, Mengelkamp H-T, Werrmann N (1995) The non-hydrostatic mesoscale model GESIMA. Part II: Parameterizations and applications. Contrib Atmos Phys 68:15-41

Famiglietti J, Wood EF (1991) Evapotranspiration and runoff for large land areas: Land surface hydrology for atmospheric General Circulation Models. Surv Geophys 12:179-204

Foster J, Liston G, Koster R, Essery R, Behr H, Dümenil L, Verseghy D, Thompson S, Pollard D, Cohen J (1996) Snow cover mass intercomparison of General Circulation Models and remotely sensed datasets. J Climate 9:409-426

Friedrich K, Mölders N, Tetzlaff G (2000) On the influence of surface heterogeneity on the Bowen-ratio: A theoretical case study. Theor Appl Climatol 65: 181-196

Fröhlich K, Mölders,N (2002) Investigations on the impact of explicitly predicted snow metamorphism on the microclimate simulated by a meso-beta/gamma-scale non-hydrostatic model. Atmos Res 62:71-109

Giorgi F, Avissar R (1997) Representation of heterogeneity effects in earth system modeling: Experience from land surface modeling. Rev Geophys 35:413-438

Graham LP, Bergström S (2000) Land surface modelling in hydrology and meteorology – lessons learned from the Baltic Basin. Hydrol Earth Syst Sci 4:13-22

Hagemann S, Dümenil L (1998) A parameterization of the lateral waterflow for the global scale. Clim Dyn 14:17-31

Holzbecher E (1998) Modeling density-driven flow in porous media. Springer Publ, New York, p286

Holzbecher E (2000) Characterisation of heat and mass transfer in porous media. 16th IMACS World Congress, Lausanne

Imboden DM, Lerman A (1978) Chemical models of lakes. In: Lerman A (ed) Lakes - chemistry, geology, physics. Springer Publ, New York, pp341-356

IPCC (Intergovernmental Panel on Climate Change) Assessment report climate change 2001: The scientific basis. Cambridge University Press, p980

Jørgensen SE (1995) State-of-the-art management models for lakes and reservoirs Lakes & Reservoirs: Res Manage 1:79-87

Kinzelbach W (1986) Groundwater modelling. Elsevier, Amsterdam, p333

Kramm G, Dlugi R, Mölders N, Müller H (1994) Numerical investigations of the dry deposition of reactive trace gases. In: Baldasano JM, Brebbia CA, Power H, Zannetti P (eds) Air pollution II Vol. 1: Computer Simulation. Computational Mechanics Publications, Southhampton Boston, pp285-307

Kramm G, Beier N, Foken T, Müller H, Schröder P, Seiler W (1996) A SVAT scheme for NO, NO_2, and O_3 - model description and test results. Meteorol Atmos Phys 61:89-106

Leung LR, Ghan SJ (1995) A subgrid parameterization of orographic precipitation. Theor Appl Climatol 52:95-118

Leung LR, Wigmosta MS, Ghan SJ, Epstein DJ, Vail LW (1996) Application of a subgrid orographic precipitation/surface hydrology scheme to a mountain watershed. J Geophys Res 101:12803-12817

Lin B-L, Sakoda A, Shibasaki R, Goto N, Suzuki M (2000) Modelling a global biogeochemical cycle in terrestrial ecosystems. Ecol Mod, 135:89-110

Liston GE, Sud YC, Wood EF (1994) Evaluating GCM land surface hydrology parameterizations by computing river discharges using a runoff routing model: Application to the Mississippi Basin. J Appl Meteorol 33:394-405

Manabe S (1969) Climate and ocean circulation. 1. The atmospheric circulation and the hydrology of the earth's surface. Mon Weather Rev 97:739-774

Maniak U (1996) Gitterpunktgestütztes Flußgebietsmodell als Abflußmodul eines Wasserhaushalts- bzw. Klimamodells. Zwischenbericht. Leichtweiß-Institut, TU Braunschweig

Mengelkamp H-T, Warrach K, Raschke E (1999) SEWAB – a parameterization of the surface energy and water balance for atmospheric and hydrologic models. Adv Water Resourc 23:165-175

Miller JR, Russel GL, Caliri G (1994) Continental-scale river flow in climate models. J Climate 7:914-928

Mölders N (1999) On the effects of different flooding stages of the Odra and different landuse types on the local distributions of evapotranspiration, cloudiness and rainfall in the Brandenburg-Polish border area. Contrib Atmos Phys 72:1-24

Mölders N (2001) Concepts for coupling hydrological and meteorological models. In: Arnold K, Raabe A (eds) Meteorologische Arbeiten aus Leipzig (VI), Wiss Mitt, Leipzig, 22:1-15

Mölders N, Raabe A (1997) Testing the effect of a two-way-coupling of a meteorological and a hydrologic model on the predicted local weather. Atmos Res 45:81-108

Mölders N, Rühaak W (2002) On the impact of explicitly predicted runoff on the simulated atmospheric response to small-scale land-use changes - an integrated modeling approach. Atmos Res 63:3-38

Mölders N, Walsh JE (2004) Atmospheric response to soil-frost and snow in Alaska in March. Theor Appl Climatol 77:77-105

Mölders N, Hass H, Jakobs HJ, Laube M, Ebel A (1994) Some effects of different cloud parameterizations in a mesoscale model and a chemistry transport model. J Appl Meteorol 33:527-545

Mölders N, Raabe A, Tetzlaff G (1996) A comparison of two strategies on land surface heterogeneity used in a mesoscale-β meteorological model. Tellus 48A:733-749

Mölders N, Kramm G, Laube M, Raabe A (1997) On the influence of bulk-parameterization schemes of cloud microphysics on the predicted water-cycle relevant quantities - a case study. Meteorol Z 6:21-32

Mölders N, Raabe A, Beckmann T (1999) A technique to downscale meteorological quantities for the use in hydrologic models - description and first results. IAHS 254:89-98

Müller E, Foken T, Heise E, Majewski D (1995) LITFASS - a nucleus for a BALTEX field experiment. DWD Arbeitsergebnisse No. 33

Nield DA, Bejan A (1992) Convection in porous media. Springer Publ, New York, p408

Orlob GT (ed) (1983) Mathematical modeling of water quality - streams, lakes and reservoirs. Wiley, Chichester, p518

Parkhurst DL (1995) PHREEQC: A computer program for speciation, reaction-path, advective transport, and inverse geochemical calculations. U.S. Geol. Surv., Water Res. Invest. Report 95-4227, Lakewood, p143

Petoukhov V, Ganopolski A, Brovkin V, Claussen M, Eliseev A, Kubatzki S, Rahmstorf S (2000) CLIMBER-2: A climate system model of intermediate complexity. Part I. Clim Dyn 16:1-17

Peyret R, Taylor TD (1985) Computational methods in fluid flow. Springer Publ, New York, p258

Plüss C, Ohmura A (1997) Longwave radiation on snow-covered mountainous surfaces. J Appl Meteorol 36:818-824

Prandtl L (1952) Essentials of fluid dynamics; with applications to hydraulics, aeronautics, meteorology and other subjects. Blackie, London, p452

Prandtl L (1965) Führer durch die Strömungslehre. Vieweg, Braunschweig, p523

Pyles RD (2000) The development and testing of the UCD advanced canopy-atmosphere-soil algorithm (ACASA) for use in climate prediction and field studies. Ph.D. Thesis, University of California at Davis

Rajar R, Cetina M (1995) Hydrodynamic models as a basis for water quality modelling: A review. In: Wrobel LC, Latinopoulos P (eds) Water pollution III - modelling, measuring and prediction. Computational Mechanics Publications, Southampton, pp199-211

Robinson DA, Serreze MC, Barry RG, Scharfen G, Kukla G (1992) Large scale patterns and variability of snowmelt and parameterized surface albedo in the Arctic Basin. J Climate 5:1109-1119

Robock A (1985) An updated climatic feedback diagram. B Am Meteorol Soc 66: 786-787

Rodi W (1980) Turbulence models and their application in hydraulics - a state-of-the-art review. IAHR publication, Delft

Saaltink MW, Ayora C, Carrera J (1998) A mathematical formulation for reactive transport that eliminates mineral concentrations. Water Resour Res 34(7): 1649-1656

Sausen R, Schubert S, Dümenil L (1994) A model of the river runoff for use in coupled atmosphere-ocean models. J Hydrol 155:337-352

Savijärvi H (1992) On surface temperature and moisture prediction in atmospheric models. Beitr Phys Atmos 65(4):281-292

Scheidegger AE (1961) General theory of dispersion in porous media. J Geophys Res 66:3273-3278

Steefel CI, MacQuarrie KTB (1996). Approaches to modeling of reactive transport in porous media. In: Lichtner PC, Steefel CI, Oelkers EH (eds) Reactive transport in porous media. Mineral Soc Am, Rev Mineralogy 34:83-129

Stieglitz M, Rind D, Famiglietti JS, Rosenzweig C (1997) An efficient approach to modeling the topographic control of surface hydrology for regional and global climate modeling. J Climate 10:118-137

van Genuchten MTh (1977) On the accuracy and efficiency of several numerical schemes for solving the convective dispersive equation. In: Gray WG, Pinder GF, Brebbia CA (eds) Finite elements. Water Res, 1. Conf., Pentech. London, pp71-102

Vasiliev OF (1987) System modelling of the interaction between surface and ground waters in problems of hydrology. Hydrol. Sci J 32(3):297-311

von Storch H, Zorita E, Cubasch U (1993) Downscaling of global climate change estimates to regional scales: An application to Iberian rainfall in wintertime. J Climate 6:1161-1171

Walko RL, Band LE, Baron J, Kittel TGF, Lammers R, Lee TJ, Ojima D, Pielke RA Sr, Taylor C, Tague C, Tremback CJ, Vidale PL (2000) Coupled atmosphere-biophysics-hydrology models for environmental modeling. J Appl Meteorol 39:931-944

Walter GG, Contreras M (1999) Compartmental modeling with networks. Birkhäuser, Boston

Yeh GT, Cheng HP (1996) On diagonalization of coupled hydrological and geochemical reaction equations. In: Wheeler MF (ed) Environmental studies. Springer Publ, New York, pp373-398

Ziemann A (1998) Numerical simulations of meteorological quantities in and above forest canopies. Meteorol Z 7:120-128

CHAPTER 4

Case Studies

4.0 Introduction

Axel Bronstert, Nicole Mölders

In this chapter, 14 examples of coupled modelling studies are presented. They cover a wide range of scientific disciplines, including climate, hydrology, soil sciences, and economic aspects. However, it is obvious that this cannot be a complete survey of recent coupled modelling studies.

For brevity, the presentations of the case studies purposely focus on the description of the couplings only. The case studies are to give an idea of how model couplings can be realised, and do not cover the entire range of recent coupled models. The examples show how models may differ with respect to detail and complexity.

The case studies have been arranged in five groups according to the main coupling issues which are addressed by each specific study: The first group contains six studies which deal with the coupling of different compartments of the hydrological cycle, such as surface and groundwater (case study No. 1, 2, and 3), permafrost and runoff (No. 4 and 6) and atmosphere and various compartments of the terrestrial hydrological cycle (No. 5). The second group comprises three case studies focusing on the linkage of water and energy cycles. Though these studies have a similar focus, they differ strongly concerning their typical scales (spatial scales: from small catchments to global; time scales: from weeks to thousands of years). The major application field addressed in these studies is the evaluation of landscape change effects on atmospheric processes. The next two case studies deal with the coupling of the hydrological cycle with chemical cycles, such as carbon and nutrients. The following group includes examples of coupled models which in addition to natural processes also take into account socio-economic processes, i.e., the economic (e.g., the income of farmers) and sociological impacts (such as demographic development and migration). They cover a regional scale (No. 12) and a global scale (No. 13). Finally, one example of a study is presented which has a strong focus on methodological aspects of coupled models, such as data assimilation and model organisation.

An important feature of the different coupled models presented here is the type of coupling achieved. In this context we speak of a one-way coupling if the results from one discipline are being used as prerequisite information for another scientific discipline, but no feedback effects are considered. A two-way coupling mode includes the consideration of feedback effects. Even within one study, different degrees of coupling may occur, if certain process feedbacks are included and others are not. Case studies 4 and 5, for instance, both apply one-way coupling with respect to the atmospheric component, but use two-way coupling for some other parts. Coupled models can be set up in a mode where two models exchange information (e.g. results from sub-models) via an external interface, i.e. the two (or more) models can also be run in a stand-alone mode, and have been programmed in separate codes. This may be termed a *weak coupling* mode. A model where all coupling has been achieved within a single and consistent computer code, and the exchange of information is organised in time steps and spatial units specifically adjusted to the dynamics and spatial variations of the coupled processes, may be termed a *strong coupling* mode. If a coupled model accounts for feedback effects and at the same time is organised in a strong coupling mode, we may term this an *integrated* mode.

As discussed before, the case studies cover a very wide range of spatial and temporal scales. Some investigate processes at the hillslope scale (No. 10), and others (e.g. No. 9, 13) at the global scale. It is worth noting, however, that most case studies deal with a spatial scale which is termed meso-scale in hydrology, or regional scale in atmospheric sciences. The timescale of the case studies spans from days and weeks (e.g. No. 7, 8) to several thousand years (No. 9). All case studies have in common that the degree of coupling is determined by the question to be addressed by the study, i.e. the temporal and spatial scales and the feedback processes that are relevant at these scales. In this context the same temporal scales may be combined with different spatial scales. Case studies 9 and 10, for instance, aim at a coupling on the local and global spatial scales, respectively, but both on the long term with respect to the temporal scale.

All case studies are presented in a similar way in the following pages. First, the motivation and the objectives of the study are summarised, then the study set-up and the study region are introduced. The next sub-section discusses the complexity addressed in the study and finally an appraisal of the experiences gained with the coupled model is given. Table 4.0-1 lists all the case studies and gives a classification with respect to their type of coupling as discussed above.

Table 4.0-1 Case studies and their classification with respect to the type of coupling as discussed above.

Case study No.	Authors	Name of the model	Topic	Study region	Main spatial scale	Main time scale	Main coupling issue	One-way mode	Two-way mode	Integrated mode
1	Vázquez-Suñé et al.		ground-/ surface water/ salinity	Barcelona (Spain)	meso / regional	seasons - decades	different compartments of the hydrological cycle	X		
2	Holzbecher		groundwater / lakes	Lake Dagow (Germany)	meso / regional	seasons - decades	different compartments of the hydrological cycle		X	
3	Holzbecher	FAST-C	salinity	Nile Delta (Egypt)	meso / regional	--	different compartments of the hydrological cycle		X	
4	Motovilov & Georgiadi	ECOMAG	permafrost & climate change	Kenkeme basin (Russia)	meso / regional	days - seasons	different compartments of the hydrological cycle	X	X	
5	Halldin et al.	MIUU; ECOMAG	boreal region	South Sweden; North Finland	meso / regional	hours - days	different compartments of the hydrological cycle	X	X	
6	Kuchment et al.		runoff in permafrost	Kolyma (Russia)	meso / regional	seasons - decades	different compartments of the hydrological cycle	X		
7	Mölders		land-use change impacts	East Germany	meso / regional	minutes - weeks	water & energy			X
8	Pielke et al.		weather	Florida (USA)	meso / regional	seasonal	water & energy		X	
9	Ganopolski	CLIMBER-2	Earth system modelling	the globe	global	thousands-millions of years	water & energy			X
10	Band & Tague	RHESSys	water, carbon and nutrients	Pond Branch (USA)	hillslope catchment	days - decades	water & biogeochemistry			X
11	Hall	BOREAS	boreal ecosystems	North Canada	local - regional	diurnal - centuries	water & biogeochemistry	X	X	
12	Krol et al.	SIM	regional impacts of climate changes	Northeast Brazil	regional	decades - centuries	water, climate, economy, sociology			X
13	Eickhout & Leemans	IMAGE-2	global impact of climate changes	the globe	global	decades - centuries	water, climate, economy, sociology			X
14	Leavesley & Springer		"virtual watershed"	--	meso / regional	hours - years	methodological aspects	X	X	

4.1 Groundwater Modelling in the Urban Area of Barcelona

Enric Vázquez-Suñé, Jesus Carrera, Xavier Sánchez-Vila, Elena Abarca, Ramón Arandes

4.1.1 Motivation and Objectives

The study of urban groundwater is motivated by the strong interaction between city socio-economic development and groundwater environmental impacts. Urbanisation clearly affects both availability and quality of groundwater resources: seepage from sewage system pollutes groundwater, this leads to abandoning existing wells, which causes groundwater heads (actually, recovery), leading to flooding of basements and other underground infrastructures; and the cycle goes on. These effects have significant social, environmental and economic implications. This kind of problem is being faced by many cities worldwide, but unfortunately, researchers and municipal managers have tended to address them separately, rather than within an integrated framework.

Barcelona groundwater has suffered, and is still suffering, many of the problems associated to urban growth: over-pumping, sea-water intrusion, water-level rises, flooding of subway, pollution (Custodio 1997; Vázquez-Suñé et al. 1997).

The objective of this case study is to present some of the models prepared to address these issues. Specifically, we discuss the city-wide Barcelona model, as shown in Fig. 4.1-2, and a small-scale model, the Besós model, shown in Fig.4.1-4.

4.1.2 Groundwater in Barcelona

Barcelona is located between the mountain range Serra de Collcerola and the Mediterranean Sea. The city is bounded by two rivers, the Besós and Llobregat. There are several aquifers. In the upper parts is a Palaeozoic aquifer, consisting of shales and granites, and Quaternary aquifers formed by the alluvial and deltaic sediments of Llobregat and Besós rivers; in between, are aquifers consisting of piedmont cones and coarse alluvial sediments.

Groundwater levels in Barcelona have followed the typical evolution of an industrial city. Water extracted for industrial uses has increased since the early 20th century and reached a maximum of 60 – 70 million m^3 per year by the early 1970s. Drawdowns in different parts of the city ranged from a few metres to more than 15 m in several areas. Pumping has been severely reduced since the late 70s because of contamination and industrial migration. Water levels have almost recovered to their natural levels except in particular zones. An important part of the subway system and many of the buildings that exist today in Barcelona, were built during the period 1950 – 1975, coinciding with the period of maximum depletion. At that time, neither designers nor constructors were aware that groundwater levels could revert to the original levels. As groundwater levels recover, those constructions face severe flooding and pollution problems. For example, the subway system is suffering seepage problems in many areas, having to pump around 10 -15 million m^3 per year from inside the subway tunnels (Fig. 4.1-1).

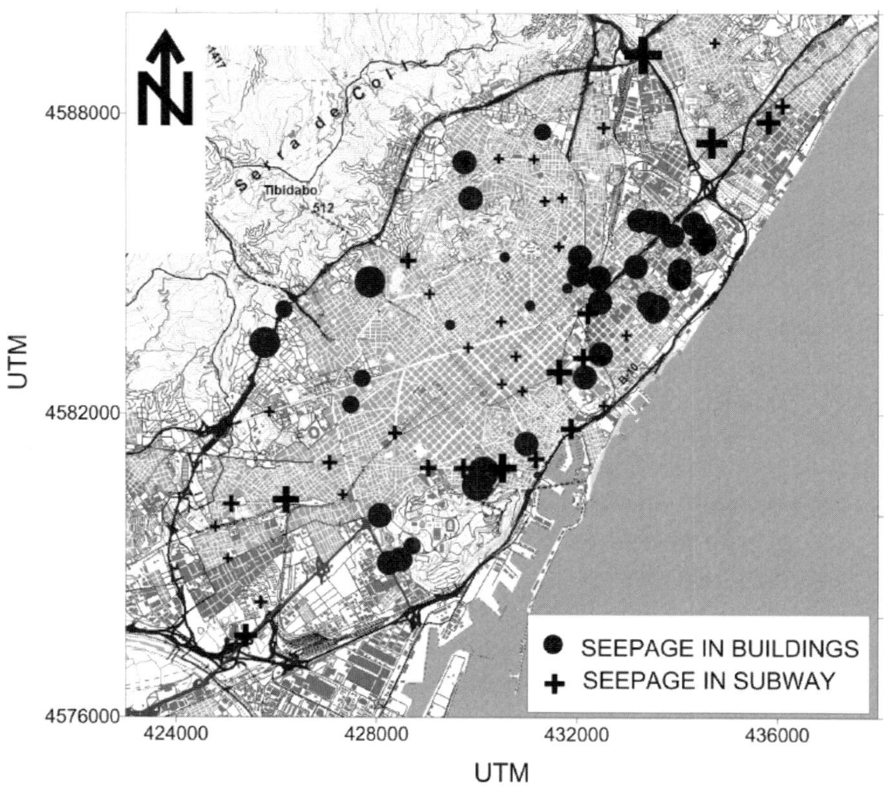

Fig. 4.1-1 Seepage problems in Barcelona city (symbol size is proportional to amount of seepage)

At present, the economic impact of reducing seepage by maintaining groundwater below the foundations of structures is very high. The impact of allowing the levels to stay high (by taking no external actions) is also very significant: To the need for drainage works, such as impermeabilisation, pump installation and water disposal, one must add the energy cost of continuously pumping the rather high flows that have to be evacuated. Moreover, the city sewage system is often used for evacuating the seepage. This is not a good solution because by mixing with wastewater, the seepage becomes polluted, causing an increase in treatment costs. Moreover, portions of the sewage network become saturated, which affects the proper functioning of the system during periods of critical need, such as during storms.

4.1.3 Complexity of the Groundwater Model

Several numerical models have been built with various purposes: integrating different historical data, accounting for the present situation, validating the conceptual model, quantifying groundwater flows (water balance) and predicting the future evolution of groundwater under different scenarios. Additionally, integration efforts helps to reduce uncertainty in hydraulic parameters. The Barcelona groundwater model is unique in several aspects that are relevant to the nature of this case study:

- It takes into account the historical evolution of the urban area. This is achieved by changing over time the sink/source terms associated with changes in soil use (i.e. changing from agricultural or industrial to urban use).

- Areal recharge depends basically on specific factors (notably water and sewage network density), directly related to the density of population and to the time the area was developed.

- Extractions are difficult to evaluate and they are often underestimated. A specific study was made to evaluate the evolution of pumping.

- Urban structures are singular features in a model and they should be addressed individually. In fact, small-scale specific models are needed to address them. One such model is described below.

- Interactions with subway and railway tunnels are modelled by means of appropriate sink/source terms. As head rises, so does the pressure behind the tunnels (and, hence, the flux) and the length of section affected.

- Seawater intrusion is addressed by means of an appropriate boundary condition.

Fig. 4.1-2 Model domain, including street network (grey) and highways (blue), and finite element mesh (red)

From a modelling viewpoint, all these interactions are treated as one-way but in reality, some are two-way interactions. For instance, pollution caused some of the wells to be abandoned. However, this has been incorporated externally into the model. All interactions are subsumed in the flow equation. For the city-wide model, this was solved using the finite element method with the grid shown in Fig. 4.1-2. An output of the model is the quantification of the terms involved in the groundwater balance (Fig. 4.1-3) and its evolution over time. Quantification of the different terms involved in the water balance is important for two reasons: a) knowing which are the main points to be stressed when managing the groundwater resources, and b) suggesting corrective measures.

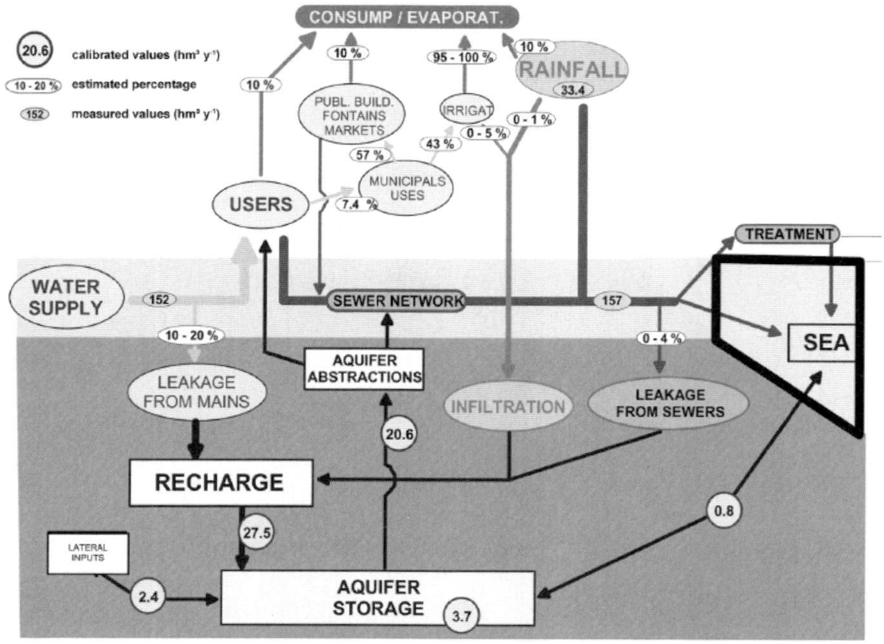

Fig. 4.1-3 Quantification of the terms involved in the urban groundwater balance in Barcelona

Three main sources of pollution have been identified in Barcelona: losses from the sewers system, runoff water infiltration, and seawater intrusion. Regarding seawater intrusion, its growth has been related to rising groundwater levels, and this is correlated directly with the industrial development of the city. In the eastern part of the city, seawater intrusion was detected as far as 2 km inland from the seacoast during the 1970s. The recovery of groundwater levels has reversed the direction of flow and some of the wells that were highly saline are now back to fresh water.

Discrimination among pollution sources can probably be best achieved by performing a mass balance computation involving different chemical species. Many authors compare solute concentrations in different water sources as the way to decide the chemical species that can be used for discrimination purposes (Lerner and Yang 2000). This methodology has been applied to the city of Barcelona to evaluate the total amount of water coming from each source. This has been implemented in the numerical model to compute the percentage of losses in mains, sewers and the contribution of direct infiltration, the river, or the sea.

4.1.4 Specific Experiences

The overall model is too large for use in a local framework. Specific models may be needed for specific applications. These include (1) the on-going work to explain the evolution of water quality by means of a coupled flow, transport and chemical reactions model; (2) models to anticipate the effect of infrastructures (e.g. new underground tunnels); or (3) to define local pumping areas aimed at groundwater as an additional resource and at reducing groundwater levels. An example of the latter is presented below.

The selected case is related to a Fluvial Park that is being constructed by the Besós River (Fig. 4.1-4). The project involves recovering the riverbed, the quality of surface water, and building an urban park in an area with a historical deficit in green zones.

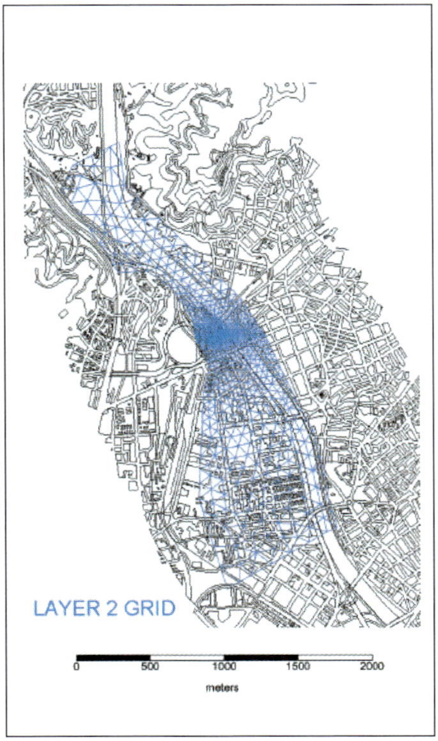

Fig. 4.1-4 Areal extent and finite element grid of the Besós model

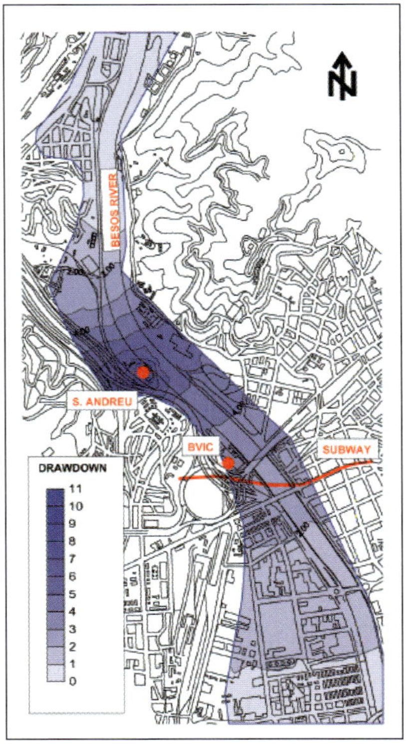

Fig. 4.1-5 Calculated drawdowns in the Besós model. The red line represents the subway

We focus here on water management for this new area. Approximately 1 $hm^3 y^{-1}$ must be supplied to the area to irrigate the green areas and to maintain an artificial lake. The first idea was to use water from the local drinking water supply system. An alternative is to use groundwater from wells located nearby. Using water from the river was discarded for quality reasons and to maintain river flow during the summer. The final decision was set within a global management concept where the model presented previously was a key tool.

The numerical model presented previously was used to analyse a number of pumping scenarios. The main scenario studied was the compatibility of pumping 15 $hm^3 y^{-1}$ for water supply plus 1 $hm^3 y^{-1}$ from an abstraction well located beside the Baró de Viver (BVIC) subway station (see Fig.4.1-5). Compatibility is measured in terms of admissible drawdowns. As a further outcome from the model we could gauge the effect of pumping upon the groundwater levels along the subway line.

Groundwater flow in this model is coupled to interaction with the subway tunnel and with the river. Results are shown in Fig.4.1-5. Both interactions are one-way, in the sense that neither river stage nor water level in the tunnel is assumed to be affected by the groundwater level.

4.1.5 Summary

This case study illustrates that urban groundwater displays significant feedbacks. It is strongly influenced by urban growth, which affects recharge and drainage. In turn, groundwater affects social developments. By creating problems, it forces the administration to react by changing water management practices, which in turn affect the urban environment (pumping for park irrigation, reduction of demand over conventional water sources, maintenance of river flow, etc.). The resulting behaviour is complex and rich and can hardly be represented by a model. Yet the models are essential to help in organising knowledge and in understanding the levels of interactions.

4.2 The Lake Dagow Coupled Model for Groundwater and Surface Water

Ekkehard Holzbecher

4.2.1 Motivation and Objectives for the Coupled Approach

Lake Dagow is a sub-catchment of another groundwater lake, Lake Stechlin (Fig. 4.2-1). Since the construction of the Polzow canal between 1745–1750 connecting the lakes Dagow, Stechlin, Nehmitz, and Roofen with the River Havel, the region drains to the North Sea. As a consequence, the groundwater level and the water levels of several lakes sank considerably and the littoral belts were reduced. For Lake Dagow, a further major influence on the water level may be attributed to groundwater withdrawal when water utilities began operating wells in the vicinity of the lake shore in the 1990s. However, groundwater had certainly been withdrawn for private use at a similar rate before this period, although there is a complete lack of data about pumping by land owners in the two villages Dagow and Neuglobsow located on the southern shore.

One of the aims of the model approach presented here is to reconstruct former water levels. Such an aim can be achieved only by a coupled model connecting groundwater and surface water compartments.

4.2.2 Short Description of the Study

Lake Dagow (Dagowsee) is located near the border between Brandenburg and Mecklenburg-Vorpommern, 130 km north of the city of Berlin. The interaction with groundwater is especially important because Lake Dagow is a groundwater lake with no inflow from creeks or rivers, typical of the north German lake district.

Lake Dagow and its catchment (see Fig. 4.2-1) is part of the Northern (Baltic) Land Ridge formed in the latest glacial period – the Weichselian. As a result of the relatively fast melting of glaciers, dead ice was left in the grooves and depressions formed earlier below the ice sheet. The depressions filled with dead ice were later covered with sand and gravel, leading to the present landscape with its numerous lakes and wetlands.

In the region studied, two aquifers are separated by boulder clay 10-30 m thick. The uncovered aquifer consists of fluvioglacial deposits; the sandy aquifer consists of fluvioglacial deposits classified as highly permeable. Both the uncovered aquifer and the surface water comprise one hydraulic system, with the lakes fed from the groundwater of the uncovered aquifer.

4.2.3 Complexity of the Study

A steady-state groundwater model, covering the entire catchment of Lake Stechlin and Lake Dagow, was described in detail by Holzbecher and Ginzel (1998). Holzbecher (2001, 2003) presents quasi steady-state computations, in which the model region is extended to the lake shores of the neighbouring lakes Glietzen, Peetsch and Nehmitz. These models dealt with the groundwater compartment only, for which measured lake levels were used as boundary conditions.

Here, a model variant is outlined which tackles a smaller region, including the entire sub-catchment of Lake Dagow (Fig. 4.2-1). The simulations are based on measured time-series for surface water and several groundwater

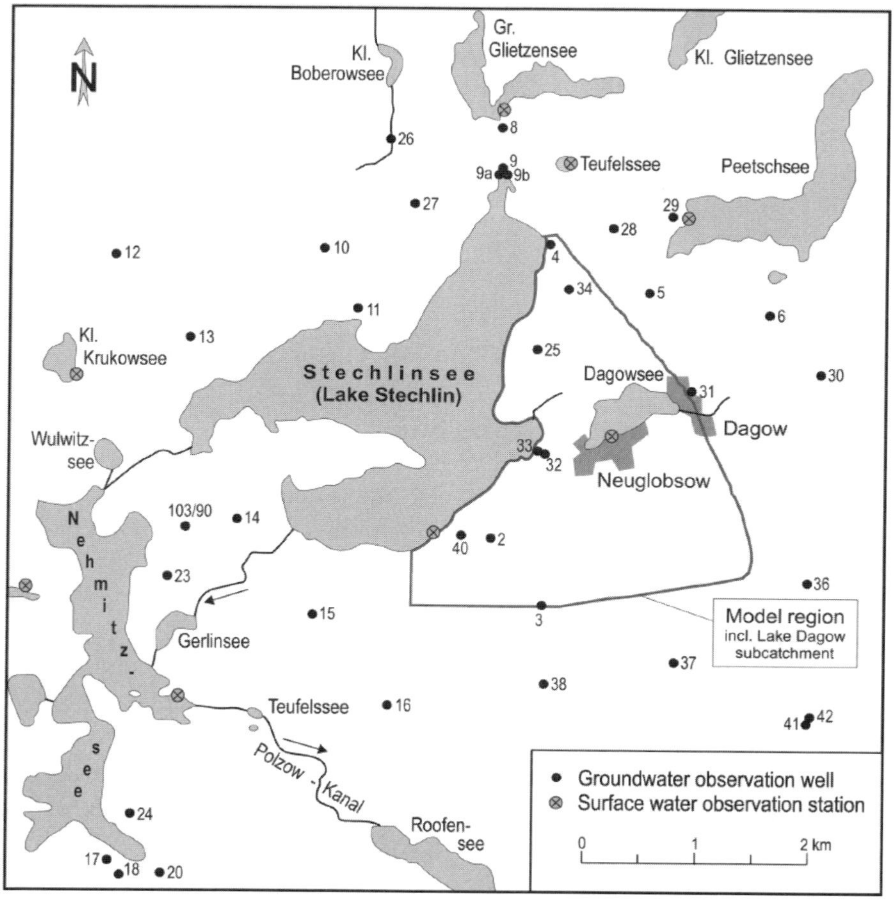

Fig. 4.2-1 Map of Lake Dagow and vicinity showing observation points and model region

observation wells recorded monthly since at least 1959. Moreover, meteorological data, as precipitation, evaporation and evapotranspiration, are also available as monthly time-series (Richter 1997). The measurements of water levels for lakes Stechlin and Dagow serve as boundary conditions for the groundwater model which in turn yield water exchange fluxes between groundwater and surface water. These fluxes are then used in the flux balance for lakes, according to the hydrological balance equation:

$$\Delta S = P - E + I - 0 + R \tag{4.2-1}$$

where ΔS denotes change in water storage within a given time Δt, P the precipitation and E evaporation from the lake surface. I, the sum of inflowing surface water, can be neglected for Lake Dagow (Holzbecher and Nützmann 1999). Also, the amount of outflowing surface water O is negligible. R, denoting the exchange between aquifer and lake, is the only variable on the right-hand side of the balance equation which may take negative values, as happens when the outflow from the lake as groundwater exceeds the amount of fluid inflowing from the aquifer.

For a lake, the water storage change can also be approximated by

$$\Delta S = A \cdot \Delta h \tag{4.2-2}$$

with the mean lake surface denoted by A and the water level change by Δh. This formula may have to be replaced by a more complex one when there are relevant changes in the lake surface area.

The modelling concept for Lake Dagow and its catchment is as follows: balance Eq. 4.2-1 is computed using measured values for P and E, as well as the value for R derived by the groundwater model. Eq. 4.2-2 is then utilised to calculate the change Δh in lake water levels. Head values calculated in this way can be compared to measurements and thus used to improve the groundwater model.

Various input data in the transient model are based on assumptions. Some parameters can only be estimated, like the specific yield and the hydraulic conductance between lakes and aquifer. The temporal development of several variables is unknown, particularly that concerning groundwater withdrawal for private use and groundwater recharge through the unsaturated zone. The values for these parameters and variables can be improved by a calibration or parameter estimation procedure based on the comparison of measured and calculated head values as described above.

In the following, results of the steady state version of the model are reported. Storage ΔS is zero in long-term studies under the assumption of no climatic changes. Thus, the groundwater model was calibrated to produce $\Delta S=0$ in the surface water balance Eq. 4.2-1. The unknown parameter in the groundwater model is groundwater withdrawal for private use. Using such a loosely coupled groundwater–surface water simulation, the inconsistency in the first Stechlin model (Holzbecher and Ginzel 1998) could be corrected.

4.2.4 Experiences

In the beginning it had been assumed that groundwater withdrawal could be neglected because there are no data available from 1959 to 1992, i.e. before the water works went into operation in the vicinity of Lake Dagow with a permitted maximum allowed pumping rate of 200 m^3d^{-1}. A mass balance check for Lake Dagow in a post-processing run for the first groundwater model delivered a value for long-term recharge R of $132 \cdot 10^3$ m^3y^{-1}, while according to the meteorological record, the difference $E - R$ amounts to $17 \cdot 10^3$ m^3y^{-1} only (compare Eq. 4.2-1). The difference of almost one order of magnitude revealed that a relevant term of the water balance was not considered. As groundwater has been pumped for private use during the period under consideration, it is most likely that these water losses from the aquifer represent the missing term. A parameter estimation run showed that the best match between observed and modelled head values were obtained for a groundwater withdrawal rate slightly above the maximum pumping rate of the water works.

Another run of the coupled groundwater and surface water procedure was started, to reconstruct the water level of Lake Dagow for a scenario without groundwater withdrawal. The lake water level was used as an optimisation parameter with the goal (function) given by Eq. 4.2-1. It turned out that the water level would lie more than 0.6 m above the current datum. The procedure had to be extended to take in the local topography, available in a 25 m grid. It then transpires that the surface of Lake Dagow would be extended on the northern and southern shores and, moreover, Little Lake Dagow, the presence of which is reported in historic sources but is nowadays merely a wetland, would re-appear (Fig. 4.2-2).

Feedback between lakes and groundwater is surely the rule and not the exception. Inflow from or outflow to the aquifer is one component of the lake budget while on the other hand hydraulic gradients in the vicinity of the lake – and thus the quantity of water flux – is determined by the water level in the lake.

Various model approaches can be followed to deal with coupled hydrological compartments. In the case study presented, both groundwater withdrawal rate and the lake water level have been used as free parameters in the sub-models for groundwater and surface water; both have a significant effect on the net groundwater exchange with the lake. In most situations, the lack of data will lead to a situation in which not all components of Eqs 4.2-1 and 4.2-2 can be determined using either measured values or the results of a computer model. When only one of the contributions on the right-hand side of the Eqs. 4.2-1 and 4.2-2 is unknown, it may be calculated from the other terms. When, as for example the outflow *O* from a lake is unavailable, it can be computed from the lake balance Eq. 4.2-1 where the change in storage is calculated first as output from a groundwater model. Such an approach was used by Richter (1997) with a simple balance model for groundwater.

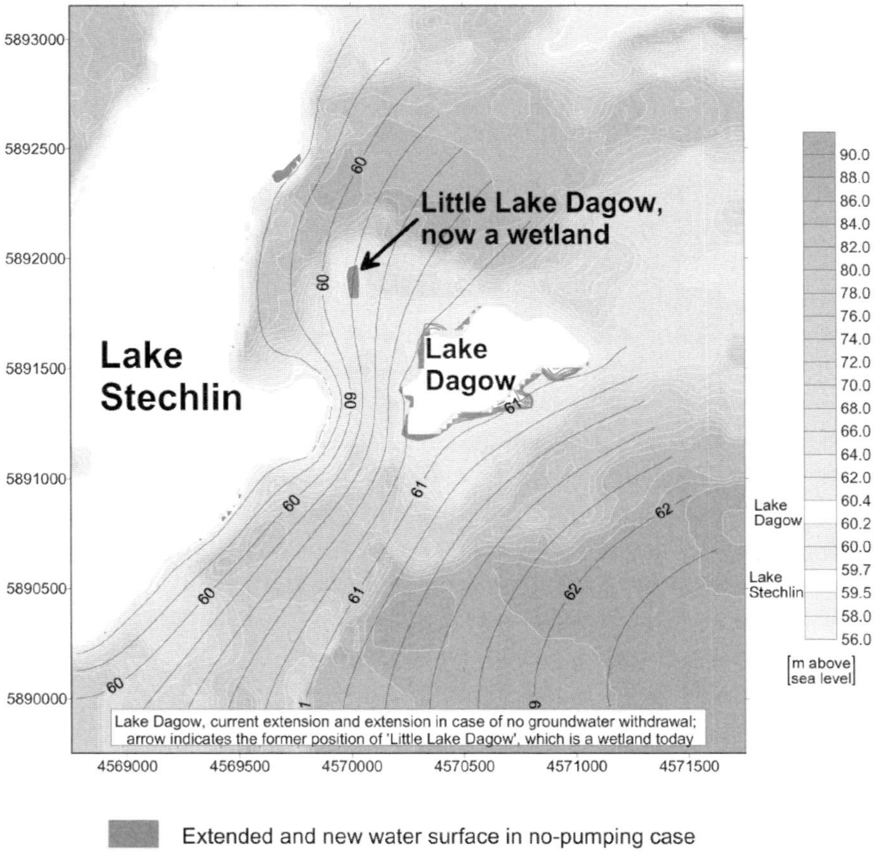

Fig. 4.2-2 Lake Dagow catchment and vicinity in a no-pumping scenario; topography (white contours), groundwater hydraulic head (black contours), water surfaces

4.3 Modelling the Hydrology of the Nile Delta

Ekkehard Holzbecher

4.3.1 Motivation and Objectives for the Coupling Approach

The connection between the hydrological compartments considered in the study is simplified, with no inter-compartmental feedback. The influences on groundwater from surface water, the atmosphere and the sea are taken into account, but effects in the reverse direction are not included in the model. The case chosen here is an example of intra-compartmental two-way coupling. In problems dealing with salinisation, a two-way coupling of flow and transport is usually necessary, while in hydrology there are good reasons to neglect the link between transport (salinity) and water flow.

4.3.2 Short Description of the Study

The Nile delta is the major supplier of the increasing food demand in Egypt. Although Egypt is the outflow recipient of the Nile drainage basin, which covers almost 10% of the African continent, the water requirements of the fast growing population has begun to outstrip its allocated water supply. There are problems of both water quantity, as water is a scarce resource in an arid region and also with water quality, particularly concerning the salt content.

The Nile delta is located around and between the two major branches of the River Nile, the Damietta and Rosetta, before the Nile reaches the Mediterranean Sea. Covering an area of more than 23.000 km^2 and stretching more than 150 km from east to west at the coast, it forms a giant oasis within the Sahara desert. The underlying Pleistocene aquifer, consisting of permeable sands, has a north–south extension of more than 100 km, reaching from the Mediterranean to the city of Cairo. The aquifer is limited on both sides by the desert. The main source of water into the region is the Nile river, which bifurcates into the two delta arms, Damietta and Rosetta, 20 km north of Cairo. The depth of the aquifer, from a maximum of 1000 m, decreases gradually with the distance from the sea.

In an interdisciplinary project managed by the Technical University Berlin (TUB), a hydrological model was set up which takes into account the main components of the water and salt balance. The project was coordinated by TUB within the Special Research Project "Arid and Semiarid Areas"

(SFB 69) and included hydrogeologists, sedimentologists, hydrologists and hydromechanics. A major aim of the project was an evaluation of the saltwater intrusion into the Nile delta aquifer.

4.3.3 Complexity of the Study

The two main compartments of the model are the surface model and the aquifer model. Aside from the hydrosphere, the surface model also includes the biosphere, as different crop patterns and time periods are considered. The aquifer model covers the saturated fluid phase within the geosphere. The coupling between the two compartments is one way only, as described below. The coupling term is groundwater recharge and a two-way coupling between the water and salt balance is implemented. Using the terms introduced above, there is intra-compartmental feedback but no inter-compartmental feedback.

The surface model considers surface water inflow from the Nile and groundwater withdrawal. There is outflow through drainage channels into the Mediterrranean. Outflow from the Nile river to the Mediterranean is relevant only in the Rosetta branch; outflow through the Ismailiya canal is even smaller. Precipitation and evaporation observations were taken from 27 climate stations distributed throughout the model region (Arlt 1995). The CROPWAT code (FAO 1991) was used to compute evapotranspiration, depending on climatic conditions, crop pattern and agricultural practice, once for the winter season and then for the summer season. Groundwater recharge was calculated as a missing variable from the fluxes mentioned and used as input for the groundwater model (Baumann 1995)

The aquifer model considers the following sources: direct inflow from the Nile river and from the Ismailiya canal, groundwater recharge and inflow from the Nile river aquifer. Aside from groundwater withdrawal, groundwater is lost from the aquifer into the adjacent Wadi Natrun and the Moghra aquifers and to the Mediterranean (Baumann 1995). Fig. 4.3-1 shows the flux balance of the aquifer.

The vertical cross-section in the north-south axis through the aquifer was selected as representative for the set-up of a two-dimensional model. The model region reaches from the Mediterranean for 100 km (approximately two-thirds the distance to Cairo) to the south (along the symmetry axis of the delta) and has a maximum depth of 1 km at the northern boundary. The permeability of the Pleistocene aquifer has been obtained on the basis of numerous pumping tests reported in former publications (Zaghloul 1959; Farid 1980). The hydraulic conductivity is $6 \cdot 10^{-4}$ m s^{-1}. Some semi-permeable

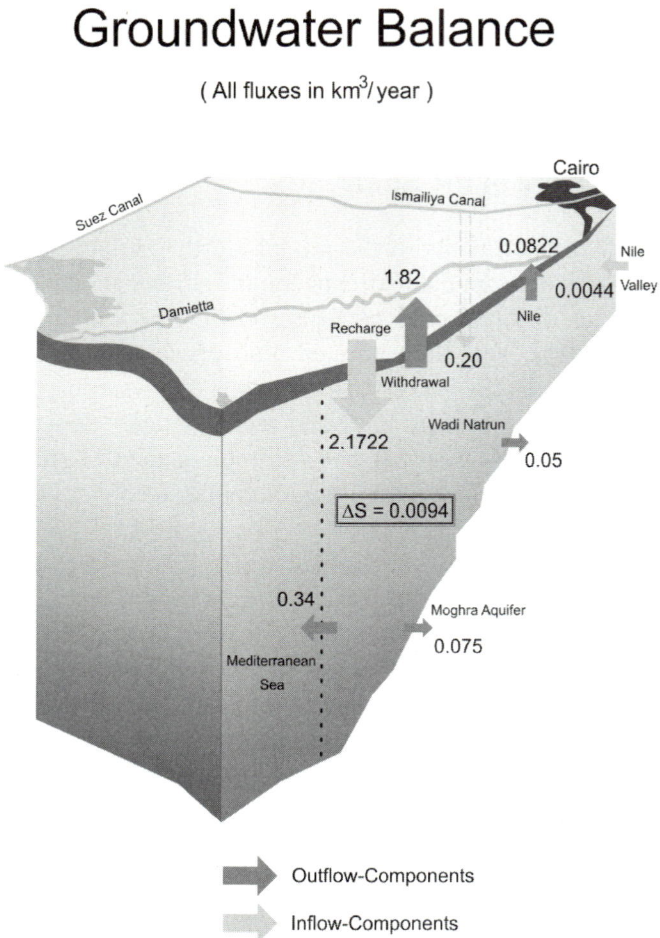

Fig. 4.3-1 Fluid balance terms of the Nile-delta aquifer according to Baumann (1995)

lenses with a conductivity of $1 \cdot 10^{-7}$ m s^{-1} were taken into account. In most parts the aquifer is confined by a clay cap with a conductivity of $2.8 \cdot 10^{-9}$ m s^{-1}. The clay cap originates from Nile river sediments in the Holocene and its depth reaches from a few metres in the south up to 75 metres below today's coastline. There is groundwater recharge through the semi-permeable clay in the southern part and discharge in the northern part of the delta.

The computer model is based on the mixing approach, i.e. on the fact that saline and fresh waters mix and that a transition zone is built up. Water flow and salt transport are calculated in a coupled approach. Boundary conditions for both flow and salinity need to be given. The flow boundary conditions at the surface and at the inflow boundary in the south are determined from the hydrological budget of the Nile-delta region. Of relevance to the aquifer model is the difference between groundwater recharge (2.17 $km^3\ y^{-1}$) and groundwater withdrawal (1.82 $km^3\ y^{-1}$). Although of minor importance, exchange between different bodies of surface water and groundwater is considered. Top boundary flow conditions are specified according to the discharge/recharge characteristic in the clay cap. At the bottom of the model the no-flow condition is required. At the vertical boundary below the seaside, there must be no vertical component of the velocity.

The FAST-C(2D) code used for the aquifer model is designed for modelling saline or thermal, steady or transient convection in a porous media in two-dimensional regions. The code was tested on several cases, including Henry's classical saltwater intrusion set-up (Henry 1964; Holzbecher 1998) and the transient movement of a saltwater front in a saline disaster in Japan (Holzbecher and Kitaoka 1993). Various scenarios – transient and steady state – were simulated in order to evaluate the potential consequences of increased use of groundwater resources (Baumann 1995).

Within the project, extensive field trips in 1991 and 1992 were carried out. Water analyses were taken from new drilled boreholes. The analyses show a gradual change from NaCl-type waters in the north to $CaHCO_3$-type in the south of the delta. The field measurements down to a depth of 330 m indicated seawater concentrations (35 000 ppm) only along the coastline. Moreover, they show that the transition zone reaches about 40 km inland. Freshwater concentrations (less than 1000 ppm) were found in the middle of the delta only, using measuring electrical conductivity (Baumann 1995; Arlt 1995). The output of the FAST-C(2D) model corresponds well with these measurements.

4.3.4 Experiences

The results from the computer model simulated a total saltwater intrusion of no more than 40 km for water with a total dissolved solids (TDS) concentration of ca. 1000 mg l^{-1}. Variations in grid spacing and in the discretisation method for first-order derivatives demonstrate the extreme influence of numerical dispersion on the computational results. Saltwater intrusion in the Nile Delta has been overestimated in previous models because grids had been used which were too coarse. The Ghyben-Herzberg relation (Herzberg 1901; Holzbecher 1998), which relates the subsurface seawater-freshwater interface to freshwater head above the mean sea level, was found to be invalid for the Nile Delta aquifer: this is probably due to a question of scale (the classical rule can be questioned in long shallow aquifers with high aspect ratio e.g. length/height ≥ 100). The numerical results agree well with new measurements of salt concentrations (Arlt 1995). A severe and deeply penetrating salinisation problem in the Nile Delta aquifer was not discernable from the data obtained and evaluated.

Not addressed in the project was the problem of rising groundwater tables due to increased irrigation in some parts of the delta – a problem that is well-known in other regions of the world. Aside from other problems caused by rising groundwater tables, there is an additional loss of water by evaporation from the soil and by evapotranspiration. Increasing salinity in near-surface ground and soil water indicates that the water balance could be affected significantly in these regions. The feedback from the subsurface compartment to the surface compartment should not be neglected, at least locally, as it has been in the reported project.

4.4 Modelling the Changes in Hydrological Cycle Processes for Small and Middle River Basins in Conditions of Permafrost under Climate Change

Yuri G. Motovilov, Alexander Georgiadi

4.4.1 Objectives and Motivation

The objective of the model ECOMAG is the analysis, mathematical description and modelling of hydrological cycle processes of river basins in conditions of permafrost. This includes research into the effects of climate change and human activity on hydrological systems with, in particular, the development of parameterisation of the subgrid effects of hydrological cycle processes in climatic models.

Intensification of human activity, connected with using and transforming natural resources, causes essential changes to geosystem status and frequently promotes intense ecological changes: the quality of air, water in the rivers and reservoirs is worsened, degradation of soils and forests is amplified, all of which change hydrological cycle processes on the land and in the atmosphere. The hydrological cycle is an important component of the geosphere and biosphere, dependent upon global changes and also influencing these changes. In this connection, creating a logical and uniform chain of the diverse processes within geosystems requires a mathematical description of hydrological cycle processes which uses the same approaches and conceptual language applied in other geophysical sciences. This has resulted in the development of physically-based, space-distributed models of the hydrological cycle on land. At present, such models are most suitable for investigating processes in catchments on a regional scale. However, in the near future, further improvement of such models will be necessary for a more detailed description of hydrological processes, such as the influences on water quality, thermal condition of soils and biological processes. Necessary improvements will include estimation of changes in the hydrological characteristics through human influence or as a result of climate changes, as well as estimation of the exchange processes operating at the land-surface/ atmosphere interface for better hydrological representation in atmospheric models. Such research is very important, in particular for the permafrost zones such as in Siberia or the Far East, where human activity can result in irreversible changes in the hydrological behaviour of territories, resulting in catastrophic economic and ecological consequences.

4.4.2 Short Description of the Study

Distributed hydrological models allow the determination of the water balance and its variation across river basins. Several such models are in common use (i.e. SHE-model, Abbott et al. 1986; TOPMODEL, Beven and Kirkby 1979; WATBAL Knudsen et al. 1986) but none of them explicitly contains components reflecting important characteristics of the boreal and permafrost landscape. The hydrological model ECOMAG (Motovilov and Belokurov 1996; Motovilov et al. 1999) has been developed for application to boreal and permafrost conditions.

ECOMAG (ECOlogical Model for Applied Geophysics) is a space-distributed, physically-based model of hydrological cycle and pollution transformation on a catchment at a regional scale.

The model consists of two main submodels:

1) The hydrological submodel describes the main processes of the terrestrial hydrological cycle: infiltration into the soil, evapotranspiration, thermal and water regime of soil, snow cover formation and snowmelt, formation of surface, subsurface, groundwater flow and runoff in the river network (Fig. 4.4-1).

2) The geochemical submodel describes the processes of contaminant accumulation on the earth surface, dissolution and penetration of soluble pollutants into the soil, their interactions with soil solution and soil matrix, biochemical degradation, transport of pollutants by surface and underground waters, transformation through the river channels.

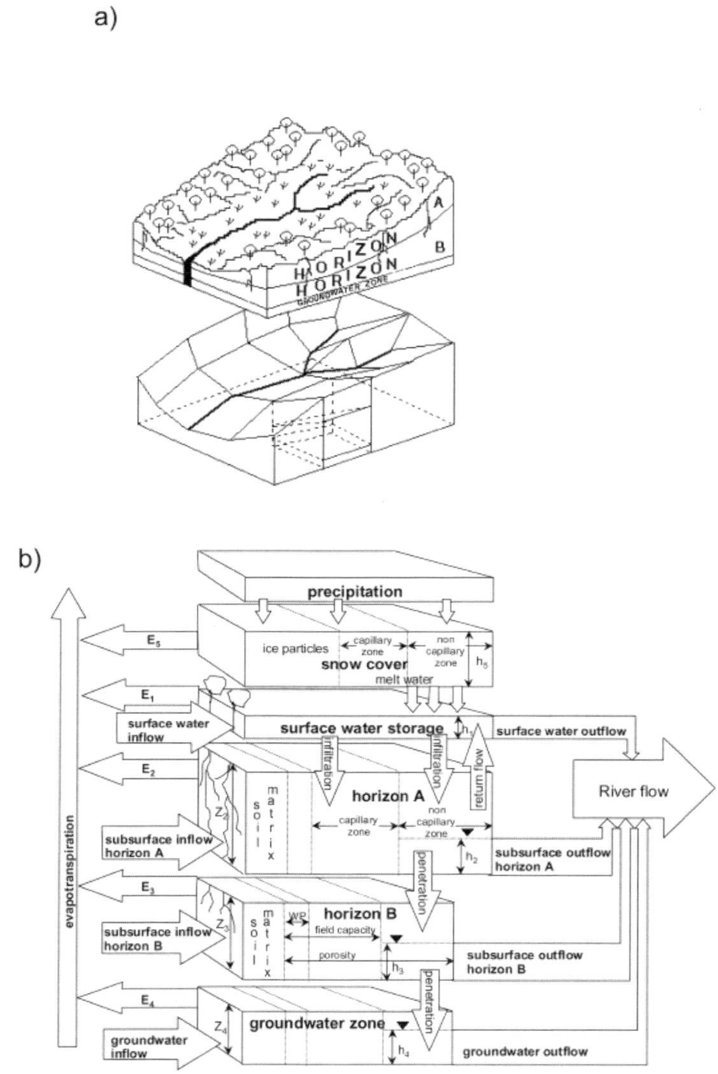

Fig. 4.4-1 Block-scheme of the ECOMAG model; a) Schematisation of a catchment; b) Vertical structure of ECOMAG for a landscape element

The model has been applied to hydrological cycle processes under different scenarios of climatic change for the Kenkeme river basin near Yakutsk (Yakutia-Sakha Republic of the Russian Federation, Fig. 4.4-2). The Kenkeme river basin extends to about 10,000 km^2 in a region of permafrost.

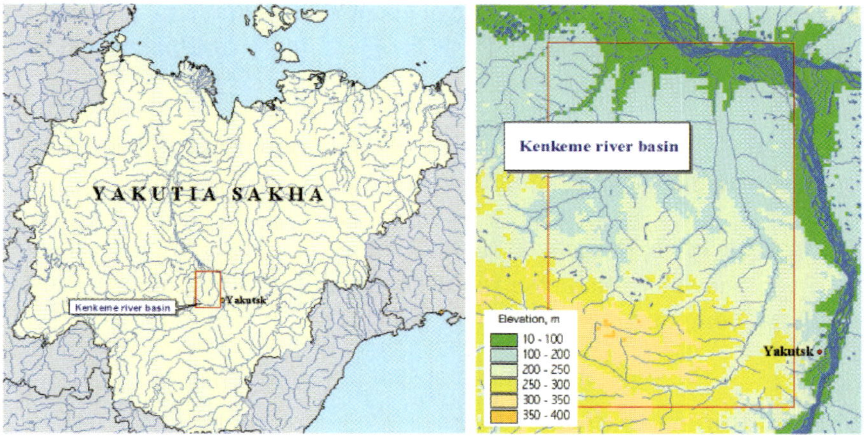

Fig. 4.4-2 Location of Kenkeme river basin

ECOMAG uses daily time steps resolution: the necessary minimum information needed for hydrological calculations includes standard meteorological observations as daily precipitation, temperature and humidity. Data on discharges, characteristics of snowcover, soil moisture, evapotranspiration, groundwater level, soil frost and thawing depth may be involved in calibration and validation of the model.

Space schematisation of the river basin (the allocation of river network, sub-catchments of tributaries, slope elements) is executed on the basis of thematic maps using GIS techniques. A drainage basin is approximated by regular square elements (grid cells from 100 m × 100 m to 4 km × 4 km, depending on the type of task and size of catchment), taking into consideration the peculiarities of topography, soil, vegetation and land-use types. The model has been applied to river basins in a range from 4 to 70,000 km^2.

The ECOMAG model is intended to help decision-makers with a wide range of hydrological and environmental river basin monitoring problems. In particular, it allows one to:

- Investigate the sensitivity of hydrological systems to the natural variations in climatic conditions as well as anthropogenic loads on the catchment;

- Simulate migration and transformation of polluting substances in river basins from point and non-point sources of industry and agriculture.

The formulation of the ECOMAG model for a specific river basin includes the following main stages:

- Collection of data necessary for calculation and calibration of model parameters (electronic thematic maps of a region, hydrological and meteorological data);

- Estimation of sensitivity, calibration of parameters and validation of the model on archival series of observations;

- Simulation of environmental characteristics on various levels (surface, soil, groundwater zone, river runoff) under different hydrometeorological conditions and anthropogenic loads.

4.4.3 Complexity of the Study

Hydrologists are studying climate-change effects on water resources using sensitivity analysis. The main approach for assessing the patterns of climatic change for the future is through modelling the climate system with General Circulation Models (GCMs). Then standard methods are usually used for generation of meteorological input from the results of GCMs into hydrological models. The mean monthly climatic differences in temperature and precipitation from GCM simulations are added to the daily values of each climatic variable during the hydrological simulation period.

Many studies of climate change impact on the terrestrial hydrological cycle and river runoff have been carried out using different hydrological models, scenarios of climate changes, river basins, geographic zones, continents. These sensitivity analyses contain two main types of uncertainty. The first depends on the adequacy of the hydrological model applied under the assumed climatic scenario: different types of hydrological models have been used to estimate climate change impact on river runoff, from regression equations to detailed physically-based models. Unfortunately, most of them were not validated by the Klemes hierarchic test-system (Klemes 1986) and, therefore, care should be taken with the conclusions made on the basis of such models. The second type of uncertainty is connected with the climatic scenario, which forms the basis for runoff simulations: GCMs give the climatic projection for the whole century but the precision of such forecasts are not known.

In the coming years, a further improvement in hydrological models and GCMs is necessary for a more detailed description of hydrological processes, especially in permafrost regions. There is a need to estimate the feedbacks and exchange processes at the land-surface/atmosphere interface to gain better hydrological representation in atmospheric models as well as better estimation of changes in the hydrological characteristics under the influence of human activity or as a result of climate changes.

4.4.4 Experiences

Primarily, the ECOMAG model was constructed and applied in Russia for regional environmental monitoring (Motovilov and Belokurov 1996). Since 1995, this version of the model has been improved for boreal and permafrost conditions in co-operation with L. Gottschalk and his group at Oslo University, Norway. The model was applied to regional simulation of the terrestrial water cycle during the international projects NOPEX (NOrthern hemisphere climate Processes land-surface Experiment; Gottschalk et al. 1998, 2001; Motovilov et al. 1999) and GAME-Siberia (GEWEX Asian Monsoon Experiment; Motovilov and Belokurov 1997). The studies included among others:

- Climate change impact on the hydrological cycle of river basins (Motovilov 1998);
- Estimating the influence of diffuse agricultural pollution on water resources in the Baltic Sea;
- The estimation of environmental damage around the new Oslo International Airport;
- Environmental monitoring in areas where rocket stages return to Earth;
- Simulation of inflow into reservoirs of Volga river basin.

4.5 Modelling Atmospheric and Hydrological Processes in the Boreal Region

Sven Halldin, Lars Gottschalk, Sven-Erik Gryning

4.5.1 Motivation and Objectives

Hydrological model applications have traditionally been done by setting parameter values and boundaries. When computer constraints alone are no longer a limitation on modelling capability, it has become clear that the real predictability constraint is not model-structure detail but place-specific characteristics and processes. New developments in computer technology will allow implementation of integrated models of large areas of the landscape, even if not whole countries. When climate change is a possible threat to our societies and the world-water scarcity is emerging as a major international problem, the integration of atmospheric and hydrological properties over different scales becomes a key modelling issue.

Studies of elements for such new models for the boreal landscape are one of the focal points for the NOPEX project (Northern Hemisphere Climate-Processes Land-Surface Experiment; Halldin et al. 1999, 2001). The estimation of regional fluxes of sensible and latent heat are central. Flux estimations are made with mesoscale hydrological and atmospheric models. It has been a challenge to compare flux estimates critically and evaluate them in a systematic way. Such a comparison has helped to identify problems in coupling atmospheric and hydrological approaches for flux estimation, related to critical processes and differences in scale.

4.5.2 Description of the Study

Area

The NOPEX project is carried out in two regions. The southern NOPEX region, around Uppsala, central Sweden, is characteristic for the southern edge of the boreal zone. The northern NOPEX region is situated at the northern boreal edge around Sodankylä, north Finland. The NOPEX project is based on two experimental activities. Intensive, time-limited Concentrated Field Efforts (CFEs) focus on comprehensive studies of the boreal-forest climate system, especially the spatial variation in fluxes and states over a region of size similar to a GCM grid cell. CFE1 (summer 1994), CFE2 (summer 1995), and CFE3 (winter/spring 1997) will be followed by CFE4 (winter) to finish NOPEX. Long-term monitoring of the land-surface/atmosphere climate and its components has been carried out since May 1994 in the Continuous Climate Monitoring (CCM) programme in order to follow climatic trends, capture seldom-occurring but important events, and to allow initiation, validation and comparisons of models and analyses at different time scales. The NOPEX regions and main scientific results are described in the two NOPEX Special Issues of *Agricultural and Forest Meteorology* (Halldin et al. 1999) and *Theoretical and Applied Climatology* (Halldin et al. 2001). Both issues include a CD with measurements from the campaigns.

Modelling approach

Three modelling approaches have been used to estimate regional fluxes and to identify critical processes at the land-surface/atmosphere interface. The mixed-layer-evolution model, the atmospheric mesoscale MIUU model, and the hydrological ECOMAG model (see also Sect. 4.4).

The mixed-layer-evolution model (Batchvarova et al. 2001) is based on a zero-order mixed-layer growth model for near-neutral and unstable atmospheric conditions and is applicable when the mixed-layer height is well above the height where individual surface heterogeneities are felt. The model requires wind speed and temperature profiles obtained from radio soundings, remote-sensing, or radio-acoustic systems extending above the blending height.

The MIUU model (Enger 1990) solves prognostic equations for horizontal wind components, potential temperature, specific humidity and turbulent kinetic energy. A terrain-influenced coordinate system is used to introduce topography in the model. The second-order turbulence closure is level 2.5 whereas the advection scheme is of third order, both in time and space, and has been corrected for numerical diffusion. Each grid is divided into a vegetated and a non-vegetated part. The latent and sensible heat fluxes from these are calculated separately and the total flux to the atmosphere is obtained as a weighted sum. The land-surface parameterisation uses a bulk-canopy model which calculates the ground and canopy energy balances separately. The soil is modelled with a two-layer force-restore method. The canopy parameterisation includes influences of both ground and canopy albedos and emissivities, leaf area index, stomatal resistance and roughness characteristics. The grid resolution is 1.5 km × 1.5 km in the central part of the model domain and expanding towards the boundaries.

The distributed ECOMAG model (Gottschalk et al. 2001) was developed for boreal conditions and describes soil infiltration, evapotranspiration, thermal and water regimes of soil – including freezing, surface and subsurface flow, groundwater and river flow, and also snow accumulation and snowmelt. In its original version, a drainage basin is approximated by irregular triangular or trapezoidal elements, depending on topography (size, form and slope), soil types (peat, clay, sand, till, bedrock), vegetation and land use (open area, forest, lake, mire, urban area). This version has been widely applied in Russia. A modified version of the model was developed within NOPEX that uses a regular grid (2 km × 2 km) to allow coupling with the MIUU model and radar-derived precipitation data. The basic assumption of the model is that a river basin can be divided into a mosaic of irregular or regular elements, each to be viewed as a landscape hydrological unit. The soil is divided into a top layer, an intermediate layer, and a capillary zone just above the groundwater surface. The vertical extension of these layers is a dynamic variable controlled by the position of the groundwater. The quotient of actual to potential evaporation equals unity under wet conditions and decreases linearly with soil moisture below a threshold. A kinematic wave equation is used for water transport in rivers, which constitutes the only water transport between elements.

4.5.3 Complexity of the Study

Three specific scales can be identified in a boreal landscape. The first is related to the hillslope flow dynamics, which determine the rate of runoff formation and flow into the permanent river network, lakes and bogs. The second is connected to the dynamics of a river basin in a representative landscape mosaic of hillslopes, lakes, bogs, etc. The third scale is defined from the soil–vegetation–atmosphere interactions in a representative landscape mosaic of different soils, vegetation and land use. The two latter scales might be related, but the boundary for the first one is set by the watershed or basin edge.

The wind near the ground changes ceaselessly over the various surfaces, trying to reach a balance with each one. The size of the dominant eddies increases with height, so the signatures of individual surface features are blended efficiently at the so-called blending height. The vertical influence of individual roughness elements, such as tree tops, is confined to the *roughness sub-layer*. The trees are organised in forested areas, separated by agricultural fields, mires, and lakes on the scale of kilometres or more. The influence of such a heterogeneity extends up to the *blending height,* loosely defined as the level above which the flow becomes horizontally homogeneous. Fluxes of surface momentum and heat are often modelled with similarity theory and require roughness lengths for momentum, z_{0m}, and for heat z_{0h}. Because momentum transport is enhanced by pressure fluctuations, z_{0m} can be greater than z_{0h}. Topography and isolated objects also enhance momentum flux through form drag but this hardly contributes to area-averaged heat flux. The thermal roughness length can be very small compared to z_{0m} when using surface radiation temperatures. The use of similarity theory to estimate the heat flux over a terrain with heterogeneous radiation temperature is complicated because heat flux can be counter to the temperature gradient. Spatially-averaged radiation temperature can, for example, be dominated by snow-covered open fields with negligible heat fluxes, while considerable positive daytime heat fluxes are typical for the northern boreal forest. Transfer coefficients for counter-gradient flux are negative and z_{0h} is not defined.

The hydrological response in a landscape with till deposits depends on the soil moisture in the unsaturated zone and the depth of the groundwater table. The soil-moisture and groundwater conditions in the NOPEX region varies between hydrological response units with different surface deposits or topographical characteristics, whereas the dynamics in units with similar characteristics are often identical. Soil-moisture and groundwater conditions in environments with till deposits are dominated by variations in topography, vegetation and soil characteristics at a length scale of less than 2 km. This means that there are no gradients in soil-moisture content or groundwater levels between sub-catchments at this spatial scale, and consequently, all lateral water transfer occur as runoff in the river network. It has been suggested that this scale should be used as a common spatial denominator for work with coupled atmospheric–hydrological modelling in the NOPEX regions.

MIUU and ECOMAG model parameters are "effective" since they are assigned a value for each grid cell according to land-use and soil type, and eventually adjusted by calibration. The direct relation to scale-aggregated values is therefore lost. An important use of the NOPEX process-scale-identification results is in the parameterisation of hydrological and atmospheric models. If the grid size is chosen in accordance with the hillslope scale, the parameterisation process implicitly includes the hillslope runoff formation processes down to its entrance to the river net and the transport between grid cells is limited to river flow. A smaller grid size would require the modelling of groundwater transport between the cells. A distinction between hillslope and catchment scales for the Scandinavian boreal zone is still problematic. The sizes of lakes, mires and bogs vary considerably. Even when a hillslope scale has been chosen as the proper size of a fundamental unit, each such element may still contain mires, bogs and lakes. A proper parameterisation at this scale needs to consider the interaction between hillslope dynamics and lakes, mires and bogs. There will also be cells occupied totally by larger lakes, mires and bogs. An obstacle for choosing the catchment scale as a base for defining a proper size of a computational element is that, from a hydrological point of view, it should correspond to the watershed boundary. This hampers the possibilities for coupling with meteorological models based on regular grid cells. A way out of this dilemma is to use a grid of fundamental units as a common denominator for catchments in the hydrological model component and another larger grid in the meteorological component.

Conditions at the northern NOPEX region during late winter are characterised by low solar elevation angles and days and nights of equal length. Measurements have revealed (Gryning et al. 2001) that forest-induced sensible heat flux could reach more than 100 W m^{-2} under these conditions (Fig. 4.5-1). The trees absorb short-wave radiation efficiently whereas the snow-covered ground reflects it. The heat flux above the forest comes mainly from tree warming, while that from the forest floor is small.

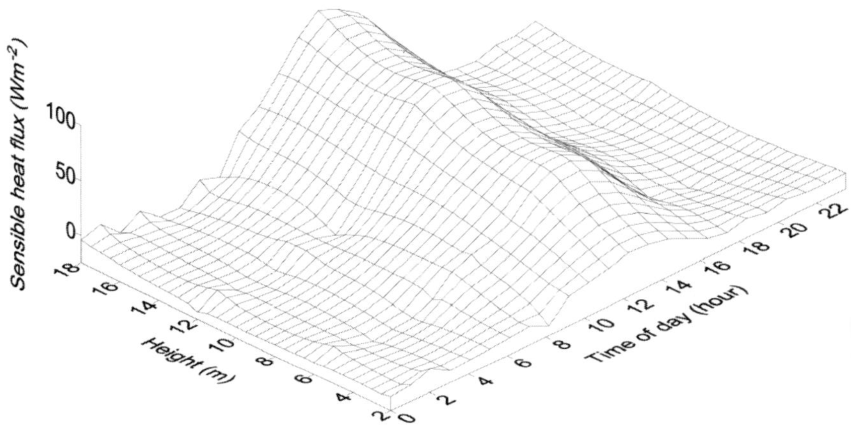

Fig. 4.5-1 Sensible heat flux in and above a sparse 9-m forest as function of height and time of day. The plot is based on half-hour measurements at 2, 6, 12 and 18 m height on 13-24 March 1997 in the northern NOPEX region

The latent heat flux, which originates mainly from snow sublimation, is ten times lower, typically 10–20 W m^{-2}. Simulations with a new parameterisation scheme showed that the forest heat flux under cloud-free conditions was controlled primarily by tree shading and not forest coverage, contrary to the situation for overcast conditions. Satellite- and aircraft-derived albedos are, thus, not relevant when modelling short-wave radiation absorption under cloud-free conditions during the boreal winter.

The mixed-layer-evolution model was used to estimate regional heat flux over the southern NOPEX region for different summer days. The regional flux was lower than the flux over the forest and higher than over agricultural fields, and agreed reasonably well with the land use-weighted average. The model was used in a similar way for winter days over the northern NOPEX region. In this case, the forest controlled regional fluxes of momentum and heat in a different way. The regional momentum flux was 10-20% smaller than the measured forest flux, and the regional sensible heat flux was 30-50% of the forest values. Good agreement was found (Fig. 4.5-2).

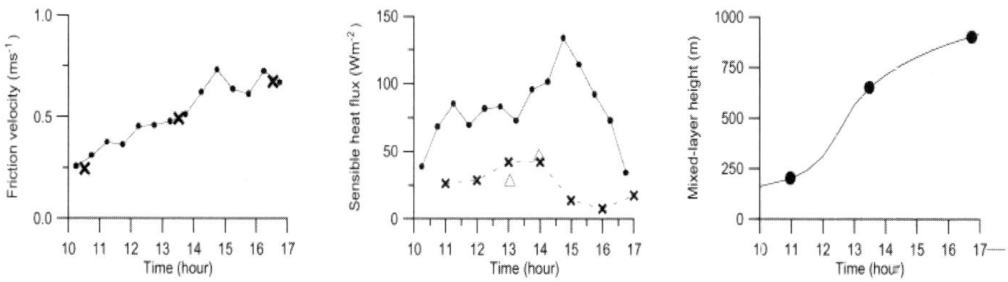

Fig. 4.5-2 Late winter conditions (CFE3) in local standard time. Left: Measured local friction velocities over forest at the northern NOPEX region (•, full line), and regional friction velocities from the mixed-layer-evolution model (×). Middle: Measured local sensible heat flux over forest (•, full line), derived regional sensible heat flux (×, broken line), and flight-track mean values (Δ). Right: Mixed-layer heights from radio soundings (•)

The ECOMAG and MIUU models were run in parallel for separate days during the NOPEX CFE1 (Fig. 4.5-3). A fundamental obstacle with this comparison is the fact that the meteorological model can only be run for 24-hour time slots and not continuously for the time period of the hydrological model. Changes in state variables of the hydrological model during 24-hour slots are minor, especially during summer months. Their role in initialisation of the meteorological model can, therefore, not be evaluated realistically. Initialisation of hydrological model states is, on the other hand, in its infancy and is presently done by "tuning in" the model for a preceding period of several months.

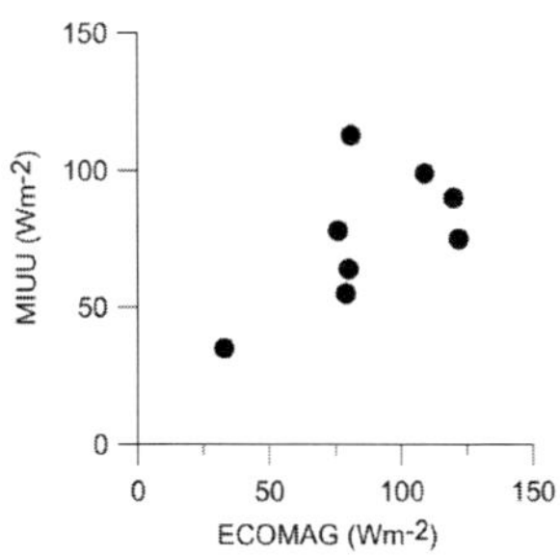

Fig. 4.5-3 Estimated latent heat fluxes from the atmospheric MIUU and the hydrological ECOMAG mesoscale models over the southern NOPEX region for selected summer days (CFE1/2)

4.5.4 Experiences

The boreal region is one of the most heterogeneous landscapes in the world and models developed elsewhere are not always applicable. New experimental studies, especially in wintertime, have revealed the importance of so far unrecognised processes, such as the effect of low solar angles on sensible-heat production. It is likely that further experiments will be needed to identify all critical processes needed to model the land-surface/atmosphere system at northern latitudes.

Results of NOPEX studies have highlighted issues that should be resolved before coupled hydrological and meteorological models can be applied successfully in the boreal region:

- Most hydrological and meteorological models define their grids and boundaries differently. They are normally only meaningful if operated with different time spans and temporal resolution. These incoherencies hide fundamental process-related problems, different for different landscapes and regions of the world;

- The selection of scale is intimately linked with the parameterisation requirements for subgrid-scale processes. A grid scale of 2 km seems suitable for both atmospheric and hydrological models in the boreal region but this conclusion needs to be confirmed with data from more sites;

- The ECOMAG model concept has been applied successfully to the southern NOPEX region (Gottschalk et al. 2001) but the parameterisation of physical processes at the hillslope and catchment scale requires further consideration;

- New procedures for parameter-value estimation, traditionally by calibration in hydrology, must be developed, based on remote-sensing data. Surface albedo, e.g. cannot be used in its traditional form for winter conditions. Retrieval of parameter values will be crucial, especially when developing macroscale hydrological models to be coupled to regional-scale atmospheric models.

4.6 A Distributed Model of Runoff Generation in the Permafrost Regions

Lev S. Kuchment, Alexander N. Gelfan, Victor N. Demidov

4.6.1 Objectives and Motivation

The permafrost regions cover approximately one-quarter of the land surface of the world, more than 60% of Russia and one-half of Canada. Because of low population density, expensive access and limited human activity, these regions have a sparse and extremely unevenly distributed hydrometeorological network. The extensive collection of field-measurements during recent decades has increased the available information on the hydrometeorological processes in the cold regions considerably but most of these data are too fragmentary. As a result, the peculiarities of permafrost hydrology has been investigated but weakly. At the same time, the comprehensive physically-based models of runoff generation (for example Abbott et al. 1986; Kuchment et al. 1983, 1986) developed for the temperate latitudes cannot be applied to the permafrost river basins because of differences in the main processes and the lack of adequate hydrometeorological data and basin characteristics.

4.6.2 Short Description of the Model

A Runoff Generation Model in Permafrost Region (RGMPR), a physically-based distributed model of snowmelt and rainfall runoff generation, has been developed for the permafrost regions (Kuchment et al. 2000). It is based on experience with modelling runoff generation processes in different geographical zones as well as on the use of the unique hydrometeorological data collected at the Kolyma water balance station (Fig. 4.6-1). Runoff generation observations have been made at this station since 1948 for seven representative basins with drainage areas from 0.3 to 21.2 km^2. The data collected include the standard meteorological measurements of solar radiation, snow-cover characteristics, soil moisture and snow evaporation, temperature and soil moisture, groundwater levels, runoff, chemical content of runoff. The model describes snow-cover formation and snowmelt, thawing of the ground, evaporation, basin-water storage dynamics, overland, subsurface and channel flow. The main difference between the RGMPR model and models of runoff generation for regions with moderate climate is the small role given to infiltration of water into soil and a larger dependence of runoff losses

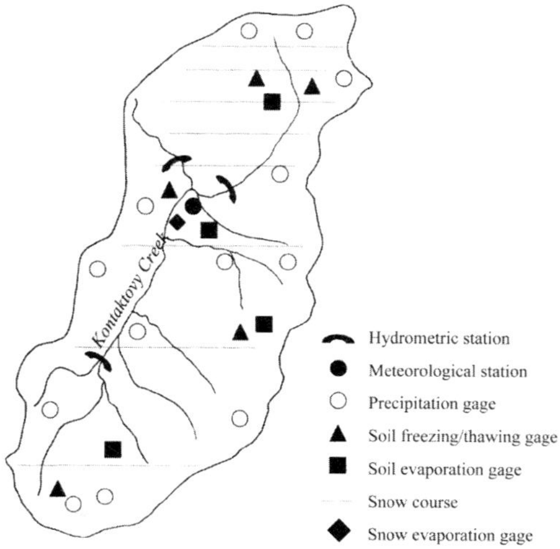

Fig. 4.6-1 Plan of the Kolyma water balance station

on the depth of thawed ground. The thaw of the frozen ground increases significantly the water input and the water storage capacity, changing the ratio between surface and subsurface flow. The choice of the structure of the model is based on the analysis of the long-term observations of the runoff generation processes at the Kolyma water balance station and is orientated towards the standard hydrometeorological information available in the cold regions.

4.6.3 Complexity of the Study

A case study with the model has been performed for the Upper Kolyma River basin (99,400 km^2) in which the Kolyma water balance station is situated (Fig. 4.6-2). The river basin is in the zone of continuous permafrost, interrupted only by patches of unfrozen ground under the river beds. The dominant soils are coarse-grained mountain-tundra podzols with large gravel content. The peatlands occupy about 2% of the basin area. The depth of the active layer is controlled by elevation, exposure, vegetation and presence of rivers and lakes. On the slopes of northern exposure, the average depth of the active layer is 0.2 - 0.8 m; on the slopes of southern exposure, the average depth of the active layer reaches 1.5 - 3.0 m. On the

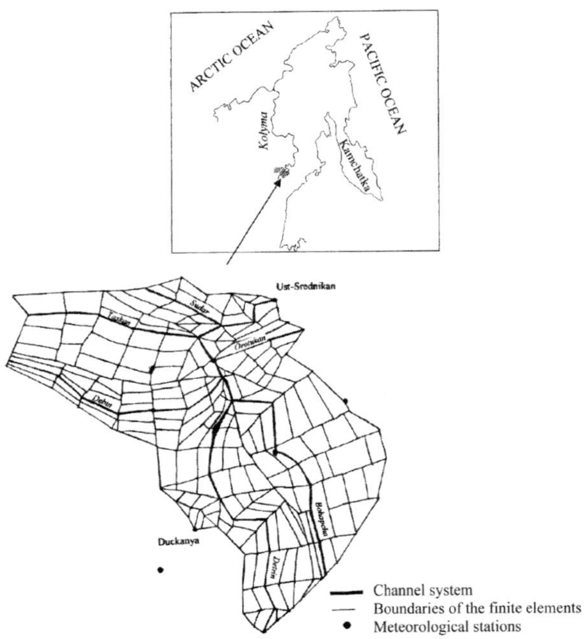

Fig. 4.6-2 Finite-elements schematisation of the Upper Kolyma River basin

basis of hydrometeorological data for the Upper Kolyma River basin and the literature on permafrost hydrology, the following general scheme for runoff generation is proposed (Fig. 4.6-3). First, the snowmelt water fills up the free storage capacity in topographic depressions in the peat mats and the ground,

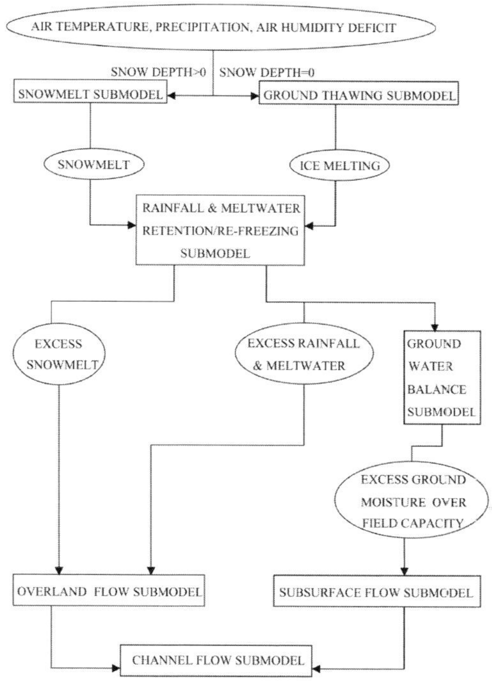

Fig. 4.6-3 Schematic diagram illustrating the major components of the model

where this water freezes. It is assumed that the basin storage capacity is distributed statistically over the basin and the mathematical expectation of this distribution before snowmelt depends only on the wetness of the ground in the antecedent summer-autumn period (before snowmelt the ground is deeply frozen). During the warm period, the storage capacity increases with increasing depth of thawed ground. Excess snowmelt and rainfall water over the free storage capacity forms overland flow. Ice melting and evaporation of soil moisture begin at the snow-free areas of the river basin. The melting of ice in the ground and in the depressions produces the subsurface flow and increases the basin storage capacity. The subsurface flow occurs above the frozen layer of the ground. The water retained by the capillary forces does not take part in subsurface flow. The infiltration of rainfall into the ground is quick and does not depend on the ground moisture conditions. It is assumed that the horizontal hydraulic conductivity decays exponentially with the depth. To calculate the characteristics of snow cover during snowmelt, the system of vertically averaged equations of snow processes at a point has been applied. This system includes the description of temporal change in snow depth, ice and liquid water content, snow density, snowmelt, sublimation, re-freezing melt water, snow metamorphism.

To take account of the spatial mosaic variability of the snowpack characteristics before melting, the Upper Kolyma River basin is divided into 16 approximately equal sub-areas, corresponding to the available snow course network. It is assumed that the spatial stochastic variations in the snow water equivalent within each sub-area can be described by log-normal statistical distribution. The movement of the front of ground thawing is described by the system of equations accounting for heat transfer in snow, frozen and unfrozen ground. It is assumed that the spatial distribution of the free storage capacity of the river basin can be described by exponential law. The evaporation rate is determined as a function of the air humidity deficit.

To model overland, subsurface and channel flow, a finite-element schematisation of the Upper Kolyma River basin and kinematic wave equations have been applied. The channel river system is represented by the main channel of the Kolyma River and by six tributaries (Fig. 4.6-2). The channel system is divided into 44 reaches (finite elements) taking into account the topography and the river network structure; the basin area is separated into strips, adjacent to the channel finite elements and along which one-dimensional flow to river channels is presupposed. The strips are also divided into finite elements with different topography, soil and vegetation characteristics.

The RGMPR model includes 12 constants. A set of numerical experiments had estimated the sensitivity of the runoff hydrographs to changes in different model parameters before calibration. It established that the porosity and the evaporation coefficient is the most important. These two parameters control accumulation and discharge of water in the ground during the entire period of runoff generation. The parameter controlling snowmelt has a significant influence on overland flow but during a short period only. The free basin storage before snowmelt strongly influences the form of the runoff hydrograph at both snowmelt and rainfall input.

The influence of the routing parameters of the roughness coefficients for overland river channel flow as well as the horizontal hydraulic conductivity is relatively small because of large variability of the slopes. Most of the used parameters variate in relatively narrow ranges and may be considered as regionally general. Some of these parameters can be determined using the

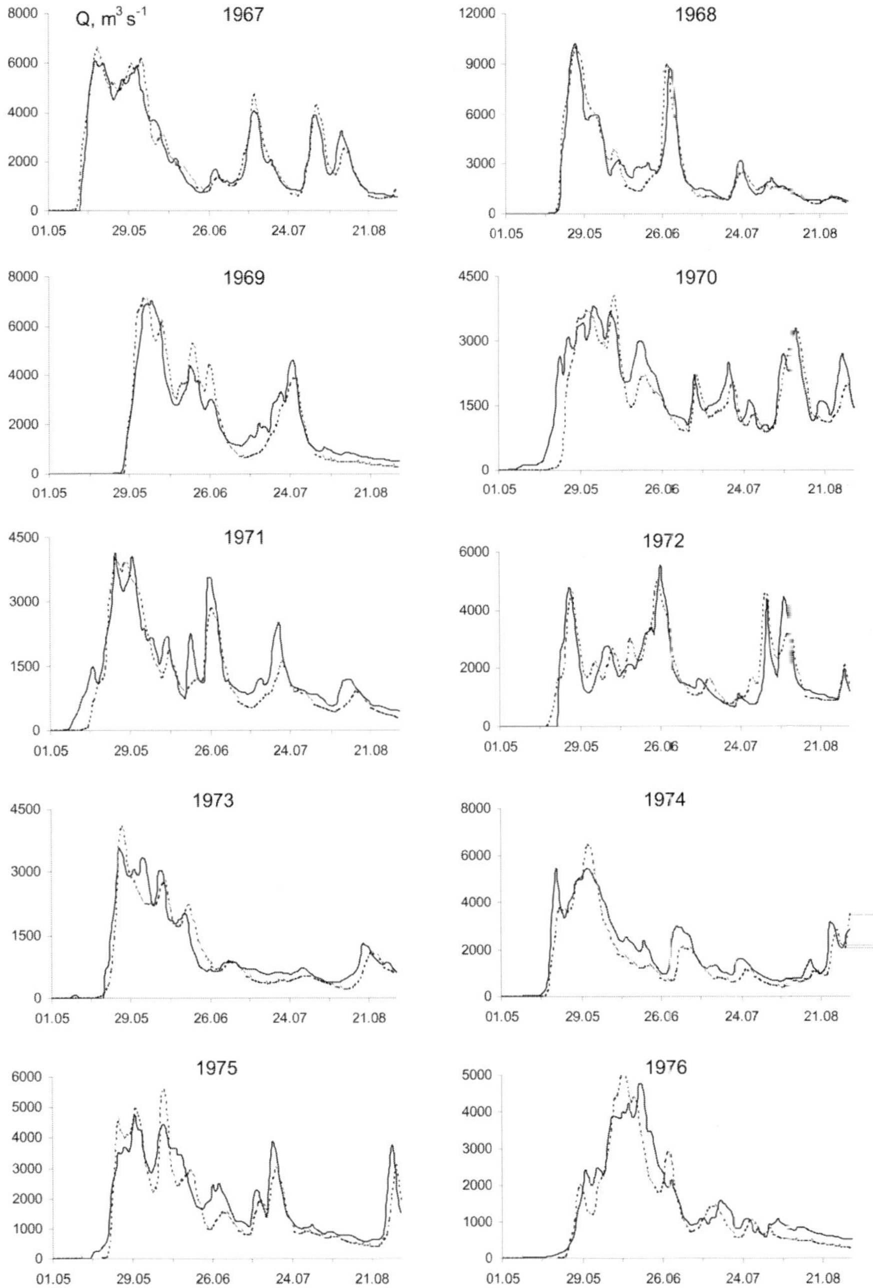

Fig. 4.6-4 Comparison between measured (continuous line) and calculated (dashed line) hydrographs

measurements on a small area within the river basin or on the basis of data taken from the literature.

To calibrate and to verify the RGMPR model for the Upper Kolyma River basin, daily hydrometeorological measurements from 1 May to 31 August during 10 years (1967–1976) were used. The meteorological data include records at eight sites. Five constants have been assigned on the basis of measurements of these constants at the Kolyma water balance station: two constants have been calibrated, using the hydrometeorological observations at this station and five have been calibrated with the aid of the hydrographs of the Kolyma River.

4.6.4 Experiences

The validation of the RGMPR model has shown that the model simulates satisfactorily the main peculiarities of runoff generation in cold regions (Fig. 4.6-4). The standard error of the calculated runoff hydrographs is 482 $m^3 s^{-1}$ (the standard deviation of the measured hydrographs is 1434 $m^3 s^{-1}$). The standard error of the calculated peak discharges is 490 $m s^{-1}$. The largest relative error of the peak discharges is 18% (for 1974); the largest relative error of the runoff volume are 14% (for 1971) and 13% (for 1970).

An important part of the model is the interaction between heat transfer processes in the snow and the soil and basin characteristics that control water movement. Basin storage is directly affected by melting and freezing water at the surface and at different depths in the soil. The thaw of the frozen ground increases runoff losses significantly and changes the ratio between surface and subsurface flow.

4.7 Investigations on the Impact of Land-use Changes using an Integrated Hydrometeorological Model

Nicole Mölders

4.7.1 Motivation and Objectives

With an increasing world population as well as changing climate and land-use conditions, the availability of water becomes a central question of research interest. Water availability, amongst other things, depends on precipitation, infiltration, surface runoff, as well as on the exchange of heat, matter and momentum at the earth-atmosphere interface. Because they depend on both the land and the atmospheric parts of the water cycle, studies on water availability require a detailed consideration of the water cycle to guarantee the applicability of models for future planning tasks. As an example, the role of surface runoff in the regional water cycle and its meaning for water availability under changed land-use conditions is examined.

4.7.2 Short Description of the Study

Area

Simulations with and without surface and channel runoff are performed for northern Saxony and southern Brandenburg, Germany. These simulations and their results are referred to hereafter as *ho* and *so*. In addition, these simulations are performed for a modified landscape wherein the water meadows along the rivers are changed to willow-forest, denoted as *hm* and *sm*, respectively.

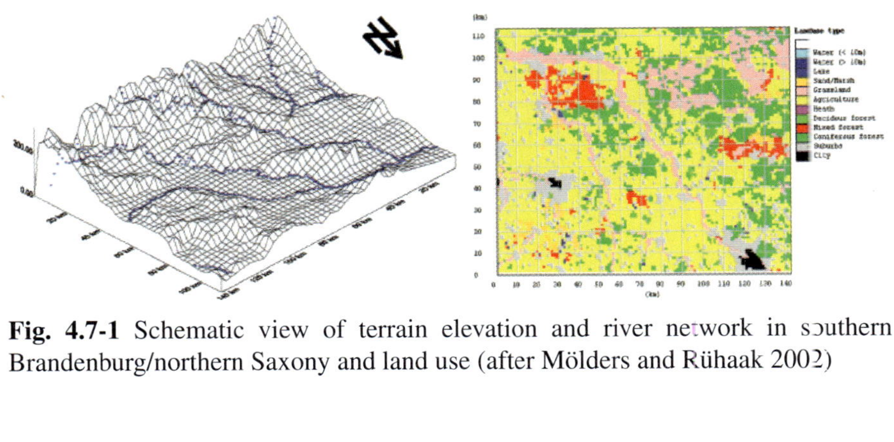

Fig. 4.7-1 Schematic view of terrain elevation and river network in southern Brandenburg/northern Saxony and land use (after Mölders and Rühaak 2002)

Coupled processes and compartments

A module to explicitly predict surface and channel runoff was developed (Mölders and Rühaak 2002) and coupled to the hydro-thermodynamic soil-vegetation scheme (HTSVS; Kramm et al. 1994, 1996) that was coupled with a mesoscale-β meteorological model as a hydrometeorological model (see Chapter 3.3).

Surface runoff and river runoff are described by the St.Vernant equation. Every subgrid cell may receive surface runoff from its neighbouring subgrid cells. Rivers are fed by surface runoff. There is no recharging of soil moisture by the rivers. Darcy's law and the Richards equation are used in the vertical direction, i.e. no horizontal transport within the soil is considered. The heat- and moisture-processes occurring within the soil are described by balance equations for soil-temperature and volumetric water content. The Ludwig-Soret-effect (i.e. a temperature gradient is able to generate a change of soil volumetric water content) and the Dufor-effect (i.e. a moisture gradient may alter soil temperature) are taken into account. The volumetric heat capacity of moist soil, soil albedo and hydraulic conductivity are functions of soil volumetric water content. Infiltration is determined by an explicit formulation of the Green-and-Ampt approach. The parameterisation of vegetation includes a mixture approach to consider bare soil and/or vegetation simultaneously within the grid cell. This heterogeneity at the microscale is relevant for the near-surface stratification of the atmosphere (stability) and the atmospheric fluxes of sensible and latent heat. Transpiration of water by plants is calculated by a bulk-stomatal resistance approach.

Spatial and temporal scales

The time step is 10 s as required by the atmospheric model part to fulfil the Courant-Levy-criteria. The horizontal grid resolution of this part of the model is 5×5 km^2. Surface and channel runoff are simulated on a subgrid having a horizontal resolution of 1×1 km^2. The river network is divided into pieces of different flow direction, length, slope and boundary conditions for inflow and outflow (Fig. 4.7-1). Each subgrid cell has its own in- and outflow conditions. An explicit subgrid scheme is applied for aggregation and disaggregation (e.g. Mölders et al. 1999).

4.7.3 Complexity of the Study

Type of feedbacks involved

Primary differences between the simulations with and without the runoff module result from runoff. Whenever or wherever the precipitation rate exceeds the infiltration rate, water is ponded on the surface and will contribute to runoff in *ho* or *hm*. Neglecting surface runoff means that the water will be ponded if precipitation rate exceeds infiltration rate, and the ponded water will be infiltrated later in *so* or *sm* than in *ho* or *hm*. Consequently, in the areas of heavy precipitation, the total infiltrated water amounts of *so* and *sm* exceed that of *ho* or *hm*, respectively. Conversely, in the latter simulations, infiltration also occurs in regions without precipitation that are downhill of the areas receiving precipitation because of the surface runoff of ponded water. Furthermore, since water flows downhill, it can reach areas where the precipitation rate is less than the possible infiltration rate. Thus, here infiltration is higher in simulations incorporating runoff than in those without (see Mölders and Rühaak 2002).

The modified infiltration yields to altered distributions of soil volumetric water content, soil temperature, surface temperature and surface moisture as well as water availability. Consequently, neglecting runoff also affects evapotranspiration (Fig. 4.7-2). Thus, in the simulations with and without the runoff module, different amounts of water vapour are supplied to the atmosphere for which cloud- and precipitation-formation differ. This means that the simulated local recycling of previous precipitation is affected by the inclusion or omission of surface runoff. The altered precipitation again yields to differences in ponded water, infiltration, evapotranspiration, water availability and groundwater recharge. In general, the differences between the simulations with and without the runoff module grow with increasing simulation time.

Simplifications

The interaction between rivers and groundwater, lateral soil water fluxes as well as evaporation from the water surface of rivers are all neglected. A constant water flux into the rivers that flow into the model domain is assumed at its boundaries. The temporal change of plant physiological parameters (e.g. albedo, emissivity, root depth, LAI, aerodynamic and hydraulic roughness length, stomatal resistance, etc.) and the possible variation of soil type with depth are also neglected.

Internal processes and external factors

The land-use changes mean a change in plant physiological and surface parameters (e.g. stomatal resistance, aerodynamic and hydraulic roughness, albedo, emissivity, etc.). During daytime, surface moist static energy flux will not be altered by the parameter modifications if moist static energy does not change. The partitioning between the sensible and latent heat fluxes may be affected and, hence, the Bowen-ratio may change. An increase in Bowen-ratio means a decrease in moist static energy that potentially reduces convection. In the case of a change from agriculture to deciduous forest, albedo is enhanced. For the change from grassland to deciduous forest the opposite is true: reducing albedo increases the moist static energy.

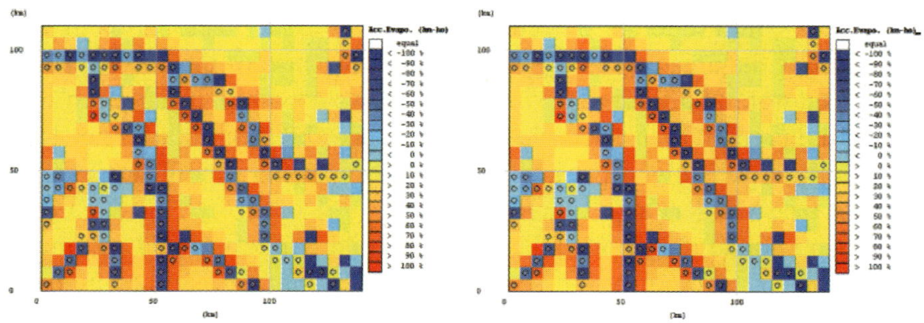

Fig. 4.7-2 Percentage difference of 24 hours accumulated evapotranspiration as obtained by *hm* – *ho* scaled by *ho* (left), and *sm* – *so* scaled by *so* (right). Modified after Mölders and Rühaak (2001)

Fig. 4.7-2 shows the percentage difference of 24 hours accumulated evapotranspiration. First, differences in evapotranspiration occur in the areas of the land-use changes. Later on, in the unchanged leeward-side regions of the land-use changes, further difference in evapotranspiration result from the modified wind, temperature and moisture states of the surface layer, the altered recycled precipitation and insolation due to differences in cloud distribution. As a consequence of the changed evapotranspiration, the surface moisture distribution and water availability differ. In addition to the altered water vapour supply to the atmosphere, the modified vertical mixing and heating affect cloud- and precipitation-formation. The changed pattern, rates and amounts of precipitation again affect infiltration, ponded water, evapotranspiration, runoff, water availability, and groundwater recharge.

Neglecting runoff can yield an underestimation of the secondary effects of land-use changes in the areas downwind and/or downwards of the land-use changes. In the simulation incorporating surface runoff, the land-use changes contribute to a stronger change in infiltration, soil water fluxes and groundwater recharge than in that without surface runoff. Comparing the percentage differences in ponded water with and without consideration of runoff, for instance, shows that neglecting surface runoff can lead to an underestimation of the effect of land-use changes. The absolute percentage difference in infiltration caused by the land-use changes will be less than about 10%, on average, if runoff is neglected, but will be much higher when runoff is taken into account.

Coupled natural cycles

The processes of the hydrosphere, biosphere, geosphere and atmosphere are coupled, as depicted in Fig. 3.3-4. The anthropogenic impact on these processes is examined by assuming land-use changes to willow forest along the rivers. The sensitivity of bulk-stomatal resistance in the transpiration process to photosynthetic active radiation, water vapour deficit between leaf and ambient air, leaf temperature, soil water deficit and CO_2 concentration of the air is considered by correction functions that range from 0 to 1.

Treatment of uncertainty

Various studies were performed to examine the sensitivity of the integrated model (or parts of it) on the choice of plant physiological and surface physical parameters, parameterisations (e.g. Mölders 2001) as well as boundary conditions.

4.7.4 Experiences

Difficulties

Detailed data sets of river characteristics (e.g. Manning coefficients, form, width and depth of the river, etc.) are scare for small rivers. At the boundary of the model domain, river inflow has to be defined by observations or else prescribed reasonably. Furthermore, for long-term simulations the meteorological model has to be nested into a meteorological (global) model with coarser resolution for which assumptions are required on the nesting (one-way, two-way) at its boundaries.

Conclusions

The results of the simulations with and without runoff confirm the general impact of runoff on the regional water cycle found by Mölders et al. (1999) when applying a tightly two-way coupled system of a conceptual hydrological and meteorological model. Note that although they carried out their studies with the same meteorological model, they also applied another SVATS (force-restore method for soil moisture, heat diffusion for soil temperature) than that used in this case study.

The sensitivity of water availability to land-use changes was found to be greater when surface runoff is taken into account. Since very many rivers are of subgrid-scale with respect to the grid- or subgrid-resolution of mesoscale atmospheric models, one has to expect that the impact of land-use changes on water availability predicted by mesoscale atmospheric models without consideration of surface runoff may be underestimated if the patch-size of the land-use changes is assumed realistically. Secondary differences in water availability result from the altered surface moisture and evapotranspiration and, hence, different local recycling of previous precipitation. Based on the findings of this study, one may conclude that investigations on the regional water availability and its modification by land-use changes should include surface and channel runoff.

Further applications

In the future, studies will have to use measurements of river inflow at the boundaries of the hydrometeorological model domain. The hydrometeorological model should be driven by data from a coarser atmospheric model nested into a global model to allow for long-term studies. Doing so ensures capturing large-scale atmospheric phenomena (e.g., cyclones) and propagating them into the hydrometeorological model domain. As these phenomena build on scales much larger than those covered by mesoscale-β models, these models can only simulate time scales of a day or so when run on their own.

4.8 The Influence of Anthropogenic Landscape Changes on Weather in South Florida

Roger A. Pielke Sr., Curtis Marshall, Robert. L. Walko, Louis T. Steyaert, Pier-Luigi Vidale, Glen E. Liston, Walter A. Lyons, Thomas N. Chase

4.8.1 Motivation and Objectives

Model simulations in two studies were used to assess the extent to which land-use change in south Florida may have affected local precipitation during the summer season. A more detailed discussion of this work appeared in Pielke et al. (1999) and Marshall et al. (2004). Previous studies in other geographic areas demonstrated that landscape patterns can generate local atmospheric circulations due to contrasting surface properties that can be as strong as a sea breeze caused by a land-water contrast (see Pielke 2001, for a review).

4.8.2 Description of the Study

The Regional Atmospheric Modeling System (RAMS) Version 4.3 was used for the numerical simulations presented in this study (Marshall et al. 2004). All of the simulations were performed on a nested grid configuration with an outer grid of 42 × 48 points at a 40 km interval covering the southeast Atlantic and Gulf Coast states, southward to the latitude of the Yucatán Peninsula. An inner grid with 42 × 50 points at 10 km spacing was nested to cover central and south Florida and adjacent coastal waters. Both grids extended over 30 vertical levels, with the lowest level nearly 100 m above ground level. The vertical grid spacing was geometrically increased with height to a maximum of 1 km at the model top (20 km). Initial conditions and outer grid lateral boundary conditions were provided by the National Centers for Environmental Prediction-National Center for Atmospheric Research (NCEP-NCAR) global analysis dataset (Kalnay et al. 1996). During the integration of the simulations, the reanalysis data were updated every 6 h and nudged over the five outer grid points at each time step.

4.8.3 Complexity of the Study

The experiments differed solely in the definition of land-use category and initial soil moisture. The observed land-use change is illustrated in Fig. 4.8-1.

The land-use change is prescribed (a one-way feedback to the atmosphere), with the biophysical effect of sensible and latent turbulent heat fluxes two-way interactive between the soil, vegetation and the atmosphere.

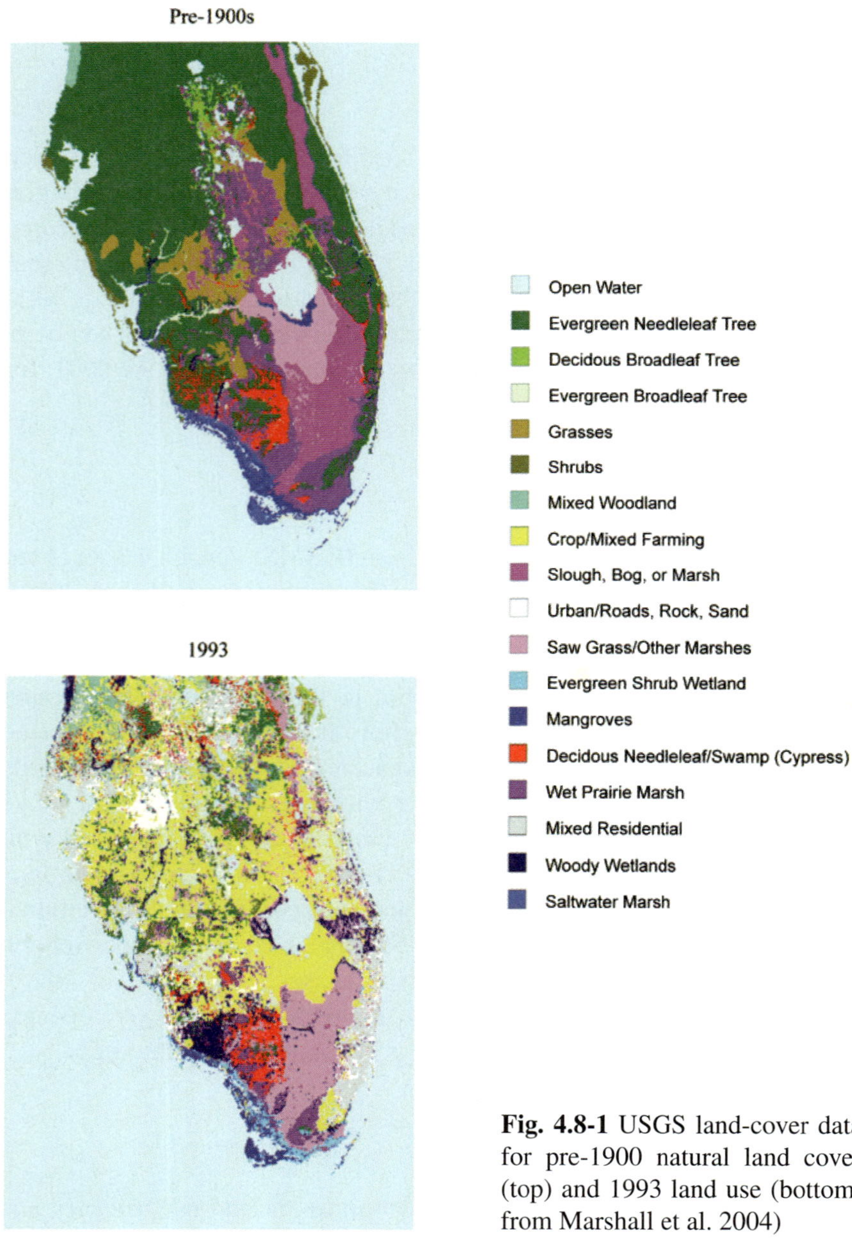

Fig. 4.8-1 USGS land-cover data for pre-1900 natural land cover (top) and 1993 land use (bottom; from Marshall et al. 2004)

The representation of the surface heat energy and moisture budgets used the LEAF-2 scheme (Walko et al. 2000). LEAF-2 provides algorithms to partition incoming net radiation into sensible heat fluxes, physical evaporation and transpiration based on landscape type and more rapidly varying land-surface conditions such as soil moisture. Sea surface temperatures were obtained from the global climatological files at the National Center for Atmospheric Research (NCAR). The July and August average values are used.

4.8.4 Results

Fig. 4.8-2 shows the accumulated two-month deep cumulus precipitation for July-August 1973 using pre-1900 land cover and 1993 land use from the Marshall et al. (2004) study. The difference between these two model simulations is shown in the right panel. The differences exceed 100 mm and more in places, along with an altered spatial pattern. The human-caused landscape change has apparently had a major effect on the hydrology of the region. The spatial pattern of the simulated rainfall for the 1973 model run is consistent with the climatological pattern of rainfall. The area-averaged rainfall decrease is over 10%. This reduction in rainfall reduces the amount of water that reaches the Everglades areas. The average maximum surface air temperature occurring for the entire period has increased in the model by about 2° C for many inland areas.

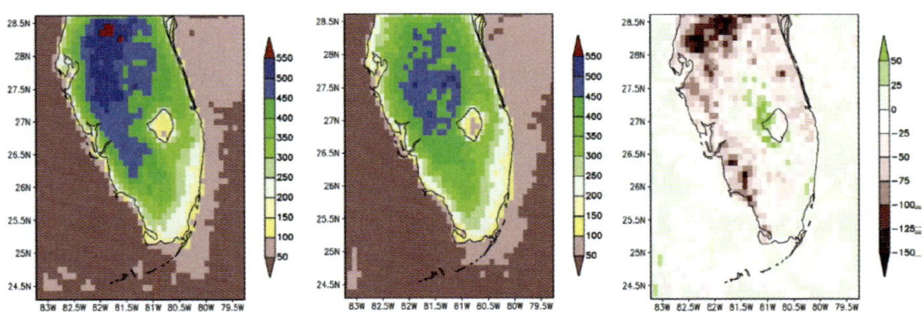

Fig. 4.8-2 Accumulated convective rainfall (mm) from the model simulations of July-August 1973 with pre-1900 land cover (left), 1993 land use (middle), and the difference field for the two (right; 1993; minus pre-1900 case). From Marshall et al. 2004

There are, unfortunately, only limited data available with which to evaluate temporal trends over south Florida, which makes comparison with the model results difficult. Fig. 4.8-3 shows regionally-averaged precipitation data during July and August for at least part of this past century. The model results indicate that if land-use change was the only factor influencing rainfall, we would expect a general decrease in rainfall over the interior, although there is considerable grid-point to grid-point variability near Lake Okeechobee and the coastlines. Some locations along the coast show a modest increase in rainfall.

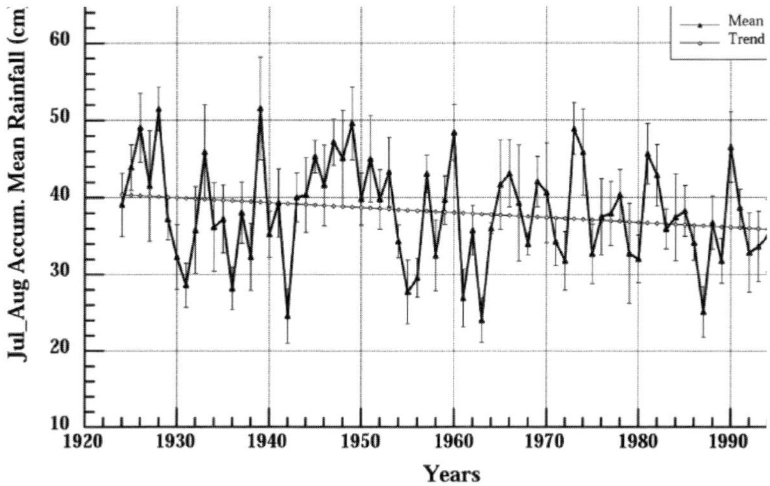

Fig. 4.8-3 Regional average time series of accumulated convective rainfall (cm) from 1924 to 2000, with corresponding trend based on linear regression of all July-August regional average amounts. The vertical bars overlain on the raw time series indicate the value of the standard error of the July-August regional mean (from Marshall et al. 2004)

The differences in precipitation in the model are due to the alteration in the spatial pattern of transpiration and physical evaporation due to the land-cover change. The reduction in this water flux to the atmosphere results in less precipitation in the interior of the peninsula from thunderstorms and a spatial displacement in the rainfall pattern.

Fig. 4.8-4 presents regionally-averaged maximum temperature trends. The observed increase in maximum averaged temperature for July and August is consistent with the model-simulated warming in response to land-use change.

4.8.5 Conclusions

Over south Florida during the past 100 years, there has been a widespread conversion of natural vegetation to urban and agricultural land, and into grassy shrubland. These landscape changes are likely to have altered the

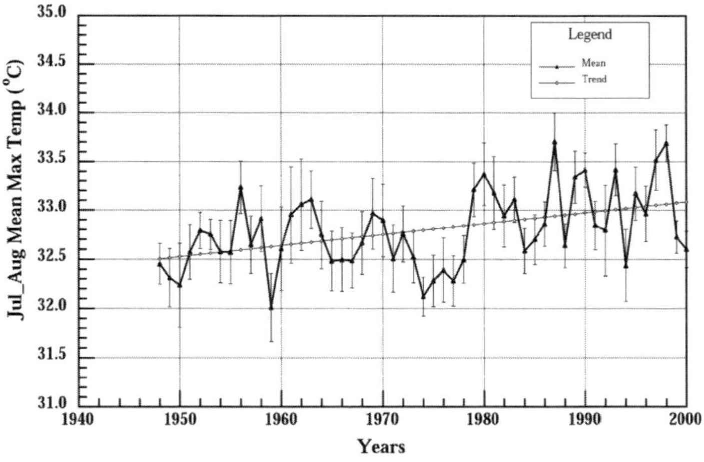

Fig. 4.8-4 Same as Fig. 4.8-3 except for daily maximum shelter-level temperature in degrees C

local weather patterns, with average summer rainfalls for south Florida decreasing by 10% or more. This reduction in precipitation due to land-cover changes would be in addition to precipitation variability (which is also evident in Fig. 4.8-3) caused by year-to-year variability and long-term trends in synoptic weather features.

Along with the reduction in deep cumulus rainfall, the model indicated that surface temperatures should warm in response to the landscape conversion. A warming is observed in the regional averaged maximum temperatures. These results indicate that unless land-use change effects on weather are included in climate trend analyses, the reasons for climate change can be concluded erroneously. In addition, because of the permanent landscape changes in the regions around the Everglades, it will be impossible to restore the climate in this region to what it was prior to the landscape change, thereby making the restoration of the Everglades ecosystem even more difficult. Rainfall, for example, is likely to be permanently less in the absence of changes in larger scale climate influences.

Acknowledgements

Portions of this work were originally reported in Marshall et al. (2004). This research was supported by USGS Contract #1434-CR-97-AG-00025, Task 7, and NASA Grant No. NAG5-11370.

4.9 CLIMBER-2: An Earth System Model of Intermediate Complexity

Andrey Ganopolski

4.9.1 Motivation and Objectives of the Study

The growing understanding that the Earth system functions in a complex, strongly nonlinear way, where separate components (atmosphere, hydrosphere, biosphere, etc.) are closely interlinked, stimulates development of a new class of computer models: Earth System Models of Intermediate Complexity (EMICs). Among the crucial questions which can be addressed with such models is the physical and geochemical interaction between the geosphere and biosphere on global and long-term scales. Due to the complexity of the Earth system and the large range of temporal scales of the different components of the system (up to a thousand years and longer) it is quite natural to synthesise our knowledge of the Earth system through a relatively simple computer model and with coarse spatial resolution. This strategy is behind a hierarchy of EMICs under developments at the Potsdam Institute for Climate Impact Research. CLIMBER-2 (CLIMate-BiosphERe model; see detailed description in Petoukhov et al. 2000) represents the medium level of this hierarchy, and it has been used for a variety of studies of past and future climate changes, as well as for the sensitivity and stability analysis of the Earth system.

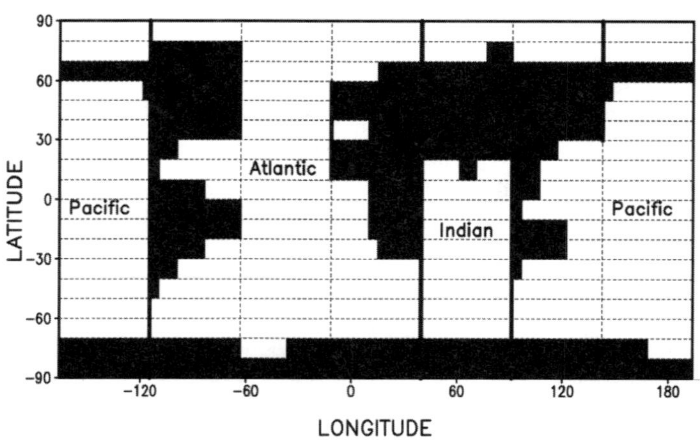

Fig. 4.9-1 Representation of the Earth´s geography in the model. The black area represents the areas of land. Dashed lines show the atmospheric grid, solid lines separate ocean basins

4.9.2 Short Description of the Study

The basic strategy in the development of EMICs is to achieve a high level of integration of the components of the Earth system and high computational efficiency by simplification in description of individual components (e.g. atmosphere or ocean) and by using a relatively coarse spatial resolution. The CLIMBER-2 model has a spatial resolution (Fig. 4.9-1) which resolves only individual continents or subcontinents and ocean basins: latitudinal resolution is 10° in the atmosphere and land models (2.5^{o} in the ocean) and in the longitudinal direction the Earth is represented by seven equal sectors. CLIMBER-2 encompasses six modules, as shown in Fig.4.9-2:

1) an atmospheric module;

2) an ocean and sea-ice module;

3) a vegetation module;

4) an inland-ice module;

5) and modules of marine biota and

6) oceanic biogeochemistry.

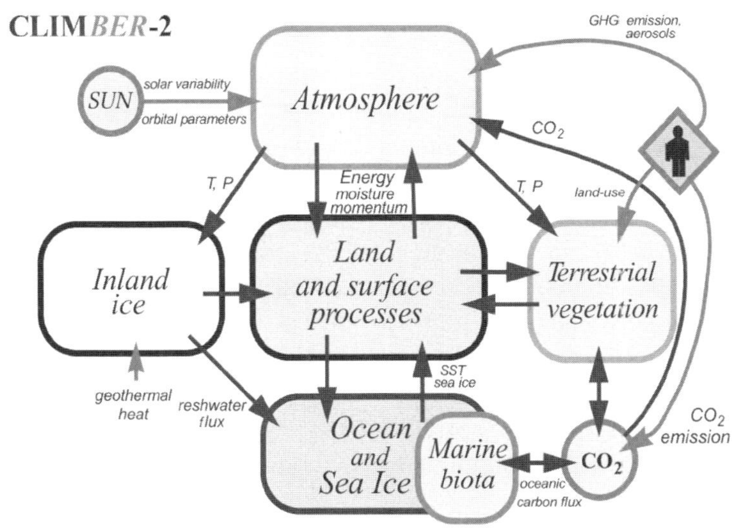

Fig. 4.9-2 Principal scheme of the CLIMBER-2 model. Grey arrows show external forcings, black arrows show exchange of information between individual modules

All components are fully interactively and bi-directionally coupled via fluxes of energy, moisture, momentum and carbon without using any explicit information about present day climate state (e.g. flux-correction). This allows us to use the model for simulations of the past and future climates which differ considerably from the modern one. Anthropogenic activities, such as land use and emission of greenhouse gases, are given as external boundary conditions. Due to low computational costs the model allows us to perform long-term (multi-millennial) simulations and numerous sensitivity experiments needed to facilitate understanding of the role of model uncertainties and limitations.

4.9.3 Complexity of the Study

Individual components of the Earth system are represented in CLIMBER-2 by relatively simple models, but in respect of the total number of the processes and feedbacks our model is comparable with the comprehensive models. In particular, the atmosphere module of CLIMBER-2 is a 2.5-dimensional[1] statistical-dynamical model. This model is based on the assumption that the general structure of the atmosphere can be expressed in terms of large-scale, long-term fields of the main atmospheric variables and ensembles of synoptic eddies and waves. The latter are parameterised in terms of their statistical characteristics. The atmosphere module describes atmosphere dynamics, moisture transport, precipitation and cloudiness, and radiative processes. The ocean module describes the ocean hydro- and thermodynamics, sea ice and the ocean carbon cycle. It is based on the multi-basin zonally-averaged approach. Simulations of terrestrial vegetation are based on a continuous description of plant functional types. For each continental grid cell, the vegetation module computes fractions of vegetation cover and desert. The vegetation module includes a simple carbon cycle model in which allocation of carbon to four pools (leaves, stems and roots, humus, soil) is evaluated. Vegetation adapts to climate on a time scale proportional to the turn-over time of carbon in the pool of slowly varying biomass, i.e. stems and roots. Hydrology and interaction between climate and vegetation are described using Atmosphere-Surface Interface (ASI). ASI is based on BATS (Biosphere-Atmosphere Transfer Scheme) (Dickinson et al. 1986), simplified and modified for the level of spatial

[1] The term "2.5-dimensional" reflects the fact that unlike atmosphere GCMs the vertical profiles of main atmospheric characteristics - wind speed, temperature and humidity - are not computed using prognostic 3-D equations but rather diagnosed based on an assumption on the universal vertical structure derived from climatology. This is why our atmosphere model is placed between the simplified 2-D and comprehensive 3-D models.

and temporal aggregation and degree of complexity of CLIMBER-2. The ASI distinguishes six surface types, including two vegetation types - forest and grassland. The presence of vegetation affects interaction between land surface and atmosphere via surface albedo (especially for snow-covered conditions), evapotranspiration (different root distribution in soil and LAI) and surface roughness. In turn, the natural distribution of vegetation is controlled by temperature and precipitations.

4.9.4 Experiences (Lessons Learned)

CLIMBER-2 has been successfully tested against modern climate data, as well as for a variety of palaeoclimates (Last Glacial Maximum, Holocene Optimum, etc.). Furthermore, the model has been extensively compared with comprehensive climate models (GCMs) in respect to the sensitivity to CO_2, solar insolation and perturbations in freshwater flux. Special emphasis in our research has been given to climate-biosphere interaction on different spatial and temporal scales. These studies demonstrate the importance of realistic description of climate properties of vegetation cover-surface albedo, roughness, transpiration. Results of our experiments also indicate that interaction with the ocean (primarily via the hydrological cycle) affects considerably the climate sensitivity to changes in vegetation cover (Ganopolski et al. 2001; Claussen et al. 2001).

Using CLIMBER-2, we have studied separately the impact of deforestation/aforestation on climate via biogeophysical processes (energy and hydrology) and the combined effects of biogeophysical and biogeochemical (carbon dioxide) processes. With respect to the biogeophysical effect only, our results demonstrate the important role which the ocean plays in amplification and modification of the climate response to changes in vegetation cover. Fig. 4.9-3 shows that in the case of interactive ocean, cooling caused by deforestation is much more pronounced and wide-spread. The same is true for the reduction in precipitation. One of the prominent features of climate response to boreal deforestation is a strong reduction in summer monsoon precipitation in southern and south-eastern Asia. An important result of our study is that in the coupled climate model both boreal and tropical deforestation leads to global cooling, while the atmosphere-only version of the model predicts the opposite sign of temperature changes for boreal (cooling) and tropical (warming) deforestation. The reason for global cooling due to tropical deforestation in the coupled model is a strong reduction in downward long-wave radiation, caused by the reduction of water content in the atmosphere. The interactive ocean reacts to the decrease of water content in the atmosphere by cooling and a decrease in evaporation that amplifies the decrease in the water vapour greenhouse effect.

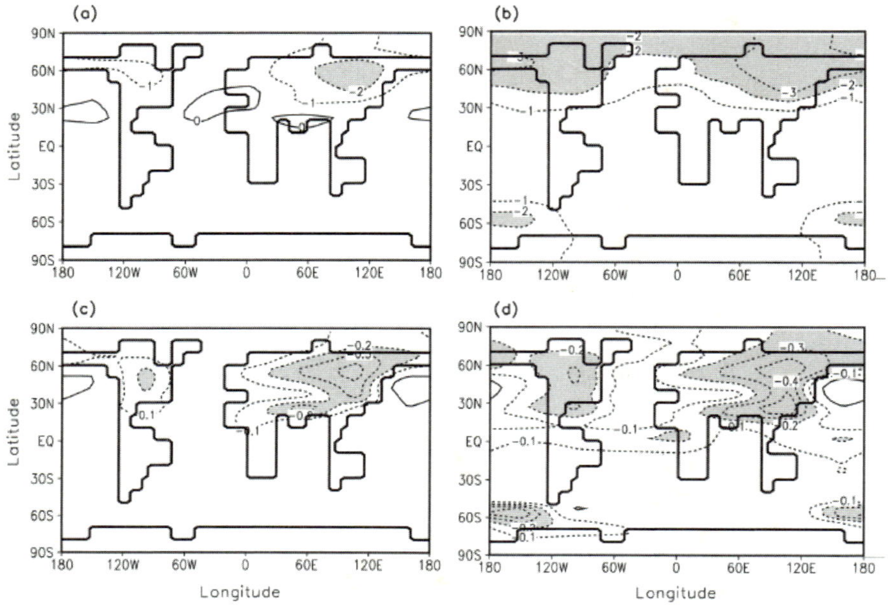

Fig. 4.9-3 Differences in **a, b** annual near surface air temperature (in ^{o} C) and **c, d** precipitation (in mm d^{-1}) between "boreal deforestation" and control experiments. **a, c** correspond to experiment with fixed ocean characteristics; **b, d** to experiment with interactive ocean (from Ganopolski et al. 2001)

Release of carbon due to deforestation leads to increased greenhouse effect and, thus, works in the opposite direction to the biogeophysical effect. When both effects are taken into consideration (Claussen at al. 2001), our sensitivity studies show that in the tropics the climate impact of deforestation via changes in CO_2 outweighs the biogeophysical effects, while in boreal latitudes biogeophysical processes play a dominant role. These sensitivity studies have a practical implication: they show that afforestation in high latitudes is not an efficient way to curb global warming via uptake of CO_2, since the net effect of boreal afforestation is warming rather than cooling.

Our current analysis is focused on the large-scale patterns of climate response to different external or internal perturbations, in particular the question of stability and synergism in climate systems. As the first results are quite promising, further progress needs considerable improvements in the description of climate-biosphere interactions by increasing spatial resolution and the number of considered processes and feedbacks. It is crucially important to achieve this by preserving relatively low computational costs – otherwise the model cannot be used for the kind of studies which have been performed with the CLIMBER-2 model. This is why the new model, CLIMBER-3, currently under development at the Potsdam Institute for Climate Impact Research, is based on the combination of simplified (highly parameterised) and state-of-the-art components and is specially designed for a high-performance parallel-processor computer.

4.10 Feedbacks and Coupling between Water, Carbon and Nutrient Cycling at the Hillslope Scale

Lawrence E. Band, Christina L. Tague

4.10.1 Motivation and Objectives for the Coupling Approach

The motivation behind the coupling methods presented here is to capture mechanistically the impacts of hillslope level lateral water flow and redistribution on nutrient biogeochemical cycling and export from catchments. Much of the cycling and transport of nutrients from terrestrial to aquatic ecosystems represents a complex response to the distribution and flux of water within the length scale of a hillslope. With the exception of deep groundwater that may follow flow paths independent of surface topography, the drainage divides of the hillslope represent no flux boundaries – and a stream at the hillslope base is an absorbing boundary. The average hillslope length is the inverse of the drainage density, and will typically range between tens to at most hundreds of metres. Note that this is well under the resolution of standard regional to global simulations but contains much of the land-surface heterogeneity and process dynamics.

Within this domain, in situ organic matter decomposition, mineralisation and uptake by plant and microbial biomass are mediated by the soil water content in the upper soil horizons, which in turn are mediated by local soil and canopy structure as well as topographically driven catenary patterns. The feedbacks between water, carbon and nutrient cycling and transport occur at time scales ranging from individual storm events and interstorm periods through decadal or centennial canopy growth, biomass accumulation and succession. Spatial coupling occurs at scales ranging from within and between canopy gaps through a full topographic gradient, including hillslope wetness and topoclimatic patterns. Efforts to explain dynamics at one of the space/time scales may be subject to prescription of system state or dynamics at the other(s). Therefore, we seek to develop operational models to investigate feedbacks between dominant space/time variables within hillslope systems. An extension to the terrestrial phase of catchments is made by treating the catchment as a population of hillslopes organised around the drainage network.

4.10.2 Short Description of the Study

Area

Simulations of integrated water, carbon and nutrient cycling and transport are carried out for a small catchment that is instrumented as part of the Baltimore Ecosystem Study (BES), one of the NSF-funded LTER (Long-Term Ecological Research) sites in Maryland, USA. Pond Branch is in the Oregon Ridge County Park in Baltimore County, draining 34 ha of upland Piedmont topography with a dominant oak-hickory canopy cover. The canopy has been growing for about 80 years, and last had selective cutting in the 1950s. Typical of headwater Piedmont catchments in this area, the upper portions of the drainage area are characterised by gentle slopes with broad valley bottoms, which narrow and become increasingly incised downstream (Fig. 4.10-1). Silt loam soils overlay deep saprolite on gentle interfluves,

Fig. 4.10-1 Colour infrared image superimposed on the Digital Elevation Model (DEM) for the Pond Branch catchment in Baltimore County, Maryland, USA

which thin into patchy bedrock exposures of the underlying schists in parts of the lower slopes. The broad bottomland in the upper catchment is wet and organic-rich, with disorganised drainage at low flows characterised by anastomosing, discontinuous channels. Streamwater sources include a combination of diffuse seepage from the hillslopes as well as discrete spring sites. Continuous flow monitoring and weekly stream chemistry sampling at the outlet have been ongoing since the autumn of 1998. A more detailed description of the site and an earlier version of the model application to this catchment can be found in Band et al. (2001b).

Coupled processes and compartments

A key approach that we have taken is to attempt to describe a formal geomorphic framework for the spatial nesting of surface attributes and processes. We have produced this approach in successive versions of RHESSys (Regional HydroEcological Simulation System; Band et al. 2001a,b). Within this framework (Fig. 4.10-2), the primary processes of biogeochemical cycling and fluxes of water, carbon and nutrients are

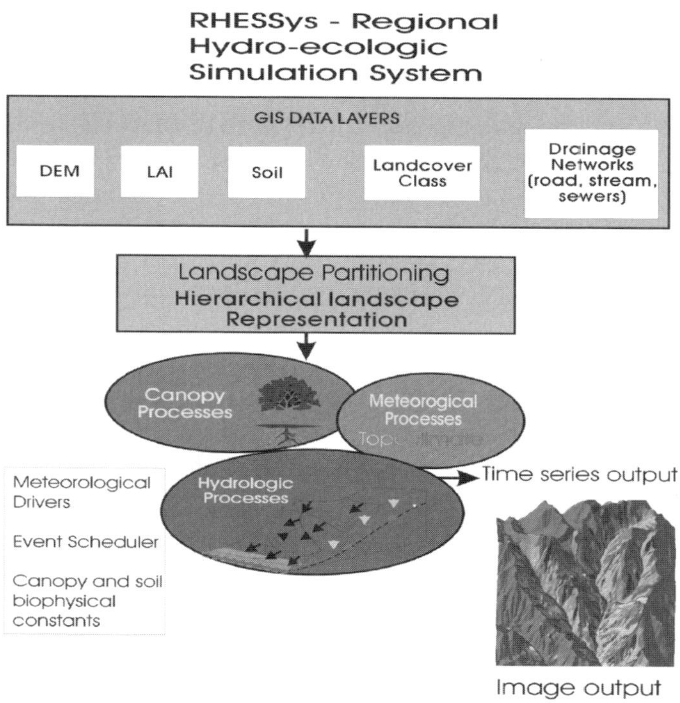

Fig. 4.10-2 RHESSys model structure showing ingestion of spatial catchment data, partitioning into hierarchical landscape representation and hydroecological process modelling driven by meteorological forcing

distributed through the canopy and soil, solving for flux divergence at the surface patch level, and building these components hierarchically into a full basin representation. Using the geomorphic framework, the catchment is partitioned successively into a containment hierarchy with component hillslopes arranged around the stream network (one hillslope draining to the left and right side of each stream link). Each hillslope in turn is partitioned into a set of contiguous patches based on topography and canopy cover, as well as the requirements of solving subsurface flow equations. Grids can serve as patches as a special case, although we often prefer to vary resolution and shape to characterise small, but essential, elements of the topography in hollows or along riparian areas. Each patch in turn is partitioned into canopy elements which can have fractional cover and multiple vertical (i.e. overstorey and understorey) layers within the patch. Canopy layers are sorted by height to resolve radiation and precipitation interception.

Water, carbon and nutrients are subject to mass conservation with respect to in situ cycling and flux divergence over the network of patches and into the stream at the base of each hillslope. All biophysical processes are associated with a specific element of the catchment representation hierarchy. Transfer between compartments occurs both within a level of the hierarchy (e.g. patch-to-patch transport of water and nutrients) and between levels of the hierarchy by aggregation or disaggregation procedures (litterfall or uptake between a patch and its component canopy strata). Table 4.10-1 lists specific processes and the compartment level with which they are associated.

Spatial and temporal scales

Time steps for different processes range from sub-daily (e.g. canopy radiation, infiltration) up through seasonal (canopy carbon allocation). However, most of the carbon, nutrient and water transport and cycling processes are run at the daily time step. The daily time step is used as the model is operated typically over temporal domains, ranging from annual to centennial in order to develop feedbacks and dynamics of the more slowly varying stores, such as the canopy and soil development.

Maintaining the dynamics of the slowest responding variables in the hillslope system places an additional burden in model initialisation as imbalanced system components may lead to very long transients. Therefore, it is necessary to be able either to prescribe realistic spatial patterns of carbon and nutrient pools or to run very long (e.g. 10^3 years) spin-up simulations to derive balanced stocks. Once model spin-up has produced a stable distribution of carbon and nitrogen stores, this initialised state can be used to run multiple scenarios at shorter time scales.

Table 4.10-1 Simulated processes for water, carbon and nutrient cycling associated with different levels of the watershed hierarchy. Note that the patch represents all soil processes and the strata are considered well mixed within the patch (share a common soil pool)

	Hydrologic Processes	**Biogeochemical Cycling**	
	Water	**Carbon**	**Nitrogen**
Basin	Stream routing		Nitrogen export
Hillslope	Lateral water flux		Nitrate transport
Patch	Soil evaporation	Decomposition	Decomposition
	Infiltration	Soil respiration	Mineralisation
	Vertical unsaturated,	Litter accumulation	Immobilisation
	saturated flux	Humus formation	Nitrification
	Root uptake		Denitrification
			Root uptake
			Deposition
Strata	Interception	Photosynthesis	Uptake
	Transpiration	Respiration	Rubisco dynamics
		Allocation	Rubisco dynamics
		Litterfall	Litterfall,
			retranslocation

RHESSys has been operated over catchments varying by four orders of magnitude in area, with the largest application being the South Platte River Basin (Baron et al. 1998) with 63 000 km^2. However, more recent applications coupling nutrients or resolving roads in a fully distributed, explicit flow routing and cycling have been carried out on much smaller experimental catchments. These smaller scale applications have shown that the feedbacks and interactions between lateral water transport in the terrestrial phase and ecosystem water/carbon/nutrient dynamics exist within the domain of the hillslope, requiring a representation of system patterns evolved at these length scales. A current goal is to find methods that can capture the effects of flow routing on net nutrient retention, treating these as subgrid processes with much lower computational expense by using a combination of spatial distribution and explicit routing approaches.

4.10.3 Complexity of the Study

Type of feedbacks involved

The mix of rapidly and slowly varying stores produces a complex response in nutrient dynamics which, at short time scales, are controlled strongly by variations in temperature and moisture content that influence root uptake, nitrogen transformations, and soil immobilisation that vary with landscape position. At progressively longer time scales, variations in the canopy cover, maturity and carbon balance of the canopy, and the distribution of the amount and quality of soil organic matter, are important long-term controls. These latter characteristics build a long memory into the ecosystem such that previous land uses may still exert an influence on system structure and dynamics. Interactions between soil water state and in situ carbon and nutrient cycling are represented at patch and canopy strata by process modules drawn from the Biome-BGC (Biogeochemical Cycling) model (Running and Hunt 1993) and the NGAS (Nitrogen Gas) module of CENTURY (Parton et al. 1996). In terms of nitrogen cycling, denitrification rates increase non-linearly with soil water as values approach saturation, while for decomposition and mineralisation, rates follow parabolic trajectories with maximum values at intermediate moisture conditions. While a set of other factors and processes influence in-situ N cycling, including temperature, carbon substrate availability and net immobilisation, soil moisture is a dominant factor controlling space–time variation of the transformation and flux of soluble carbon and nitrogen constituents over landscape level gradients.

Soil water state is developed by solving patch and canopy level interception, infiltration, evapotranspiration, unsaturated and saturated zone dynamics, with lateral divergence computed with routing techniques drawn from modified forms of either TOPMODEL (Beven and Kirkby 1979) or the Distributed Hydrology Soils Vegetation Model, DHSVM (Wigmosta et al. 1994). Note that TOPMODEL does not actually route water, but redistributes saturation deficit by use of hydrological similarity measures, and therefore does not account for transformations of transported constituents along a flowpath. Therefore, in the examples given here, we make use of the DHSVM routing modules.

At time scales of daily to seasonal, simulated soil water patterns in Pond Branch vary between states of showing strong or weak coupling with topographic gradients. During the winter and early spring, near-saturation conditions are frequently exhibited in the broad valley bottoms, hollows and lower slopes, while growing season conditions show saturation levels declining in the interfluves with drying trends extending into the valley bottoms during the summer (Fig. 4.10-3). These space–time soil water patterns interact with nitrogen cycling by restricting nitrification and promoting denitrification during the wet period, both in the uplands and bottomlands, while shifting to a greater rate of net nitrification during the drier period. In the bottomlands, the organic-rich soils which accumulate due to a combination of fluvial/colluvial transport and low decomposition rates, may shift from acting as nitrate sinks to becoming small nitrate sources. Full field verification of this shift is still in progress. However, nitrate concentrations sampled at the weir show values below the detection limit for most of the dormant period, but show consistently higher values during the growing season, with greater concentrations corresponding to the drier summers (out of three summers sampled so far).

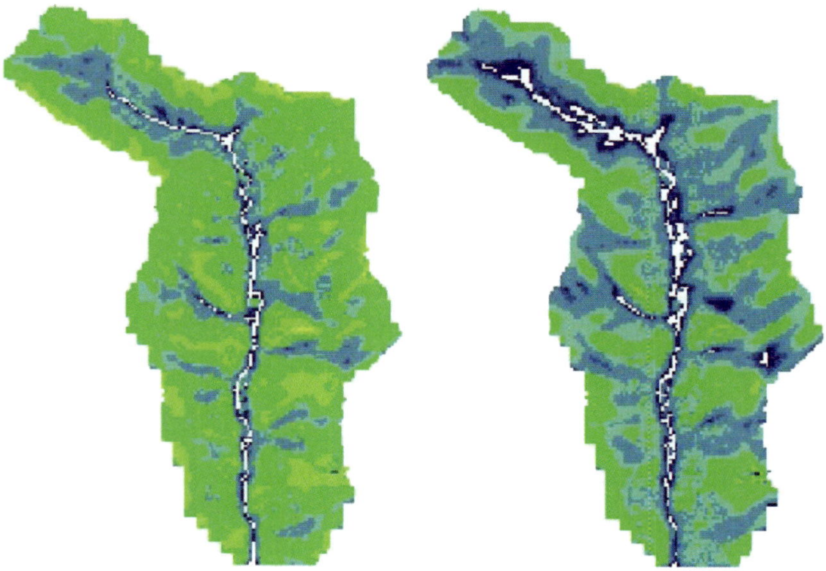

Fig. 4.10-3 Mean monthly saturation deficits in a catchment for August 1998 (left) and January 1998 (right), showing expansion of saturated areas in bottomland and hillslope hollows during the winter. Blue represents saturated conditions, ranging through greens and yellows (driest). Saturated and near saturated areas are restricted to the bottomland and hollows in August and expand during the winter, on right)

The results we show here are qualitatively similar to results presented in Band et al. (2001b) in terms of the temporal patterns of predicted streamwater nitrate concentrations, but are closer to the observed trends (Fig. 4.10-4).

This results from a few improvements in model structure as well as an improved, higher resolution Digital Elevation Model (DEM) that allows us to resolve the topography and particularly the critical riparian areas within the bottomland. In addition, we have used a 5 m grid for this simulation as opposed to the larger, irregular polygons used previously. Even with these improvements, we are still over-predicting NO_3-N concentrations, but would point out that we are not currently modelling within-stream sinks and transformations.

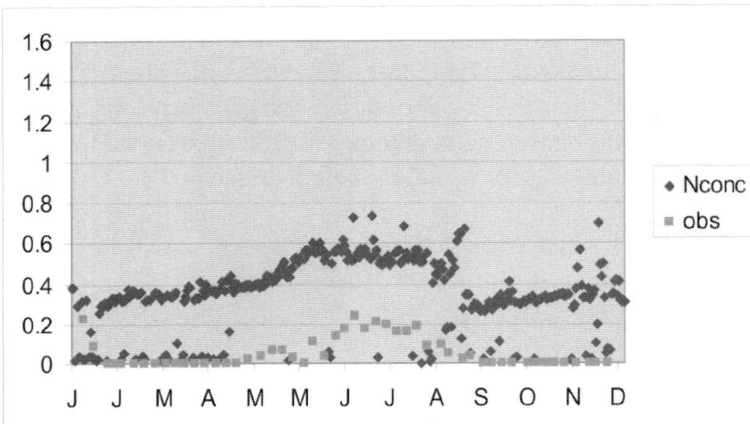

Fig. 4.10-4 Observed and simulated nitrate concentrations (mg $NO_3 - N\ l^{-1}$)at the catchment weir. Note the consistent overprediction of concentrations as in-stream processing is not incorporated, but the consistency with elevated export during the active growing season

Simplifications

Among a number of simplifications in our model approach, an important one is the parsimonious representation of unsaturated/saturated zone dynamics which includes a litter layer, unsaturated and saturated zone, following the approaches in TOPMODEL and DHSVM. The mass balance for nitrate is carried out for the full soil column, with the nitrate profile represented as an exponential distribution with depth. This produces the observed flushing effect as lateral flow in the saturated phase rises towards the surface. While a number of recent observations in both forest and agricultural sites appear to support the flushing concept, we are currently evaluating this assumption with more detailed simulations of unsaturated zone dynamics and field measurements.

We also maintain a simplified treatment of dissolved organic nitrogen loss which may lead to an underestimate of total N export from the basin and may be contributing to our overestimates of inorganic N export by maintaining higher N soil content. We are currently implementing a more mechanistic approach to dissolved organic matter evolution and export.

Internal processes and external factors

Currently, we operate spatially distributed water, nutrient and carbon mass balances with prescribed meteorological forcing, atmospheric CO_2 concentrations (no vertical gradients through the canopy) and N deposition. The nearest National Atmospheric Deposition Program (NADP) sites are some tens of kilometres away so we are using regional estimates of wet and dry deposition. Atmospheric nitrogen deposition is an important source in this region, and while retention in the ecosystem is very high (>90%), long-term growth of the canopy is enhanced by the moderate to high values. To spin-up the simulations to the last decade, we have run the model for a period of 400 years to gain stable spatial distributions of soil C and N, canopy above and below ground biomass and leaf area index (LAI). As patterns are internally generated, this provides additional information to compare with field measurements to check the consistency of model performance. Soil depth and hydraulic conductivity profiles are highly uncertain and are estimated from knowledge of the geomorphology of the area, limited pit data (four sites) with measured conductivity profiles (to a depth of two metres) and calibration to hydrographs. At present, the model operates its carbon, water and nutrient flux as a function of species-specific physiological parameters, to which the model can be very sensitive. Species composition is prescribed in the model as we do not treat succession.

Coupled natural cycles

The spatially distributed cycles of nutrient cycling (in this case nitrogen), carbon and water are coupled directly at subhillslope scales. Over long periods, water and nutrients control biomass accumulation, carbon allocation within the canopy which determines changes in canopy structure, canopy LAI and soil organic matter. These variables, in turn, are important short-term controls on water, carbon and nitrogen flux through the full system.

Treatment of uncertainty

A shortcoming of integrative process models such as that presented here is the proliferation of model parameters that can be known only with limited certainty (or high uncertainty). This is especially true of subsurface parameters, as soil biogeochemical and hydraulic characteristics are well known to be highly variable in lateral and vertical extent. This is one motivation to attempt to keep the model to as simple a representation of the key processes as is feasible, while attempting to retain the major non-linearities and feedbacks. As model complexity increases, it becomes less feasible to create adequate samples of model realisations with variable parameter sets. However, a potential advantage of an integrated model is the availability of a larger set of simulated model state and flux variables, a set of which can be observed for comparison. In our model, in addition to hydrographs, we develop patterns of soil water content, C/N content, standing biomass and LAI. The Pond Branch site has a set of terrestrial ecosystem plots in which these variables are estimated and used to develop pattern indices to describe the topographic zonation of these slowly and more rapidly developing variables with landscape position which can be compared with model output.

In addition to parameter uncertainty, the physics and chemistry controlling water and nutrient movement through hillslopes may still be poorly known. A major uncertainty in hydrology at present is whether treatment of soil water flux by Darcy matrix flow is correct, given recent demonstration of the effectiveness of preferential flow.

4.10.4 Experiences

Difficulties

The processes we describe in this paper are typically sensitive to conditions near the stream as well as channel conditions. Several researchers have shown that stream chemistry can be largely reset in the riparian zone, such that connectivity of the hillslope water to the stream, as mediated by the riparian buffer, is a critical control. However, topographic and soil information are typically poor in these areas, and standard methods of terrain interpolation tend to smooth out bottomland topography substantially. The uncertainty attached to soil information in the near stream environment, of course, can be extended to soil information through the full catchment.

Additional difficulties arise out of the problem of equifinality in complex simulations, as discussed by Beven and Freer (2001) and others. In this case, a number of different model structures and parameterisations may lead to similarly good fits of simulated to observed flow and quality data. Therefore, it is important to maximise the number of independent observational data available to assess the consistency of model behaviour.

Conclusions

Conceptual and model coupling of space-time variations in water storage and flux with biogeochemical cycling and export allows us to consider a richer set of dynamics and feedbacks to catchment behaviour. In particular, the short and long term interactions of catchment hydrological cycling with ecosystem pattern and process embedded within our modelling framework provides a set of spatial and temporal signals that can form more comprehensive testing with field observation. As examples, we can use space–time distributions of leaf area, standing biomass, streamflow and water quality to test model behaviour with observations. The hierarchical representation of the landscape also allows us to investigate system behaviour at scales commensurate with ecosystem field plots, as well as aggregate this information to resolutions commensurate with medium to coarse resolution satellite observations (Band 2000). The feedbacks of short-term flux of water, carbon and nutrients with canopy growth in mixed topographic settings also allows consideration of long-term effects of forest harvest and road development on hydrological and ecosystem behaviour (Tague and Band 2001). In short, this coupling allows us to move beyond the need to prescribe canopy and ecosystem conditions independent of hydrological regime.

Further applications

While we have concentrated our application of this modelling approach to forested catchment, we are currently extending the model approach to incorporate the set of features and processes typical of urbanising catchments. This includes the patchy nature of the surface with detailed mixtures of pervious and impervious surfaces, drainage infrastructure, and localised water and nutrient applications. These applications are being tested and supported by distributed monitoring in more urbanised catchments monitored in the Baltimore Ecosystem Study.

Acknowledgements

The Baltimore Ecosystem Study project is supported by the National Science Foundation grant GRS 0095796 to L. Band and C. Tague, and NSF Long-Term Ecological Research program, grant number DEB 9714835. We thank the USDA Forest Service Northeastern Research Station for site management, and in-kind services to the BES. In addition we thank the University of Maryland, Baltimore County for their contribution to office and laboratory space at the Research Technology Center on their campus.

4.11 The Boreal Ecosystem-Atmosphere Experiment (BOREAS)

Forrest G. Hall

4.11.1 Motivation and Objectives for the Coupling Approach

The boreal ecosystem encircles the Earth above about 48° N, covering Alaska, Canada and Eurasia. It is second in areal extent only to the world's tropical forests and occupies about 21% of the Earth's forested land surface. Nutrient cycling rates are relatively low in the cold wet boreal soils, resulting in relatively high long-term boreal carbon storage rates. These rates average roughly 30 to 50 g C m^{-2} y^{-1}, a result of relatively high root turnover from trees, shrubs and mosses with relatively low decomposition rates. Over the past few thousand years, these below-ground storage processes have created a large and potentially mobile reservoir of carbon in the peats and permafrost of the boreal ecosystem. Currently the boreal ecosystem is estimated to contain approximately 13% of the Earth's carbon, stored in the form of above-ground biomass, and 43% of the Earth's carbon stored below-ground in its soils. Meridional gradients in atmospheric CO_2 concentrations suggest that forests above 40° N sequester as much as 1 to 2 gigatons of carbon annually, or nearly 15 to 30% of that injected into the atmosphere each year through fossil fuel combustion and deforestation. Given the enormous areal extent of the ecosystem, roughly 20 million km^2, shifts in carbon flux of as little as 50 g C m^{-2} y^{-1} can contribute or remove one gigaton of carbon annually from the atmosphere. If the strong high-latitude warming trend continues, leading to warmer soils and a reduction in the extent of the boreal permafrost zone, the resultant increases in soil organic matter decomposition could switch the boreal ecosystem from a long-term carbon sink to a significant carbon source, accelerating global warming.

The boreal ecosystem also strongly influences short-term climate: (1) Its nutrient-limited conifers growing on cold, moisture-saturated peats, strongly limit evapotranspiration, creating a deep, dry atmospheric boundary layer more characteristic of an arid ecosystem, essentially "a green desert". (2) Both long- and short-term changes in its high-albedo snow and ice extent are linked to global climate variations and coupled to climate through the albedo feedback. Since 1970, climate change has had a significant impact on snow cover, which may have enhanced spring warming through feedbacks from the subsequent decrease in surface albedo. (3) Smoke aerosols produced by extensive lightning-induced forest fires modify regional climate by reducing surface insolation and influencing cloud formation and precipitation tendencies.

4.11.2 Short Description of the Study

BOREAS was an interdisciplinary, multi-scale field experiment to study the role of the boreal ecosystem in global change (Sellers and Hall 1997). 85 science teams, including atmospheric physicists, micrometeorologists, ecologists, hydrologists, biogeochemists, and remote sensing specialists were involved. During 1993 through 1996 the BOREAS science team consisted of over 300 scientists. BOREAS was originally planned to last three years, with field campaigns in the first year and disciplinary data analysis occurring in the second and third years. However, analyses performed in the second year of BOREAS showed that there were a number of gaps in the first year's data; thus a third year of data collection was proposed and funded, along with an additional fourth year to analyse the data. During the first four years of BOREAS, scientists focussed on acquiring, quality assuring, and analysing their own data. Process model work focussed mainly on analysing tower site data for model development and validation.

An additional call for proposals was issued in the fourth year of BOREAS to permit more integrative, interdisciplinary, regional-scale analyses focussed on broader scientific questions. The call resulted in the selection of thirteen guest investigations. These investigations focussed generally on modelling and analysis, extending and building on the BOREAS scientific results and database.

A major emphasis within the Guest Investigator Program was to link carbon modelling efforts closely to those of the energy, water, and hydrology modelling teams. The BOREAS nested resolution modelling approach, of tower > study area > transect > region was a cross-cutting theme for all simulation efforts. Nine ecosystem process models and seven hydrometeorological models were evaluated in three phases. Of the ecosystem process models, three ran at hourly time steps, five at daily time steps and one at a monthly time step. Model outputs such as net primary production (NPP) and net ecosystem production (NEP) as well as plant and soil respiration were cross-compared. In addition, model outputs were compared to tower-observed values for radiation, CO_2, water vapour, and heat flux. The hydrometeorological model output was also evaluated by comparison with values of runoff in gauged catchments.

4.11.3 Complexity of the Study

The objectives of BOREAS relate to two spatial scales that had to be reconciled within the experiment design: the local scale, a few centimetres to a kilometre, and the regional scale from a few kilometres to the 1000 km x 1000 km BOREAS study region in central Canada. The primary focus of local-scale experiments was to improve and characterise the performance of the process models that describe the exchanges of radiative energy, water, heat, carbon and trace constituents between the boreal forest and the atmosphere. The regional-scale experiments were concerned with applying and validating the process models over large spatial scales using remote sensing. In BOREAS, as in previous field experiments, the science team adopted a nested multi-scale measurement strategy to integrate observations and process models over the scale range.

Because BOREAS was a multi-scale, multidisciplinary experiment, BOREAS investigators were able to measure, model and validate models for a number of coupled land-atmosphere exchanges of mass, heat and energy.

For example, in the boreal ecosystem, the land surface and atmospheric boundary layer are strongly coupled through the ecophysiological control on surface evapotranspiration. Boundary layer drying feeds back to stomatal conductance through the Ball-Berry-Collatz-expression for stomatal conductance $g_c = Ah/c$, (A is the canopy photosynthetic assimilation rate, h is relative humidity and c is carbon dioxide concentration near the canopy), as humidity h drops from boundary layer warming (mentioned in Fig. 4.11-1 caption). The average seasonal diurnal effects of this are illustrated in Fig. 4.11-1. The bottom row of Fig. 4.11-1 shows the early-morning increase in stomatal conductance with increase in solar radiation (third row). The top row of Fig. 4.11-1 shows the decrease in relative humidity with increasing R_n as surface heating expands the atmospheric boundary layer. Decreasing boundary layer humidity and increasing air temperature begin to exert physiological effects on leaf stomata conductance around 0200 to 0300 UTC (8 to 9 local time), reducing stomatal conductance even though photosynthetically active radiation continues to increase. Stomatal reduction of surface evapotranspiration is rapid as indicated by the rapid drop in evaporative fraction in the second row. Restriction of surface evaporation further dries the atmospheric boundary layer, which in turn couples back to stomatal conductance throughout the day. On many days during the BOREAS observing period, high surface sensible heat flux drove boundary layer depths to a height of 3 km, more typical of a desert than a forested ecosystem.

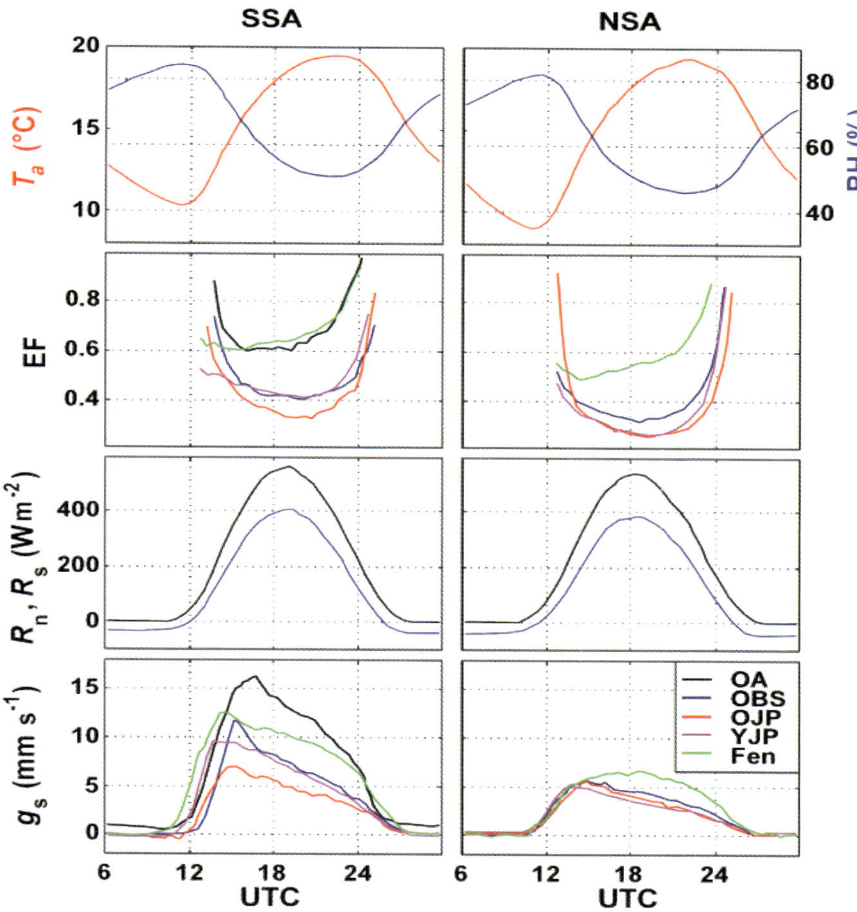

Fig. 4.11-1 Diurnal variation in air temperature, T_a (red lines), relative humidity, RH (blue lines), evaporative fraction, EF, solar radiation, R_s (black lines), net radiation, R_n (blue lines) and conductance to water vapour, g_s from BOREAS tower flux sites in the SSA (southern study area) and NSA (northern study area), averaged between 24 May and 19 Sept 1994. We first averaged the fluxes by time of day before calculating EF. The mean surface conductance was calculated from measured data only (e.g. with no gap filling) after excluding values below the 10th and above the 90th percentiles. The line style is the same for each land-cover type. Local time in the SSA and NSA is 6 hours less than UTC. (OA: old aspen; OBS: old black spruce; OJP: old jack-pine; YJP: young jack-pine)

This diurnal boundary-layer/surface feedback also couples into the seasonal atmosphere-surface exchange of carbon through temperature elevation (1) increasing canopy and soil respiration and (2) restricting photosynthetic uptake of carbon dioxide. BOREAS respiration measurements show that for wetland boreal conifer stands, below-ground respiration was ~50-70% of total ecosystem respiration. All respiration rates showed temperature dependence with Q10 values (i.e. rate increases for a 10° C temperature increase) of ~1.5 for wood, ~1.9 for roots, ~2.0 for foliage and 2.6 for soil. Chamber measurements in BOREAS peatlands showed Q10 values of 3.0 to 4.1.

Currently, respiration loss and photosynthetic uptake of carbon in the boreal wetlands are almost balanced, with a net uptake over the past few thousand years averaging about 50 g C $m^{-2} y^{-1}$. This sensitivity of the boreal ecosystem carbon exchange to temperature may couple strongly in the long-term with the observed and predicted warming at higher latitude continental interiors to accelerate high-latitude warming through increases in CO_2 input to the atmosphere.

Another important atmosphere-surface feedback that was not investigated extensively in the boreal zone is precipitation-driven. Increased storage of water in the boreal wetlands could alter carbon storage rates in these wetlands, through reduced decomposition rates below ground. Wetland spruce/tamarack communities have high root turnover, resulting in carbon storage rates ranging from 30 to 160 g C $m^{-2} y^{-1}$. Depending on the degree of soil saturation, anaerobic decomposition of organic materials in the boreal wetlands can generate methane and reduce rates of CO_2. On the other hand, drying and warming of the boreal ecosystem could switch peatlands from a net sink of carbon to a net source in exceptionally dry periods due to increased aerobic respiration.

4.11.4 Experiences

Boreal ecosystem-atmosphere exchange of carbon, water and energy from remote sensing

BOREAS thoroughly tested all aspects of computing the surface-atmosphere exchange of carbon, water and energy using a combination of surface-vegetation-atmosphere transfer models and satellite remote sensing to provide inputs to those models. These results are reported in detail in an 85-paper volume of the JGR - Atmospheres BOREAS special issue (Sellers and Hall 1997), and two subsequent issues, a 27-paper BOREAS special issue volume of JGR - Atmospheres (Hall 1999), and a 12-paper BOREAS special issue (Hall 2001). In addition there was an 11-paper volume of the BOREAS special issue of "Tree Physiology" (Margolis and Ryan 1997), a

9-paper volume of the Canadian Journal of Remote Sensing (O'Neill and Ranson 1997), and an 8-paper volume of Remote Sensing of Environment (Gamon et al. 2004).

This body of research led to a better definition and careful evaluation of the technology and analysis framework required to locate, understand the dynamics of and predict the future strengths of sources and sinks of land carbon. The analysis framework concept evolved from the BOREAS experience includes (1) inverse models, (2) coupled physical and biogeochemical process models, and (3) space-based and conventional observations. Inverse models predict the location and strength of terrestrial and ocean surface CO_2 sources and sinks, and rely on precise observations of spatial and temporal variations in atmospheric CO_2 concentrations. Process models predict carbon storage and the exchange rates of CO_2 at the atmosphere-land-ocean interfaces. Both the inverse and process modelling approaches are designed to infer regional magnitudes of net CO_2 exchange, and provide a means for cross validation. Inverse models can be used to provide a detailed analysis of what has happened to the CO_2 that has already been emitted by human activities. Physical and biogeochemical process models will provide a picture of the effects of land management and land use, terrestrial ecosystem and ocean dynamics, and other environmental factors on carbon sources and sinks over time. Importantly, these models will show how future atmospheric CO_2 concentrations might change as a result of natural occurrences, human actions, and past and future emissions. It is the synergy and interplay among advances in modelling, new observations of key Earth surface and atmospheric carbon properties, and improvements in the computational capacity that supports modelling and satellite data analysis that will enable major advances in our understanding and ability to predict climate change.

To implement this framework the major challenges for land remote sensing are (1) remote sensing of the spatial and temporal distributions of atmospheric CO_2 concentration to constrain model-based estimates of regional, continental, and global variations of land-ocean-atmosphere carbon exchange; (2) remote sensing of land biomass and biomass change resulting from land-use change, logging, fire, and changes in rates of growth resulting from climate change, CO_2 fertilisation, and nitrogen deposition; (3) more direct measurements of vegetation productivity using remote sensing to observe rates of vegetation photosynthetic activity; and (4) global satellite observations of soil moisture fields.

To accomplish these goals, an integrated and coordinated set of activities will be required. These include (1) technology development to support the new observations, (2) algorithm and model development to infer regional and global variations in atmosphere-surface carbon exchange, (3) additional coordinated field experiments of the type mounted in BOREAS for algorithm and model development and validation, and (4) data synthesis activities to provide the regional and global data sets needed to drive the carbon analysis framework.

Experiment design, coordination, data sharing and archive

To observe and quantify the many complex feedbacks regulating surface-atmosphere interactions in the boreal ecosystem, or any ecosystem for that matter, requires multi-scale, multidisciplinary research to produce the required integrated and multi-scale data sets to support research and validate models. These requirements in turn define the framework and organisation of the experiment. BOREAS involved the classic elements of any scientific study: formulation of the science questions to be addressed; definition of an analysis framework to address the questions; identification of the data required by the analysis framework; design of an approach and experiment to acquire the data; experiment execution including procurement, installation and operation of the necessary infrastructure; collection, processing and quality assurance of the data; utilisation of the analysis infrastructure and data to address the science questions; writing, review and publication of results; and finally, documenting and archiving the data. BOREAS included 85 teams involving over 300 scientists. To reduce redundancy of effort among the teams, a core staff at the Goddard Space Flight Center was formed. This staff handled many of the BOREAS-enabling tasks such as procurement, installation and operation of the experiments' core facilities (e.g. aircraft). In addition, the staff integrated the BOREAS data stream providing quality assurance, integration, documentation and archiving of the nearly 500 gigabytes of satellite imagery, aircraft imagery, and numeric data - 266 documented datasets in all.

Considering the costs involved in the planning and execution of large-scale field studies, it is advantageous to expend a bit more to create a finalised data archive that can be the legacy of the project for years to come. For a minimal additional cost (over the long-term) a centralised staff can help create a better-run project. An information system can help to publish a data product that will be useful in Earth systems science for decades. Too many projects neglect the data publication phase and in the end have no coherent data archive, limiting their usefulness to Earth systems science research. The value of these final data archives to future scientists cannot be overestimated, but is often underestimated, or worse, not even considered by project planners and sponsors. In the end, a scientific field study is not just about publishing scientific papers by the investigators, but is about creating a data archive that can be of use to scientists for decades, spawning new work and new publications far into the future.

4.12 Integrated Modelling of Water Availability and Vulnerability of Ecosystems and Society in the Semi-arid Northeast Brazil

Maarten S. Krol, Axel Bronstert, Annekathrin Jaeger

4.12.1 Motivation and Objectives for the Coupling Approach

Northeast Brazil is seriously influenced by the insufficiency and unreliability of precipitation in the semi-arid environment. The adverse natural conditions and the underdevelopment of the region means that the rural population cannot support itself in drought years.

The central objective of the project WAVES (Water Availability, Vulnerability of Ecosystems and Society, see Gaiser et al. 2003) was to improve the integrated understanding of the relationships between water availability, agro-ecosystems and society in the context of global environmental change. This understanding should assist in the assessment of climate change impacts, of the sustainability of long-term regional development and of the efficiency of various policy options.

An integrated model has been constructed covering the multidisciplinary themes involved. It has been termed "SIM" (semi-arid integrated model) and summarised by Krol et al. (2003) and documented comprehensively by Jaeger (2004).

4.12.2 Short Description of the Study

The study area shown in Fig.4.12-1 comprises the Brazilian Federal States of Ceará and Piauí, together covering 400 000 km^2. Most of this area has a semi-arid climate, with precipitation ranging from 500 to 900 mm y^{-1} and potential evapotranspiration exceeding 2000 mm y^{-1}. The climate exhibits an annual cycle with a dry period of six months (July - December); inter-annual variability is high, partly related to El Niño, with irregular severe droughts. Few rivers are naturally perennial. Dam construction, aiming at maintaining

Fig. 4.12-1 The "drought polygon" with a semi-arid climate in northeastern Brazil. The study area consists of the States of Ceará and Piauí. The border between Piauí and Maranhão is formed by the river Parnaíba

river flow and urban water supply, has a tradition of over a century. In rural areas an immense number of small reservoirs succeed in storing water to overcome shortage in normal dry periods but they fail in drought years. Groundwater availability in the interior, consisting of crystalline rock, is sparse and waters are often saline; aquifer systems exist in the coastal region and the downstream area of the Jaguaribe, the main river of Ceará. Apart from alluvial soils in the riverbed, soils are generally shallow and poor. Land use consists largely of extensive cattle holding and subsistence farming. The main crops are maize, beans, dry rice, cassava and cashew. Distribution of land property and income is very uneven, making small subsistence farmers highly vulnerable. The current population is about 10 million, increasing at a rate of 1.4 % per year. A steady rural–urban migration compensates for the rural birth excess, with migration strongly elevated in drought years. Migration to urban centres in the south of Brazil or to land-reclamation areas in the Amazon area is also an important demographic factor.

The processes described by the developed integrated model (Bronstert et al. 2000; Krol et al. 2003) cover the complete causal chain, from climate forcing to selected societal impacts.

- Climate variables are input to the model, shown in Fig. 4.12-2.

- A hydrological sub-model (Güntner and Bronstert 2003) describes runoff generation, evapotranspiration, soil humidity and deep percolation in a vertical water balance with a multi-layer soil model. Lateral flows and river flow are routed, and small water storage facilities in the landscape are accounted for. Large reservoirs in the main river network are described explicitly within the water balances.

- Water extraction is quantified for sectoral use in households, husbandry, irrigation, industry and tourism. Here, the physical availability of water is a restricting factor.

- Agricultural yields for regionally common crops are evaluated, depending on the relative sufficiency of evapotranspiration in phenological periods, regional soil characteristics and the agricultural management applied.

- Agricultural land use is represented by an agro-economic model, optimising economic returns given simulated yields, under restrictions in the availability of natural and economic resources,

- Population development is simulated, accounting for regional fertility and mortality, and for migration described by a push-pull model basing on income gradients.

Simulations are performed for the area of Ceará and Piauí (400 000 km^2) for periods of many decades (historic simulations 1921–1998 and scenarios 2000–2050). The common resolution factor in all calculations is the municipality (in average about 1000 km^2) and the year, but finer resolutions are used for various model components.

Calculations of hydrology and crop yield resolve daily time steps for terrain units within municipalities, which are homogeneous in their topographic characteristics, soils and vegetation, and average 75 km^2. Water extraction is computed at daily resolution. Model compartments are coupled directly at their finest common resolution.

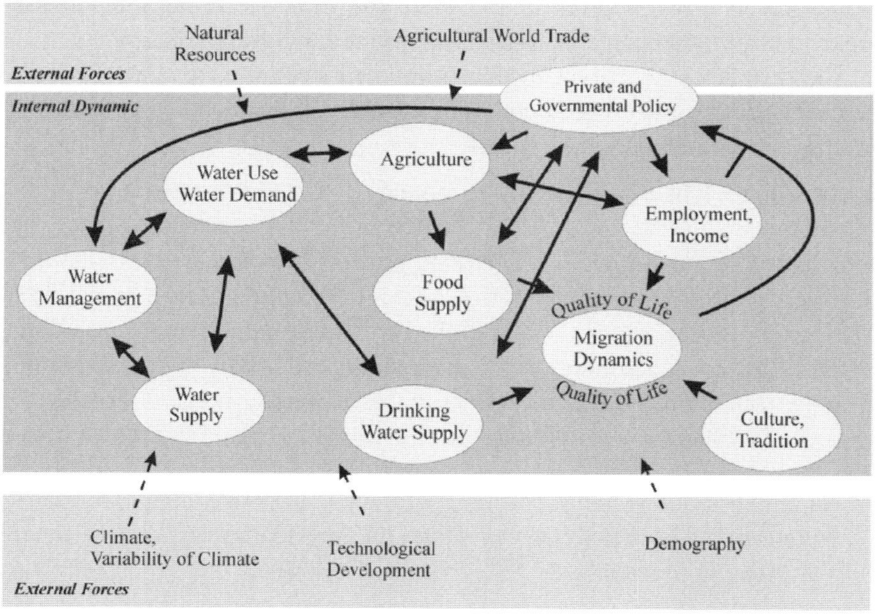

Fig. 4.12-2 Conceptual model, sketched as box-and-arrow diagram, of dynamic relationships between water availability, agriculture and migration in northeastern Brazil, as influenced by global change

4.12.3 Complexity of the Study

Feedbacks play an important role in the study. They necessitate a coupled integrated modelling, as opposed to an off-line (segmented) chain of models with one-way data transfer only. Important feedbacks in the model are, for example, the effect of water extraction for municipal water supply or irrigation on physical water availability, and the effects of land use and population development on water use.

The importance of the water extraction feedback is stressed by the fact that they become especially relevant during droughts. The land use and population feedback is especially relevant for long-term developments. Both couplings concern anthropogenic feedbacks on the natural water cycle.

Geographic couplings are of dominant relevance as well. The relative scarceness of water stresses the relevance of upstream-downstream relationships and the spatial distributions of water availability and water demands.

This aspect relates both to the level of model descriptions, which require an adequate spatial resolution, and to the level of policies. Large water storages are meant to guarantee urban water supply and small dam construction in the hinterland is presently being suppressed as the large number of small reservoirs appears to be reducing the effectiveness of large reservoirs.

The SIM Model is a concrete implementation of the result of the systems analysis shown in Fig. 4.12-2. The model is built up in a modular way, roughly representing the disciplinary contributions from different research groups, see Table 4.12-1.

Table 4.12-1 Structure of SIM: modules, content, methods applied, and scientific institutions responsible for the modules (PIK: Potsdam Institute for Climate Impact Research; FUNCEME: Fundação Cearense de Meteorologia e Recursos Hídricos (Fortaleza, Brazil); UFC: Universidade Federal do Ceará Hídricos (Fortaleza, Brazil); Uni KS: University of Kassel; FAO: Food and Agriculture Organisation; Uni HO: University of Hohenheim; FH K: University of Applied Sciences Cologne)

Module	Scientific disciplines covered by each module	methods applied	institution
climate	- historic reconstruction and scenarios of daily time series	- statistical/empirical downscaling	- PIK / FUNCEME
water	- large-scale water balance - water use and management	- process-based / deterministic - data driven budgeting	- PIK/UFC - Uni KS / UFC
land use	- crop yield - soil description - agro-economy	- empirical - distributed soil data base - mathematical optimisation	- FAO - Uni HO/ UFC/PIK - FH K
socio-economy	- quality of life - migration and demography	- empirical, data driven	- Uni KS / UFC / PIK

Many simplifications were made in constructing the different modules of the whole integrated model. The simplifications originated from lack of knowledge, lack of trustworthy data and from the focus on specific topics in the research.

Water quality issues were not addressed in the modelling, although salinisation is a serious problem for water use. The abundant small water reservoirs and reservoir operation were represented in general in a simplified, schematic way. Developments in water use and water use efficiency are only coarsely understood. Both in the agricultural economy and migration modules, only some of the well-understood processes are represented, omitting many other processes that are known only qualitatively.

The model is driven by external and internal factors, such as climate trends, market prices (external) and trends in economic growth and demographic parameters (internal).

Handling of uncertainty in an interdisciplinary integrated model poses a major challenge. In this project, only sensitivity studies were performed, estimating uncertainty effects due to uncertainties in base data and model parameters.

4.12.4 Experiences

The problems encountered in the project can be divided in those of a disciplinary and an interdisciplinary nature. Working in an underdeveloped region, reliable data over longer timescales and with reasonable spatial coverage are rarely available, posing problems for all disciplines. For interdisciplinary work, data should preferably be available with combined characteristics (e.g. water extraction both per sector and per water source) but are not found in this way; here mostly coarse assumptions have to be used. This may pose a serious problem for integrated modelling but on the other hand it illustrates the relevance of integrated studies in identifying the concrete gaps in available data and knowledge. In studying socio-economic impacts, the understanding of human behaviour in water use and strategies of adaptation to drought are only partly understood and hardly quantified. This is a problem for integrated studies connected with the state of the art in the contributing disciplines rather than of the integration itself.

Another category of problems concerns the involvement of disciplinary scientists and policy makers in integrated assessments. Both groups generally recognise the relevance of integrated studies but are not very eager to participate in modelling studies. Here, participative scenario studies and stationary integration of geographic data (in GIS) were found to have an initial attractiveness and to show the relevance of integrated modelling as a tool to support the assessments quantitatively in a way to best safeguard the consistency of scenarios.

Model applications focussed mainly on analysing regional sensitivity to climate change, on scoping pathways of regional development and assessing policy interventions related to dam construction and agricultural alternatives, see Jaeger (2004).

The integrated modelling in SIM successfully integrates the state-of-the-art understanding of regional processes. This understanding is incomplete, but consistent implications of interconnections on the systems dynamics can be shown and the most relevant knowledge gaps identified.

The integrated model proved very useful in improving the awareness of regional policy makers to the relevance of climate change for the region.

Acknowledgements

This case study concerns the project WAVES, studying the semi-arid north-eastern Brazil. German and Brazilian research institutions participated in the project. The authors thank the German Federal Ministry of Education and Research (BMBF) and the Brazilian National Council of Science and Technology Development (CNPq) and the Potsdam Institute for Climate Impact Research (PIK), Germany, for supporting this research.

4.13 Integrated Global-change Modelling with IMAGE-2

Bas Eickhout, Rik Leemans

4.13.1 Motivation and Objectives

IMAGE-2 (Integrated Model to Assess the Global Environment; Alcamo et al. 1998) is an integrated assessment model that calculates the environmental consequences of human activities worldwide. IMAGE-2 represents interactions between society, the biosphere and the climate system, to assess environmental issues like climate change, biodiversity decline and human well-being. The objective of the IMAGE-2 model is to explore the long-term dynamics of global environmental change, which requires a representation of how the world system could evolve. Future greenhouse gas emissions, for example, are the result of complex interacting demographic, technological, economical, social, cultural and political forces. The model is designed to compare business-as-usual scenarios with specific mitigation and adaptation scenarios and thus compare the effectiveness of such measures. Scenarios are "what if" representations of how the unknown future might unfold. They form an accepted and valuable tool in analysing how different comprehensive sets of driving forces may influence future emissions, concentrations, climate change and impacts. No specific likelihood can be specified to any scenario due to major uncertainties involved in the society-biosphere-climate system.

The socio-economic and energy-use calculations are performed for 17 world regions (see Fig. 4.13-1). The atmospheric and ocean components are based on globally aggregated approaches. The land-use and terrestrial-carbon calculations are performed on a high-resolution grid of 0.5 × 0.5 degrees. The innovative aspects in IMAGE-2 consisted in bringing together in a comprehensive framework these highly different resolutions and dimensions. The IMAGE-2 simulations are carefully calibrated with observed data for the period 1970–1995. Future trends in socio-economic scenario variables (e.g. demography, wealth and technology) have to be defined for each simulation. IMAGE-2 represents major global dynamic processes, including several natural interactions and feedbacks, such as CO_2 fertilisation and land-use change induced by changed climate.

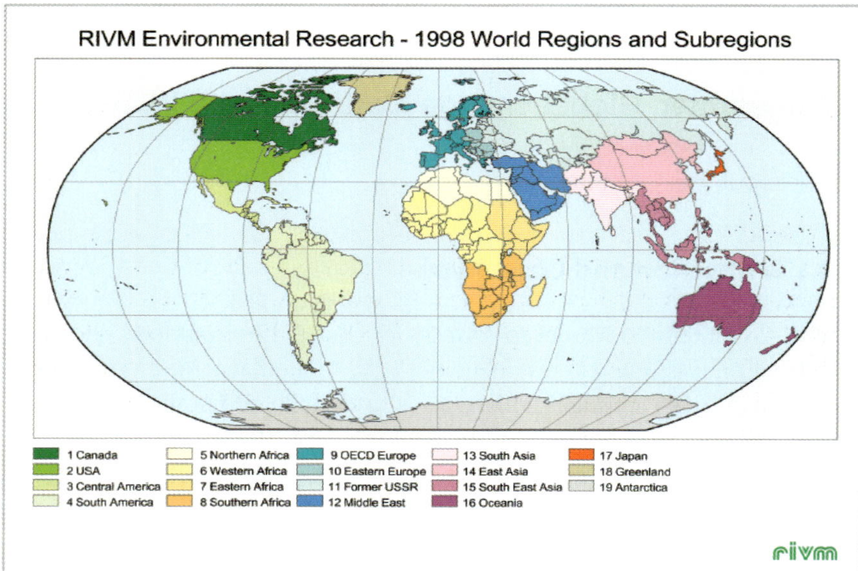

Fig. 4.13-1 The 17 world regions plus Greenland and Antarctica in IMAGE-2

4.13.2 Short Description of the Study

Recently, the Intergovernmental Panel on Climate Change (IPCC) published a set of standardised reference emission scenarios (SRES; Nakicenovic et al. 2000). A thorough review of the literature led to the development of different narrative "storylines"[1]. The quantification of these storylines was based on six different integrated models. These storylines were constructed on two axes, i.e. the degree of globalisation versus rationalisation, and the degree of orientation on material values versus social and ecological values. The resulting clusters were given neutral names: A1, A2, B1 and B2 (Fig. 4.13-2). The original SRES implementation focused mainly on the emissions from energy use and was only implemented for four regions (OECD, countries in transition, Asia and rest of the world). This aggregation concealed important regional differences in development as well as in socio-economic and environmental conditions. Also, SRES estimates emissions only: atmospheric, climatic and environmental processes that could alter emissions are neglected.

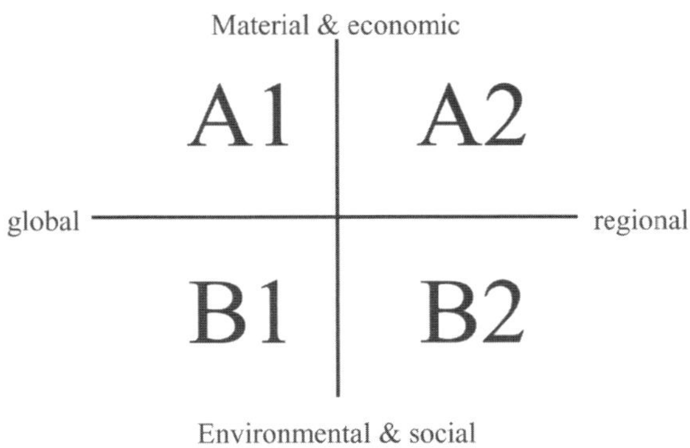

Fig. 4.13-2 The dimensions of the four SRES storylines, which define the storylines. In a globalised world, free trade and rapid conversion of technology define socio-economic developments, while in a regionalised world there is less exchange between and more diversity among regions. In a world that focuses on material and economic growth, resource use rapidly increases, while in a world with an environmental and social focus resources are used much more efficiently.

We have implemented these SRES storylines in IMAGE 2.2 on a more detailed level, taking several impacts and feedbacks into account (IMAGE team 2001). Basic demographic and economic information is derived from the original SRES data and disaggregated for 17 world regions. The data were checked for consistency by using the Phoenix model (population) and the WorldScan model (World economy) into the three linked subsystems of IMAGE-2 (Fig. 4.13-3):

[1] The innovative aspect of using narratives in the SRES storylines lies in the consistency of future trends in population, wealth, technology, equity and energy use within each scenario. The earlier IS92 emission scenarios were based on expert judgment or literature surveys for every scenario assumption independently, returning a high, middle and low scenario. In the SRES approach each narrative defines characteristic, consistent trends in the input assumptions, recognising explicitly the correlations between the different assumptions.

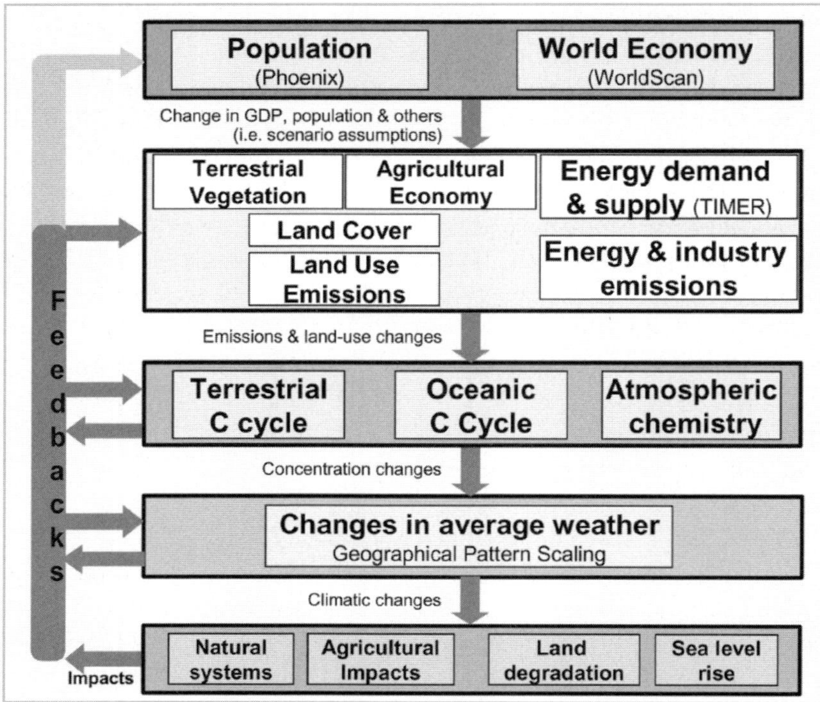

Fig. 4.13-3 The structure of the IMAGE-2 model

- The Energy-Industry System (EIS), which calculates regional energy consumption, energy efficiency improvements, fuel substitution, supply and trade of fossil fuels and renewable energy technologies. On the basis of energy use and industrial production, EIS computes emissions of greenhouse gases (GHG), ozone precursors and acidifying compounds.

- The Terrestrial Environment System (TES), which computes land-use changes on the basis of regional consumption, production and trading of food, animal feed, fodder, grass and timber with consideration of local climatic and terrain properties. TES computes emissions from land-use changes, natural ecosystems and agricultural production systems, and the exchange of CO_2 between terrestrial ecosystems and the atmosphere.

- The Atmosphere-Ocean System (AOS) calculates changes in atmospheric composition using the emissions and other factors in the EIS and TES, and by taking oceanic CO_2 uptake and atmospheric chemistry into consideration. Subsequently, AOS computes changes in climatic properties by resolving the changes in radiative forcing caused by greenhouse gases, aerosols and oceanic heat transport.

4.13.3 Complexity of the Study

The goal of simulating the complete cause-effect chain of global change requires a model with global coverage, but the multitude of components in the society-biosphere-climate system constrains the possibility to use detailed process-based models for each component. We have used simpler well-accepted models and focused on the interactions. Therefore, a more complete coverage of many feedbacks could be included in IMAGE-2 (Alcamo et al. 1998). For example, climate change induces shifts in global vegetation patterns; this changes the yield of crops and, hence, changes land-use patterns and thus influences the carbon uptake by biosphere and oceans. In other words: the strength of the IMAGE-2 model is found in the comprehensive coupling of the anthropogenic influences, the biosphere, ocean and atmosphere (c.f. Fig. 4.13-3) and not in the complexity of the different submodels.

IMAGE-2 is calibrated on historical data from 1970 to 1995. This calibration is performed on a number of data sources, concerning energy use, land use and national emission inventories. The global carbon cycle model required a longer calibration period, which is based on data for the atmospheric increase and energy and industry emissions of CO_2 for the period from 1765 to 1995.

In implementing the IMAGE-2 SRES storylines, much attention has been paid to identifying major sources of uncertainties. The different storylines result in largely different sets of energy use and energy carriers, land-use and land-cover patterns, emission trends and levels, concentrations and climate change (Table 4.13-1). Consequently, sea-level rise and other impacts also differ.

Additionally, we have run these storylines with different climate sensitivities and with or without specific feedback processes. The results of this sensitivity analysis indicate a possible global mean temperature increase, ranging from 1.6 to 5.5° C at the end of the 21st century, which is considerably larger than the range listed in Table 4.13-1. Another uncertainty stems from the pattern scaling of global-mean temperature changes to geographically explicit temperature and precipitation change by a given climate-change pattern obtained from state-of-the art climate models (GCMs). To assess such uncertainty, the pattern scaling has been carried out with different GCM climate-change patterns. In Fig. 4.13-4, the consequences of two different climate patterns for the yield of tropical cereals are depicted. These sensitivity experiments provide some insights in the uncertainty of quantification of these storylines.

Table 4.13-1 Results from the IMAGE-2 implementation of the SRES storylines after 100 simulated years (IMAGE team 2001)

	A1	A2	B1	B2
Global forest area (in Mha)	4980	3420	5750	4800
Global crop area (in Mha)	1480	2600	1260	1660
Global CO_2 emissions (in Pg C y^{-1})	13.6	29.0	2.7	10.1
Global CH_4 emissions (in Tg CH_4 y^{-1})	667	1102	491	696
Global N_2O emissions (in Tg N y^{-1})	20.7	27.4	17.5	20.8
Global CO_2 concentration (ppmv)	755	870	515	605
Global CO_2-equivalent concentration (ppmv)	1030	1315	640	820
Global mean temperature increase since pre-industrial times (in ° C)	3.4	3.7	2.3	2.9

4.13.4 Experiences

This IMAGE-2 case study provides detailed results from the SRES storylines for scientists interested in the causes and consequences of global change. Moreover, the results give a good perception on the important role of people and society in determining energy use, land use, emissions and changes in climate and biosphere. The storylines show that regionally largely different development patterns can emerge, depending on the assumptions that describe the specific SRES narrative. Especially, land-use dynamics, driven by shifts in population, diets and trade, are very important but are often neglected by the very detailed models restricted to the physical environment. IMAGE-2 gives good insight in how the terrestrial biosphere may evolve in different futures.

Uncertainty analyses with changed climate sensitivity parameters and different climate patterns show the relative importance of these factors. Especially, different climate patterns can change local and regional conclusions, as shown in Fig.4.13-4.

In the future, IMAGE-2 will pay more attention to biodiversity and adaptation of terrestrial ecosystems. For these issues, IMAGE-2 should be extended with a climate model that takes climate variability into account because ecosystems do not only respond to global-mean changes in the climate but are sensitive to local extremes. Moreover, to assess the vulnerability of ecosystems, more detailed ecosystem models will be

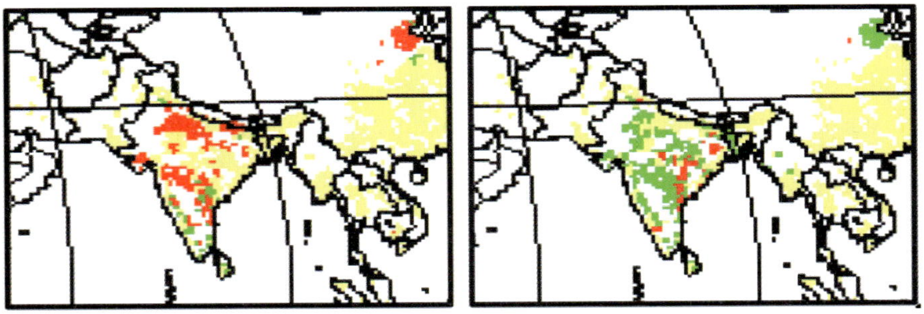

Fig. 4.13-4 Change in yield of rice in the current rice area of southern Asian, due to different climate patterns from HADCM2 and ECHAM4. Green indicates a yield increase, while red a yield decline

applied which means that the impact side of the causal chain especially has to be improved. This is policy relevant because now that climate change is accepted as a real issue, the possible impacts and adaptation have to be assessed more realistically.

4.14 The Virtual Watershed Laboratory

George Leavesley, Everett P. Springer

4.14.1 Motivation and Objectives

Hydrological simulation models have evolved from stand-alone, conceptual models that treat the entire watershed or basin as a single unit, to complex, spatially-distributed, physics-based process models. This development was driven by an improved understanding of the elements of the regional hydrological cycle, the availability of cheaper and more powerful computers, and increased societal demands for environmental decision-making tools. The National Academy of Sciences (1997), in their report "Preparing for the 21st Century", stresses the need for a scientific basis for risk assessment and the use of science to strengthen environmental policies and policymaking.

Increasingly, decision makers must balance competing demands for water and are relying on the results of strongly coupled and complex systems, for instance, by weighing the needs of riparian communities and endangered species against those of irrigated agriculture and municipal water supply, by managing the conjunctive use of surface and groundwater, or by considering the effects of climate variability on the integrated management of water. All these concerns require models in which system components are fully integrated and dynamically coupled through feedback mechanisms.

The current state of the art combines separate models of system components, operating independently and exchanging information in essentially one direction only. This does not reflect the actual dynamics of basin-scale systems whose components interact through complicated feedback mechanisms occurring on many time and space scales. Unfortunately, the development of fully integrated models is costly and time consuming. In application, the model specifics depend on the geographical location, the appropriate processes, and relevant time and space scales, all of which change from application to application. A single integrated model has limited utility.

To address these issues, we propose the development of the Virtual Watershed Laboratory (VWL), an extensible, object-oriented, parallelised, plug-and-play modelling environment with dynamic coupling of physics-based process models operating on different temporal and spatial scales. The modelling environment will consist of a data model, physics-based process models, coupling algorithms and a user interface that will facilitate both model development and analysis of model input and output. The VWL will provide a platform for the development of new, integrated models to study watershed and basin dynamics, as well as a test bed for new physical process models.

By employing information technology – developing the user interface, analytical tools, and "coupling modules" that link one physical process to another – we can advance the state of the art of basin-scale modelling without prescribing any particular computer model. Users will be free, for instance, to select a specific groundwater code from the process library or to insert one of their own choices, so long as it meets well-defined input/output conventions and software engineering requirements. The VWL project will focus initially on interactive modelling of the water and energy balance below, at, and above the land surface. However, the VWL will be a modelling environment rather than a specific model and will be designed to allow the incorporation of process models describing, for example, chemical and nutrient balances, and plant physiological and biogeochemical processes.

The VWL is a platform that will allow and encourage the development of models to simulate and predict the complex interactions taking place in the Earth system, using the latest computer tools produced by the IT community, and making them available to the scientific community, decision makers and even the public.

Specifically, the VWL will:

- Advance software engineering of coupled processes operating in distributed memory, message-passing environments;

- Develop scalable and portable tools for analysing, managing and visualising large data sets;
- Develop a graphical user environment that hides parallel programming details from users;
- Extend existing tools that support parallel workspace management and object orientation to Earth science applications;
- Implement a basic library of parallel models for groundwater, surface water, and atmospheric fluxes of mass and energy. In particular, the VWL will address the issue of parallel solvers for unstructured grids;
- Define a new generation of physics-based coupling models for linking process models operating on different time and space scales.

4.14.2 Short Description of the Study

The VWL integrates process models – the atmosphere, land-surface hydrology, subsurface hydrology, the river network, and riparian plant communities – and represents feedbacks among them in a plug-and-play user environment. We envision that the VWL will be applied to systems as small as a local watershed in eastern Massachusetts, and as large as a major river basin, such as the Rio Grande basin, with an emphasis on the basin scale. Although it is not intended to be applied at the continental scale, it could be, albeit with some modifications.

4.14.3 Complexity of the Study

Our basic goal is to apply and advance the physical science and information technology needed to model coupled processes that operate at different space and time scales, and to provide the software tools that will allow this. Although hydrological science and applications will be the direct beneficiary, the approach and many of the software tools will be transferable to other geophysical and environmental sciences, especially Earth systems science.

The plug-and-play exchangeability of software components is a widely used technique in commercial software applications, but it has not found widespread application in Earth system science. Until now, researchers have focused largely on the development and improvement of process models that simulate the water and energy fluxes within individual system components (e.g. the atmosphere, the vadose zone, the river channel). Because of the multiplicity of data formats, computer languages and software design strategies (or lack thereof), integration of these components is currently no easy task. This not only hinders the development of fully integrated models but also limits the opportunities for the developers of process models to test their models in an integrated environment and distribute their findings to a wider audience.

Experimental sciences like hydrology build on hypothesis testing and interpretation, which is based on earlier, published hypotheses and results (e.g. Reynolds and Acock 1997). However, modellers prefer to build new models from the ground up, typically because existing models are not well designed for incremental improvement by others. A plug-and-play approach to model development facilitates model process comparisons by enabling detailed analysis of individual processes and their interactions. It provides a framework in which to conduct hypothesis testing and analysis and to make incremental model improvements.

Efforts at process model integration are currently underway in a number of areas. Polcher et al. (1998) have proposed a general interface to allow coupling of land-surface schemes and general circulation models. The backdrop for their proposal is formed by the Project for Intercomparison of Land-Surface Parameterization Schemes (PILPS; Henderson-Sellers et al. 1996), which will ultimately compare different land-surface schemes coupled to the same atmospheric model, and the Atmospheric Model Intercomparison Project (AMIP; Gates 1992). Within hydrology, the only modelling environment that allows the exchange of model components to build watershed and problem-specific hydrologic models conveniently, is the U.S. Geological Survey's Modular Modeling System (MMS; Leavesley et al. 1996, 2002). MMS consists of three components: (1) The pre-processor aids in data analysis and preparation; (2) The model component consists of a process module library with modules written in C or Fortran, an interactive, graphical model-builder tool and an execution environment; (3) Finally, the post-process component provides tools to display and analyse results. Communication between the process modules is managed through a common database, accessible by all models. The USGS is porting part of the MMS into a parallel environment. The MMS is a paradigm for development of the VWL, although MMS includes a smaller set of processes and cannot handle massive data sets. VWL will be a much more powerful, flexible and evolved approach, both in terms of both information technology and Earth system science. To take advantage of the "lessons learned" by the developers of MMS, they are collaborators in the proposed project.

As an extensible, object-oriented, parallelised, plug-and-play modelling environment, the VWL will consist of the following components:

- Parallelised data management;
- Model building environment;
- Parallelised model execution environment;
- Parellelised data analysis and visualisation environment;
- A code library that will contain a hierarchy of object-oriented process models and coupling algorithms.

The VWL will be comprehensively documented. Documentation will consists of a set of web-accessible source codes, user guides and tutorials that describe and explain the VWL and its application programmer's interface.

4.14.4 Experiences

Initial experience with the VWL has been a coupling of the RAMS (Regional Atmospheric Model) (Pielke et al. 1992) with a hydrology model LADHS (Los Alamos Distributed Hydrology System) and applying this to Water Year 1993 on the upper Rio Grande Basin. This modelling approach was pursued to answer questions concerning the relative effects of climate variability versus land-use change within the upper Rio Grande. Coupling for this system is currently one-way, with meteorological variables passed from RAMS to LADHS every 20 minutes. The rapidly varying topography and shifting land use within the upper Rio Grande make it necessary to use fine grid spacing for both model components. A 5 km grid was used for RAMS and LADHS uses a 100 m grid cell size. Both codes were implemented on a parallel processor, and PAWS (Beckman et al. 1998) is used to facilitate the data transfer between components.

The one-year simulation is basically a proof of concept and also designed to answer computer science questions relative to coupling and machine performance. Fig. 4.14-1 presents soil moisture and snowpack from November 5, 1992 as an example of the data that can be produced. A key problem was the bias (over-prediction) in precipitation estimates from RAMS, especially for the higher elevation areas. Both are on a 100 m scale as the precipitation is downscaled to 100 metres from RAMS. Although this is not necessarily a problem with coupling, the precipitation bias does

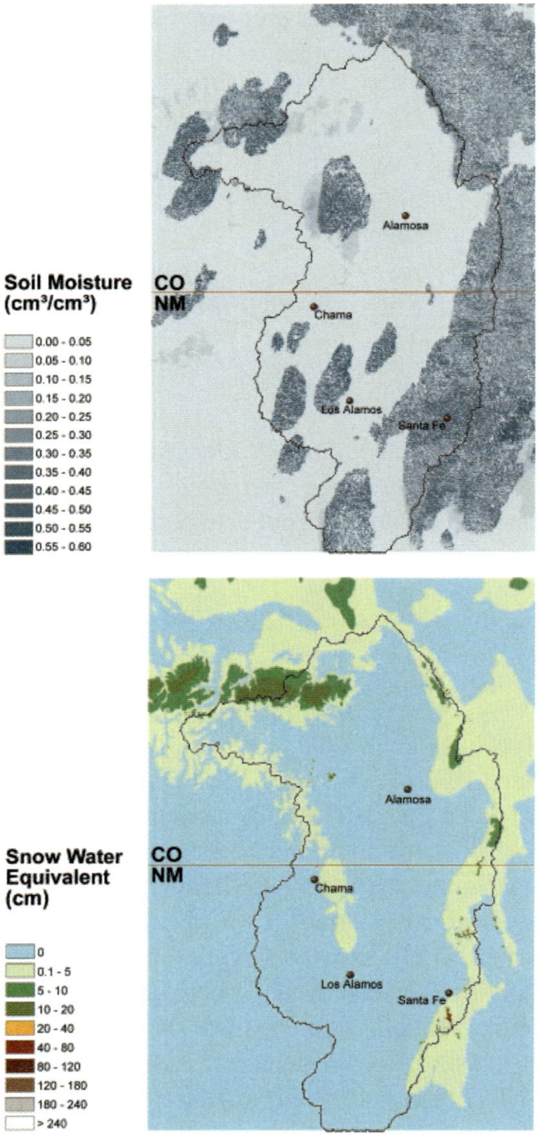

Fig. 4.14-1 Soil moisture and snow water equivalent distribution for upper Rio Grande Basin for November 5, 1992 at 00:00 UTC

represent an issue when using regional atmospheric models to address water resources scenarios for future change issues. Future modifications include developing the two-way coupling between LADHS and RAMS, which will require an energy balance to be added to LADHS. To develop the VWL further, the modular capability will need to be executed on parallel architecture using a MMS type approach. PAWS has demonstrated potential in this regard and further efforts will use the approach applied in this case study.

References Chapter 4

Abbott MB, Bathurst JC, Cunge JA, O'Conell PE, Rasmussen EA (1986) An introduction to the European Hydrological Model "SHE" – Système Hydrologique Européen. J Hydrol 87:45-77

Alcamo J, Kreileman E, Krol M, Leemans R, Bollen JC, van Minnen J, Schaeffer M, Toet AMC, De Vries HJM (1998) Global modelling of environmental change: An overview of IMAGE 2.1. In: Alcamo J, Leemans R, Kreileman E (eds) Global change scenarios of the 21st century. Results from the IMAGE 2.1 model. Elseviers Science, London pp3-94

Arlt H (1995) A hydrogeological study of the Nile delta aquifer with emphasis on saltwater intrusion in the northern area. TU Berlin, Inst. für Wasserbau und Wasserwirtschaft, Mitteilung 130, p185

Band LE (2000) Forest ecosystem productivity in Ontario. In: Perera AH, Euler DL, Thompson ID (eds) Ecology of a managed terrestrial landscape: Patterns and processes of forests in Ontario. University of British Columbia Press, pp163-178

Band LE, Tague CLBrun, SE, Tenenbaum DE, Fernandes RA (2001a) Modeling watersheds as spatial object hierarchies: Structure and dynamics. Transact Geogr Inf Syst 4:181-196

Band LE, Tague CL, Groffman P, Belt K (2001b) Forest ecosystem processes at the watershed scale: Hydrological and ecological controls of nitrogen export. Hydrol Proces 15:2013-2028

Baron JS, Hartman MD, Kittel TGF, Band LE, Ojima DS, Lammers RB (1998) Effects of land cover, water redistribution and temperature on ecosystem processes in the South Platte Basin. Ecol Applications 8:1037-1051

Batchvarova E, Gryning S-E, Hasager CB (2001) Regional fluxes of momentum and sensible heat over a sub-arctic landscape during late winter. Bound-Lay Meteorol 99:489-507

Baumann R (1995) Untersuchungen zum anthropogenen Einfluß auf die Salzwasserinvasion im Nildelta. TU Berlin, Inst. für Wasserbau und Wasserwirtschaft, Mitteilung 129, p113 (in German)

Beckman PH, Fasel PK, Humphrey WF, Mniszewski SM (1998) Efficient coupling of parallel applications using PAWS. Presented at the High Performance Distributed Computing (HPDC-7) Conference, Chicago, IL

Beven K, Freer J (2001) Equifinality, data assimilation and uncertainty estimation in mechanistic modelling of complex environmental systems using the GLUE methodology. J Hydrol 249:11-29

Beven K, Kirkby M (1979) A physically based variable contributing area model of basin hydrology. Hydrol Sci Bull 24:43-69

Bronstert A, Jaeger A, Güntner A, Hauschild M, Döll P, Krol M (2000) Integrated modelling of water availability and water use in the semi-arid Northeast of Brazil. Phys Chem Earth (B) 25(3):227-232

Claussen M, Brovkin V, Ganopolski A (2001) Biogeophysical versus biogeochemical feedbacks of large-scale land cover change. Geophys Res Lett 28:1011-1014

Custodio E (1997) Groundwater quantity and quality changes related to land and water management around urban areas: Blessings and misfortunes. In: Chilton J (ed) Groundwater in the urban environment: Problems, processes and management. XXVII Congress of the International Association of Hydrogeologists (IAH), Nottingham, pp11-22

Dickinson R, Henderson-Sellers A, Kennedy PJ, Wilson MF (1986) Biosphere-Atmosphere Transfer Schemes (BATS) for the NCAR community climate model. NCAR Techn. Note NCAR/TN-275+STR, Boulder, Colorado, p69

Enger L (1990) Simulation of dispersion in moderately complex terrain. Part A: The fluid dynamic model. Atmos Environ 24:2431-2446

FAO (1991) CLIMWAT for CROPWAT (1980). Food and Agriculture Organization of UN, Land and Water Development Division, Rome, p93

Farid S (1980) Nile Delta groundwater study. M.Sc. Thesis, Cairo University, p304

Gaiser T, Krol MS, Frischkorn H, de Araújo JC (eds) Global change and regional impacts: Water availability and vulnerability of ecosystems and society in the semi-arid Northeast of Brazil. Springer, Berlin, p428

Gamon JA, Huemmrich KF, Peddle DR, Chen J, Fuentes D, Hall FG, Kimball JS, Goetz S, Gu J, McDonald KC, Miller JR, Moghaddam M, Rahman AF, Roujean JL, Smith EA,Walthall CL, Zarco-Tejada P, Hu B, Fernandes R, Cihlar J (2004) Remote sensing in BOREAS: Lessons learned. Remote Sens Environ 89(2):139-162

Ganopolski A, Petoukhov V, Rahmstorf S, Brovkin V, Claussen M, Eliseev A, Kubatzki C (2001) CLIMBER-2: A climate system model of intermediate complexity. Part II: Model sensitivity. Clim Dyn 17:735-751

Gates W (1992) The atmospheric model intercomparison project. B Am Meteorol Soc 73:1962-1970

Gottschalk L, Grying S-E., Beldring S, Motovilov Yu (1998) NOPEX modelling activities: Towards a coupled hydrological and meteorological mesoscale model. In: Fifth Meeting of the BALTEX SSG, Riga, Latvia, April 1997, Int. BALTEX Secretariat Publ, 10:A33-A42

Gottschalk L, Beldring S, Engeland K, Tallaksen L, Sælthun NR, Kolberg S, Motovilov Y (2001) Regional/macroscale hydrological modeling: A Scandinavian experience. Hydrol Sci J 46(6):963-982

Gryning S-E, Batchvarova E, de Bruin HAR (2001) Energy balance of a sparse coniferous high-latitude forest under winter conditions. Bound-Lay Meteorol 99:465-488

Güntner A, Bronstert A (2003) Large-scale hydrological modelling in the semiarid northeast of Brazil: Aspects of model sensitivity and uncertainty. In: Servat E, Najem W, Leduc C, Shakeel A (eds) Hydrology of Mediterranean and semi-arid regions. IAH-Publications No. 278, Wallingford, pp43-48

Hall FG (1999) Introduction to special section: BOREAS in 1999: Experiment and science overview. J Geophys Res 104:27627 – 27639

Hall FG (2001) Scaling from plot to region: BOREAS III introduction. J Geophys Res 106D(24):33511-33516

Halldin S, Gryning S-E, Gottschalk L, Van de Griend AA, Jochum A (eds) (1999) Boreal forests and climate. Agr Forest Meteorol NOPEX Special Issue 98-99, p696 and appended CD with documented datasets

Halldin S, Gryning S-E, Lloyd CR (eds) (2001) Land-surface/atmosphere exchange in high-latitude landscapes. Theor Appl Climatol WINTEX/LAPP Special Issue 70, p252 and appended CD with documented datasets

Henderson-Sellers A, McGuffie K, Pitman A (1996) The project for intercomparison of land-surface schemes: 1992 to 1995. Clim Dyn 12:849-859.

Henry HR (1964) Interfaces between salt water and fresh water in coastal aquifers US Geol Survey Water-supply Paper 1613-C

Herzberg A (1901) Die Wasserversorgung einiger Nordseebäder. J für Gasbeleuchtung und Wasserversorgung 44:815-819, 842-844

Holzbecher E (1998) Modelling density-driven flow in porous media. Textbook with Software on CD-ROM, Springer Publ, Heidelberg

Holzbecher E (2001) The dynamics of subsurface water-divides. Hydrol Proces 15:2297-2304

Holzbecher E (2003) Effects on subsurface watersheds from the construction of the Polzow Canal - a case study on the effect of changes in the hydraulic system of lakes on groundwater flow. Arch Hydrobiol, Special Issue Advanc Limnol 58:25-36

Holzbecher E, Ginzel G (1998) Modeling the subsurface watershed of Lake Stechlin. 3rd Intern. Conf. on Hydroscience and Engineering (ICHE), Cottbus, Proc. on CD-ROM

Holzbecher E, Kitaoka K (1993) Saline disasters and modeling approach. XXV. Gen. Ass. IAHR (Intern. Ass. for Hydr. Research), Tokyo

Holzbecher E, Nützmann G (1999) Water and component mass balances in the catchment of Lake Stechlin. In: Leibundgut C, McDonnell J, Schultz G (1999) Integrated methods in catchment hydrology. IAHS Publication No. 258, pp37-44

IMAGE team (2001) The IMAGE 2.2 implementation of the SRES scenarios: A comprehensive analysis of emissions, climate change and impacts in the 21st century. RIVM CD-ROM Publication 481508018, Bilthoven

Jaeger A (2004): Regionale Integrierte Modellierung der Auswirkungen von Klimaänderungen am Beispiel des semi-ariden Nordostens von Brasilien. PIK-Report No.89, p156

Kalnay E, Kanamitsu M, Kistler R, Collins W, Deaven D, Gandin L, Iredell M, Saha S, White G, Woollen J, Zhu Y, Chelliah M, Ebisuzaki W, Higgins W, Janowiak J, Mo KC, Ropelewski C, Wang J, Leetmaa A, Reynolds R, Jenne R, Joseph D (1996) The NCEP/NCAR 40-year reanalysis project. B Am Meteorol Soc 77:437-471

Klemes V (1986) Operational testing of hydrological simulation models. Hydrol Sci J 31(1):13-24

Kramm G, Dlugi R, Mölders N, Müller H (1994) Numerical investigations of the dry deposition of reactive trace gases. In: Baldasano JM, Brebbia CA, Power H, Zannetti P (eds) Air pollution II Vol. 1: Computer simulation. Comp Mech Publ, Southampton, Boston, pp285-307

Kramm G, Beier N, Foken T, Müller H, Schröder P, Seiler W (1996) A SVAT scheme for NO, NO_2, and O_3 - model description and test results. Meteorol Atmos Phys 61:89-106

Knudsen J, Thomsen A, Refsgaard JC (1986) WATBAL: A semi-distributed physically based hydrological modelling system. Nordic Hydrology 17:347-362

Krol MS, Jaeger A, Bronstert A (2003) Integrated modeling of climate change impacts in Northeastern Brazil. In: Gaiser T, Krol MS, Frischkorn H, de Araújo JC (eds) Global change and regional impacts: Water availability and vulnerability of ecosystems and society in the semi-arid Northeast of Brazil. Springer, Berlin, pp43-56

Kuchment LS, Demidov VN, Motovilov YG (1983) River runoff formation (physically-based models). Nauka, Moscow (In Russian)

Kuchment LS, Demidov VN, Motovilov YG (1986) A physically based model of snowmelt and rainfall runoff genesis. Proc. The Second Science Assembly of the IAHS, Budapest

Kuchment LS, Gelfan AN, Demidov VN (2000) A distributed model of runoff generation in the permafrost regions. J Hydrol 240:1-22

Leavesley GH, Restrepo PJ, Markstrom SL, Dixon M, Stannard LG (1996) The modular modeling system - MMS: User's manual. U.S. Geological Survey Open File Report 96-151, p142

Leavesley GH, Markstrom SL, Restrepo PJ, Viger RJ (2002) A modular approach to addressing model design, scale, and parameter estimation issues in distributed hydrological modelling. Hydrol Proces 16(2):173-187

Lerner DN, Yang Y (2000) Quantifying recharge at the city scale using multiple environmental tracer. In: Dessargues A (ed) Tracers and modelling in hydrogeology. IAHS Publ 262, pp355-361

Margolis HA, Ryan MG (1997) A physiological basis for biosphere-atmosphere interactions in the boreal forest: An overview. Tree Physiology 17:491-500.

Marshall CH Jr, Pielke Sr RA, Steyaert LT, Willard DA (2004) The impact of anthropogenic land-cover change on the Florida Peninsula sea breezes and warm season sensible weather. Mon Weather Rev 132:28-52

Mölders N (2001) On the uncertainty in mesoscale modeling caused by surface parameters. Meteorol Atmos Phys 76:119-141

Mölders N, Rühaak W (2001) A runoff module for use in an atmospheric model. In: Sutmöller J, Raschke E (eds) Modellierung in meso- bis makroskaligen Flußeinzugsgebieten - Tagungsband zum gleichnamigen Workshop am 16./17. November 2000 in Lauenburg, GKSS2001/15, pp63-75

Mölders N, Rühaak W (2002) On the impact of explicitly predicted runoff on the simulated atmospheric response to small-scale land-use changes - an integrated modeling approach. Atmos Res 63:3-38

Mölders N, Raabe A, Beckmann T (1999) A technique to downscale meteorological quantities for the use in hydrologic models - description and first results. IAHS 254:89-98

Motovilov YG (1998) Simulation of climate change impacts on river runoff: Coupling of historical and modelling approaches. Proc. of the Second Int. Conference on Climate and Water, Helsinki, Vol 1, pp223-232

Motovilov YG, Belokurov A (1996) Modelirovanie gidrologicheskogo tsikla i transformatsii zagrjaznenii v rechnyh basseinah dlja zadach ecologicheskogo monitoringa (Modelling of a hydrological cycle and pollution transport in river basins for ecological monitoring, in Russian). Inst. Appl. Geophys. Publ, 81:68-76

Motovilov YG, Belokurov A (1997) ECOMAG - river basin hydrological cycle model on the basis of landscape units: Application for northern watersheds. Proc. the Third Int. Study Conference on GEWEX in Asia and GAME, Cheju, Korea, pp134-139

Motovilov YG, Gottschalk L, Engeland K, Rodhe A (1999) Validation of a distributed hydrological model against spatial observation. Agr Forest Meteorol 98-99:257-277

Nakicenovic N, Alcamo J, Davis G, de Vries B, Fenhann J, Gaffin S, Gregory K, Grübler A, Jung TY, Kram T , Emilio la Rovere E, Michaelis L, Mori S, Morita T, Pepper W, Pitcher H, Price L, Riahi K, Roehrl A, Rogner HH, Sankovski A, Schlesinger M, Shukla P, Smith S, Swart R, van Rooyen S, Victor N, Dadi Z (2000) Special report on emissions scenarios. Cambridge University Press, Cambridge, p599

National Academy of Sciences, National Academy of Engineering, Institute of Medicine, National Research Council 1997: Preparing for the 21st Century: The environment and the human future. National Academies Press, Washington D.C.

O'Neill N, Ranson J (1997) BOREAS Special Issue. Can J Remote Sens 23(2): 89-203

Parton WJ, Mosier AR, Ojima DS, Valentine DW, Schimel DS, Weier K, Kulmala AE (1996) Generalized model for N_2 and N_2O production from nitrification and denitrification. Global Biogeochem Cy 10:401-412

Petoukhov V, Ganopolski A, Brovkin V, Claussen M, Eliseev A, Kubatzki C, Rahmstorf S (2000) CLIMBER-2: A climate system model of intermediate complexity. Part I: Model description and performance for present climate. Clim Dynam 16:1-17

Pielke RA Sr (2001) Influence of the spatial distribution of vegetation and soils on the prediction of cumulus convective rainfall. Rev Geophys 39(2):151-177

Pielke RA, Cotton WR, Walko RL, TrembackCJ, Nicholls ME, MoranMD, D. A. Wesley DA, Lee TJ, Copeland HJ (1992) A comprehensive meteorological monitoring system – RAMS. Meteorol Atmos Phys 49:69-91

Pielke RA, Walko RL, Steyaert L, Vidale PL, Liston GE, Lyons WA (1999) The influence of anthropogenic landscape changes on weather in south Florida. Mon Weather Rev 127:1663-1673

Polcher J., McAvaney B, Viterbo P, Gaertner M-A, Hahmann A, Mahfouf JF, Noilhan J, Phillips T, Pitman A, Schlosser CA, Schulz J-P, Timbal B, Verseghy D, Xue Y (1998) A proposal for a general interface between land-surface schemes and General Circulation Models. Global Planet Change 19:261-276

Reynolds JF, Acock B (1997) Modularity and genericness in plant and ecosystem models. Ecol Mod 94:7-16

Richter D (1997) Das Langzeitverhalten von Niederschlag und Verdunstung und dessen Auswirkungen auf den Wasserhaushalt des Stechlinseegebietes. Berichte des Deutschen Wetterdienstes, Nr. 201, Offenbach/Main, p126

Running SW, Hunt RE (1993) Generalization of a forest ecosystem process model for other biomes, BIOME-BGC, and an application for global-scale models. In: Ehleringer JD, Field CD (eds) Scaling physiologic processes: Leaf to globe. Academic Press, San Diego, pp141-158

Sellers PJ, Hall FG (1997) BOREAS in 1997: Experiment overview, scientific results and future directions. J Geophys Res 102D(24):28,731 – 28,770

Tague CL, Band LE (2001) Simulating the impacts of road construction and forest harvesting on hydrologic response. Earth Surface Proces Landforms 26:135-151

Vázquez-Suñé E, Sánchez-Vila X, Carrera J, Marizza M, Arandes R, Gutiérrez LA (1997) Rising groundwater levels in Barcelona: Evolution and effects on urban structures. In: Chilton J (ed) Groundwater in the urban environment: Problems, processes and management. XXVII Congress of the International Association of Hydrogeologists (IAH), Nottingham, pp267-271

Walko RL, Band LE, Baron J, Kittel TGF, Lammers R, Lee TJ, Ojima DS, Pielke RA, Taylor C, Tague C, Tremback CJ, Vidale PL (2000) Coupled atmosphere-biophysics-hydrology models for environmental modeling. J Appl Meteorol 39:931-944

Wigmosta M, Vail LW, Lettenmaier DP (1994) A distributed hydrology-vegetation model for complex terrain. Water Resourc Res 30:1665-1680

Zaghloul Z (1959) Flow distribution through the groundwater aquifer of the Nile Delta. M.Sc. Thesis, Alexandria University, p106

5.0 Summary and Outlook

The primary goal of this book is to provide a current, state-of-the-art review of coupled modelling within hydrology and in integrated Earth system models at a level appropriate for graduate students entering the field. Then, on the basis of this review, to assess the value of coupled modelling, to clarify where and when their use is appropriate and beneficial, and to identify areas of opportunity and need. A systematic procedure for representing a complex, interacting system using coupled models has been attempted. In addition, the book documents examples of the application of coupled models to show how such models can be used in practice within hydrology and related sciences.

5.1 When and Why is it Appropriate to Use Coupled Models?

This book discusses many issues on the motivations for and methodologies employed when creating complex models. It is important to assess whether there is a need for a complex coupled model when addressing a specific task. Increasingly, hydrologists are asked to perform integrated analyses of environmental interactions that involve feedbacks between atmospheric, ecological and hydrological systems, as well as human society. In such cases, it is often the feedbacks between systems that are of greatest interest because they may produce unanticipated responses in a range of processes in reaction to natural or human-induced changes.

In generic terms, typical situations can be specified in which the application of coupled models is essential, for example, when:

- It is not feasible, economic, or accurate to impose a boundary between compartments in the modelled hydrogeobiosphere system and thereby isolate them;
- Observations of two or more state variables that are coupled cannot be modelled simultaneously without the processes that couple them also being included in the model;
- The science that is being addressed is beyond the normal range of activity and interest of specialists in individual coupled processes, and unforeseen whole-system behaviour may occur due to inherent non-linearity and/or non-linear feedbacks implicit in the coupling;
- Assessing the vulnerability and risk of coupled systems, as an alternative to the more conventional paradigm of forecasting.

Experiences within integrated assessments based on using coupled models have also shown that adopting a systematic procedure, or at least a well-defined coupling task, often leads to interdisciplinary cross-fertilisation and broader education of the scientists involved. This is a very high-value feature of such projects.

From a hydrological perspective, some issues may be of particular interest for integrated systems analysis based on a coupled modelling approach, including:

- Wetland eco-hydrology, including interactions of surface water bodies, shallow groundwater aquifers, vegetation dynamics, and water and energy fluxes;

- Management of river-delta systems, including interactions of surface water bodies, groundwater resources, seawater, geo-chemistry, and land-use management;

- Multi-objective water resources management has traditionally been approached recognising the coupled nature of the problem to account for the interdependency between different water resources and between the needs of different water sectors. This is of particular relevance when managing extreme water shortage situations, i.e. droughts, because the reliability of water resources systems may alter in response to local, regional and global environmental change. Hence, the corresponding impact analysis calls for a stringently integrative approach that recognises all the relevant aspects and their interactions;

- Integration of socio-economic processes into hydrological sciences has gained increasing attention recently, and the inclusion of feedback processes between nature and the anthroposphere is of particular relevance in this context.

Examples outlined in this book include the coupling of complex systems involving a range of different processes and a variety of scientific disciplines. The first six case studies focus on interactions between different compartments of the hydrological cycle. They illustrate examples of model coupling that include land-surface interactions and feedbacks at large and at catchment scale, and models that incorporate smaller scale feedbacks between canopy form, ecosystem cycling, and soil moisture content, dynamics and pathways.

Four case studies are then presented that all deal with the integration of dynamic land-surface, atmospheric and hydrological processes, including interfacial fluxes that produce non-linear feedbacks. The specific space and time scales involved in these examples vary greatly, depending on the motivation of the study and on the specific task. These examples broaden the scope of land-atmosphere-interaction studies significantly from the traditional view of the atmosphere having a one-way impact on the surface, to surface change impacts within the climate system, and the resulting two-way interactions.

Other examples in catchment ecosystem dynamics and non-point source pollution are also given that integrate biogeochemical with hydrological processes, and yield new insight into the generation and transport of non-point source pollution at catchment scale. This has been achieved by incorporating biogeochemical transformation and transport processes into hydrological models at this spatial scale.

Finally, two cases are described where the focus is on incorporating socio-economic processes and feedbacks into the natural system as part of a state-of-the-art vulnerability assessment.

However, it is important to bear in mind that there are a large range of more bounded problems in hydrology, such as, for example, storm event runoff production, flood routing, and crop water use that may be addressed more efficiently and accurately using conventional (non-coupled) analytical and modelling frameworks. In these cases, the complexity of coupled models may, indeed, provide an impediment to efficiency because of difficulties in parameterisation and model uniqueness.

5.2 Progress in Coupled Models

Over the past two decades, advances have been made in interfacing different Earth system components within hydrological models. As a result, it is becoming increasingly apparent that, as discussed above, a number of problems cannot be properly simulated separately, rather they need to be addressed by coupling subsystems together. A new paradigm is thus emerging in hydrology that emphasises the need to interact with other branches of knowledge, notably atmospheric sciences, but also soil sciences, geochemistry, biology and sociology. Several observations are apparent, including:

- Understanding the parts does not imply understanding the whole;
- Scale issues are important. Processes that dominate system behaviour at small scales may become irrelevant at large scales and new processes may emerge as being important in response to the change in scale;

- Feedbacks are ubiquitous. This, together with non-linearity and scale problems, is what causes prediction difficulties and it is one of the motivations for using coupled models;

- Spatial variability of fluxes (e.g. rainfall), properties (e.g. hydraulic conductivity) and state variables (e.g. soil humidity) often leads to unexpected system responses;

- Understanding the effects of memory may also be important. For example, land-surface and subsurface states, particularly soil moisture, affect the persistence of drier or wetter atmospheric conditions. This is of particular significance for the occurrence of large floods or droughts;

- Human feedbacks are no less important than natural feedbacks.

In part because of these and other advances in understanding, coupled models have demonstrated some successful predictive capabilities. These have been accomplished in a range of model frameworks, including those that involve loose coupling by passing data between separate model codes through a database management system (e.g. GIS), to more tightly coupled codes in which representation of media and domains are more closely integrated. Some examples of successful coupled models have been discussed as case studies in Chapter 4.

Despite these successes, there is a perceived need to be more systematic when applying coupled models, particularly when addressing problems that relate to social policies. Chapter 2 (Sect. 2.3) seeks to respond to this need. The opinion of the authors is that the very basic approaches used in model building (conceptualisation, discretisation, calibration, error analysis, model selection and validation) need to be revised. Uncertainty needs to be addressed more formally. Sensitivity analysis, error propagation and risk assessment are addressed. However, much remains to be done. Coupling may, for example, introduce such complexity that model predictions are rendered unreliable: this is discussed below. In part because of this, emphasis in coupled modelling now appears to be shifting towards risk and vulnerability assessment. This is the topic of Sect. 5.4.

5.3 How much Coupling is Needed ?

A coupled approach may not always be desirable or feasible. In addition to the labour and costs required, developing a coupled model often involves technical and conceptual constraints, e.g. the development of increasingly complex, coupled, non-linear systems usually increases model uncertainty, compromises prediction and forecasting skill, and may reduce model identifiability.

In addition, there is an interesting dichotomy in the development of interdisciplinary models that integrate different Earth system compartments in that it results in the proliferation of adjustable model parameters and more model variables that require initialisation, and in more interacting, non-linear components. The latter contributes towards the potential development of a chaotic system, which is enhanced by the uncertainties in model parameters and in the initial values of variables. Because the number and density of observations available to initialise, calibrate or validate model predictions are often small compared to the inherent degrees of freedom built into complex models, it is often possible for a number of model structures/parameterisations to be consistent with the limited observational data available. While the motivation for constructing complex models is to increase the physical realism of the simulation in terms of the feedbacks and processes that characterise the Earth system, the end result may well be a complex model with limited forecasting or predictive capability. Consequently, by analogy with the well known phrase

"Scientific approaches should be as simple as possible,
but not simpler",

in the context of coupled modelling, we recommend:

"Couple as few processes as possible, but not less".

5.4 Emphasising the Vulnerability and Assessing the Risks of Coupled Systems

As an alternative to the "forecasting" paradigm hitherto much used in hydrology, a set of research initiatives and projects investigating complex system behaviour have preferred to assess vulnerability or risk (Kabat et al. 2004; Pielke and Guenni 2004). This reverses the usual sequence of building a complex model and then using it to forecast a scenario or set of scenarios. The first step becomes to identify what attributes or potential changes in the state of the environmental system have most societal significance. There is a specific evaluation of environmental consequences so that the most critical changes can be identified. A complex modelling approach can then be used to identify stable and unstable system behaviour relative to these conditions when expected or unexpected perturbations occur, that may be natural, human-induced or, more likely, of mixed origin. Thus, the emphasis is shifted to seeking the existence of stable equilibria within the environmental system and specifying those conditions that may cause a migration from one stable mode to another.

Although with this paradigm, emphasis is taken off the development of forecasting skill, there is clearly still a need to adequately reproduce system feedbacks and non-linearity. An important task facing the scientific community is to decide the level of process specification and detail required to allow identification of regions of system stability and instability. The level of detail will, of course, depend on the nature of the questions posed, but it is clear that the complexity incorporated into model frameworks needs to carefully balance the ability to reproduce key feedbacks and non-linearities with parsimonious parameterisation, that can either be met by information that is available or that can be generated by current or developing technology. This indicates that the advancement of complex models will ultimately depend on the development of a carefully planned and maintained data infrastructure and on advancements in observational technology.

5.5 Future Research

This book has shown that there is great potential for creating new knowledge and gaining insight into and the functioning of complex systems by the development and application of integrated, coupled models. It advocates a procedure for making the development of coupled models more systematic, and describes several methods of model parameterisation to give guidance on how to deal with the several important challenges involved.

Although there is a wide range of future research that is required, our recommendation is that there should be enhanced emphasis on:

- Technical aspects of how the coupling is to be performed in an adequate manner and on how to best parameterise interface conditions;

- Identifying system conditions that tend to be stable when perturbed (i.e. damped system states) and states that are unstable and may move towards new equilibria;

- Defining the range of perturbations over which the system remains stable and/or that can generate instability;

- Assessing the importance of feedbacks so as to facilitate the removal of unnecessary feedbacks and/or addition of important but missing feedbacks;
- Incorporating societal values as part of vulnerability assessments. The examples of coupling with socio-economic processes given herein reflect the current state of the art, but defining and introducing coupling between natural and social sciences is still rudimentary in coupled models;
- Using data assimilation with coupled models;
- Establishing the credibility of coupled models by validation.

References Chapter 5

Kabat P, Claussen M, Dirmeyer PA, Gash JHC, Guenni LB de, Meybeck M, Pielke RA Sr, Vörösmarty CJ, Hutjes RWA, Lütkemeier S (eds) (2004) Vegetation, water, humans and the climate: A new perspective on an interactive system. Springer-Verlag, Berlin, p566

Pielke RA Sr., Guenni LB de (eds) (2004) How to evaluate vulnerability in changing environmental conditions. In: Kabat P, Claussen M, Dirmeyer PA, Gash JHC, Guenni LB de, Meybeck M, Pielke RA Sr, Vörösmarty CJ, Hutjes RWA, Lütkemeier S (eds) Vegetation, water, humans and the climate: A new perspective on an interactive system. Springer-Verlag, Berlin, pp481-544

Subject Index

A

B

C

E

F

G

H

I

J

K

M

N

S

T

W

Y

Z